Biogenic Synthesis of Nanoparticles and Sustainable Nanotherapeutics

Biogenic Synthesis of Nanoparticles and Sustainable Nanotherapeutics

Edited by

Malay K. Das
Department of Pharmaceutical Sciences
Dibrugarh University
Dibrugarh, Assam, India

WILEY

List of Contributors

Fazid Ali
National Institute of Pharmaceutical Education and Research (NIPER)
S. A. S. Nagar
Punjab
India

Md Ahesan Ansari
Department of Pharmaceutical Technology
Jadavpur University
Kolkata
West Bengal
India

O. T. Ayana
Sunshine College
Nekemte
Oromia
Ethiopia

Asis Bala
Pharmacology and Drug Discovery Research Laboratory
Division of Life Sciences
Institute of Advanced Study in Science and Technology (IASST)
An Autonomous Institute
under – Department of Science & Technology (Govt. of India) Vigyan Path
Guwahati
Assam
India

and

Academy of Scientific and Innovative Research (AcSIR)
AcSIR (an Indian Institute of National Importance), Sector 19
Ghaziabad
Uttar Pradesh
India

Bapan Banik
Department of Pharmaceutical Sciences
Rabindranath Tagore University
Hojai
Assam
India

Himakshi Baruah
Advanced Drug Delivery Laboratory
Department of Pharmaceutics
School of Pharmaceutical Sciences
Girijananda Chowdhury University
Guwahati
Assam
India

Giles Mary Basumatary
Drug Delivery Research Laboratory
Department of Pharmaceutical Sciences
Dibrugarh University
Dibrugarh
Assam
India

F. F. Bekere
Department of Materials Science and Engineering
Adama Science and Technology University
Adama
Ethiopia

and

Department of Advanced Materials Science and Engineering Center of Excellence
Adama Science and Technology University
Adama
Ethiopia

Ranu Biswas
Department of Pharmaceutical Technology
Jadavpur University
Kolkata
West Bengal
India

Pallabi Borah
Centre for Biotechnology and Bioinformatics
Dibrugarh University
Dibrugarh
India

Raja Chakraborty
Institute of Pharmacy
Assam Don Bosco University
Tepesia
Assam
India

Ratul Chakraborty
Pharmacology and Drug Discovery Research Laboratory
Division of Life Sciences
Institute of Advanced Study in Science and Technology (IASST)
An Autonomous Institute under – Department of Science & Technology (Govt. of India) Vigyan Path
Guwahati
Assam
India

and

Academy of Scientific and Innovative Research (AcSIR)
AcSIR (an Indian Institute of National Importance), Sector 19
Ghaziabad
Uttar Pradesh
India

Tapash Chakraborty
Advanced Drug Delivery Laboratory
Department of Pharmaceutics
School of Pharmaceutical Sciences
Girijananda Chowdhury University
Guwahati
Assam
India

Arindam Chatterjee
Institute of Pharmacy
Assam Don Bosco University
Tepesia
Assam
India

Ratan Chowdhury
Rangapara College
Rangapara
Assam
India

Malay K. Das
Drug Delivery Research Laboratory
Department of Pharmaceutical Sciences
Dibrugarh University
Dibrugarh
Assam
India

Punamjyoti Das
School of Pharmaceutical Sciences
Girijananda Chowdhury University-Tezpur Campus
Tezpur
India

Ichu Daule
Department of Pharmacy
Tripura University (A Central University)
Suryamaninagar
Tripura (W)
India

Evarin Debbarma
Department of Horticulture
Aromatic & Medicinal Plants (HAMP)
Mizoram University
Mizoram
India

Chinmoyee Deori
Department of Pharmacology
Lakhimpur Medical College and Hospital
Chowkham
Assam
India

Nikita Dey
Department of Pharmaceutical Chemistry
NEF College of Pharmaceutical Education and Research
Nagaon
Assam
India

Balisa Mosisa Ejeta
Drug Delivery Research Laboratory
Department of Pharmaceutical Sciences
Dibrugarh University
Dibrugarh
Assam
India

Barnali Gogoi
Department of Pharmacognosy
Assam Pharmacy Institute
Titabor
Assam
India

Rajlakshmi Gogoi
Drug Delivery Research Laboratory
Department of Pharmaceutical Sciences
Dibrugarh University
Dibrugarh
Assam
India

Moumita Gope
Drug Delivery Research Laboratory
Department of Pharmaceutical Sciences
Dibrugarh University
Dibrugarh
Assam
India

Ayan Goswami
Division of Pharmacology
Guru Nanak Institute of Pharmaceutical Science and Technology
Kolkata
West Bengal
India

Abhinab Goswami
Faculty of Pharmaceutical Sciences
Assam Down Town University
Guwahati
Assam
India

Hemanga Hazarika
School of Pharmaceutical Sciences
Girijananda Chowdhury University-Tezpur Campus
Tezpur
Assam
India

Emdad Hossain
Department of Pharmaceutical Technology
Jadavpur University
Kolkata
West Bengal
India

Taison Jamatia
Drug Delivery Research Laboratory
Department of Pharmaceutical Sciences
Dibrugarh University
Dibrugarh
Assam
India

Bhupen Kalita
NEF College of Pharmacy
Guwahati
Assam
India

Naba Jyoti Kalita
School of Pharmaceutical Sciences
Girijananda Chowdhury University-Tezpur Campus
Tezpur
Assam
India

Nihalini Kalita
NEF College of Pharmacy
Guwahati
Assam
India

Pratap Kalita
Pratiksha Institute of Pharmaceutical Sciences
Guwahati
Assam
India

Roohi Kesharwani
Faculty of Pharmacy
United University
Prayagraj
Uttar Pradesh
India

Harshita Krishnatreyya
National Institute of Pharmaceutical Education & Research
Guwahati
Assam
India

Parveen Kumar
NIMS Institute of Pharmacy
NIMS University
Jaipur
Rajasthan
India

Rishav Mazumder
Department of Pharmacy
Tripura University (A Central University)
Suryamaninagar
Tripura (W)
India

Jyoti Kumari Mishra
Drug Delivery Research Laboratory
Department of Pharmaceutical Sciences
Dibrugarh University
Dibrugarh
Assam
India

Junmoni Nath
Department of Pharmaceutics
School of Pharmaceutical Sciences
Girijananda Chowdhury University
Guwahati
Assam
India

Lalduhsanga Pachuau
Department of Pharmaceutical Science
Assam University
Silchar
Assam
India

Dilip K. Patel
Department of Pharmacy
Government Polytechnic
Jaunpur
Uttar Pradesh
India

Jack Patgiri
Drug Delivery Research Laboratory
Department of Pharmaceutical Sciences
Dibrugarh University
Dibrugarh
Assam
India

Padmanath Pegu
School of Pharmaceutical Sciences
Girijananda Chowdhury University-Tezpur Campus
Tezpur
Assam
India

Latifur Rahman
NEF College of Pharmacy
Guwahati
Assam
India

Abirami Raja
C L Baid Metha College of Pharmacy
OMR Thoraipakkam
Chennai
Tamilnadu
India

Bhanu P. Sahu
Department of Pharmaceutics
School of Pharmaceutical Sciences
Girijananda Chowdhury University
Guwahati
Assam
India

Anupam Sarma
Advanced Drug Delivery Laboratory
Department of Pharmaceutics
School of Pharmaceutical Sciences
Girijananda Chowdhury University
Guwahati
Assam
India

Arjun Kumar Sen
Department of Pharmaceutical Technology
Jadavpur University
Kolkata
West Bengal
India

Saikat Sen
Faculty of Pharmaceutical Science
Assam Down Town University
Guwahati
Assam
India

Shahnawaz Shameem
Faculty of Pharmacy
United University
Prayagraj
Uttar Pradesh
India

Rachna Sharma
Drug Delivery Research Laboratory
Department of Pharmaceutical Sciences
Dibrugarh University
Dibrugarh
Assam
India

L. Ronibala Singha
Drug Delivery Research Laboratory
Department of Pharmaceutical Sciences
Dibrugarh University
Dibrugarh
Assam
India

Punam Talukdar
Pharmacology and Drug Discovery Research Laboratory
Division of Life Sciences
Institute of Advanced Study in Science and Technology (IASST)
An Autonomous Institute under – Department of Science & Technology (Govt. of India) Vigyan Path
Guwahati
Assam
India

and

Academy of Scientific and Innovative Research (AcSIR)
AcSIR (an Indian Institute of National Importance), Sector 19
Ghaziabad
Uttar Pradesh
India

Preface

During the recent years, we have seen a tremendous technological progress in the field of biogenic nanotechnology for various applications in public life, including public health and well-being. The competing books published earlier have mainly focused on the applications of biogenic nanotechnology in various industrial sectors, such as agri-nanotechnology for sustainable agriculture, nanotechnology in crop protection, nanotechnology for soil conservation, advanced electronics, and energy storage devices. The current book focuses on the application of biogenic nanotechnology for public health and well-being.

This book, *Biogenic Synthesis of Nanoparticles and Sustainable Nanotherapeutics,* focuses on the biogenic synthesis of nanoparticles using various natural bioresources with emphasis on the mechanism of nanoparticle synthesis and their sustainable biomedical applications including medicine and drug delivery. Fifteen chapters cover all topics, including biogenic sustainable nanotechnology using plant extract, plant latex, egg, onion, bacteria, fungi, virus, microalgae, actinomycetes, yeast, and orange. Additionally, this book covers the characterization, safety assessment, applications, recent patents, clinical evidences, and market outlook of biogenic nanoparticles, including sustainable nanotherapeutics for application in human diseases. This book may be useful in understanding and exploring the mechanism of synthesis of biogenic nanoparticles using various natural bioresources. Further, it will give future researchers a stimulus to promote their interest in the field of green nanotechnology and sustainable nano pharmaceutical products for sustainable human health.

This book is relevant for researchers, teachers, and students in the areas of pharmaceutical sciences, pharmaceutical chemistry, pharmacognosy, pharmaceutics, pharmaceutical nanotechnology, biomedical sciences, biotechnology, environmental sciences, and pharmaceutical/biomedical/biotech industries.

The higher educational and research institutes would benefit from using this book for conducting MPharm, MS, MTech, MSc degree programs in pharmaceutical and biomedical sciences/biotechnology, including R&D activities.

Malay K. Das
Dibrugarh, India
December 2025

1

Trends in Biogenic Sustainable Nanotechnology

Taison Jamatia[1], *Malay K. Das*[1], *Evarin Debbarma*[2], *Rishav Mazumder*[3], *Ichu Daule*[3], *and Giles Mary Basumatary*[1]

[1] *Drug Delivery Research Laboratory, Department of Pharmaceutical Sciences, Dibrugarh University, Dibrugarh, Assam, India*
[2] *Department of Horticulture, Aromatic & Medicinal Plants (HAMP), Mizoram University, Mizoram, India*
[3] *Department of Pharmacy, Tripura University (A Central University), Suryamaninagar, Tripura (W), India*

1.1 Introduction

Nanotechnology is rapidly emerging as a revolutionary field with significant applications in various fields such as medicine, electronics, environmental science, and material sciences [1–3]. It involves the modification of large materials at the nanoscale, 1–100 nm, and it has unique physical and chemical characteristics [4]. Nanotechnology has a significant role in the synthesis of nanomaterials, providing sustainable and environmentally friendly solutions. It focuses on minimizing environmental impact while maximizing efficiency and product quality [5]. Nanomaterials can be synthesized by top-down and bottom-up processes [6]. Using a top-down approach, bulk materials are reduced to the nanoscale. In order to reach the required nanoscale dimensions, this method uses subtractive processes like mechanical machining, physical vapor deposition, lithography, and pyrolysis via evaporation process, which involve mechanically or physically altering larger pieces [7, 8]. Using the bottom-up method, nanomaterials are created from tiny blocks such as molecules or atoms. Bottom-up methods include chemical and biological approaches [9, 10]. The conventional techniques for creating nanomaterials, like physical and chemical methods, are more costly, need costly equipment and hazardous chemicals, require high temperatures, take longer, and are only applicable under certain circumstances, such as high

Biogenic Synthesis of Nanoparticles and Sustainable Nanotherapeutics,
First Edition. Edited by Malay K. Das.

pressures or temperatures [11–13]. The limitations of traditional methods place restrictions and highlight numerous possible risks like cytotoxicity, genotoxicity, carcinogenicity, and toxicity [11, 14–17].

Recently, environmentalist scientists have focused on the synthesis of nanomaterials, concerning the biological and environmental safety. In contrast, the biogenic synthesis of nanomaterials uses living cells and biological pathways to create nanomaterials in a way that is economical, nontoxic, and environmentally beneficial [18–20]. When it comes to synthesizing nanomaterials, this method is more environmentally friendly and sustainable than traditional chemical synthesis techniques. Biogenic synthesis of nanomaterials is sustainable, safe, and effective and utilizes renewable resources, making it a superior substitute for conventional nanomaterial synthesis techniques [5, 10, 21]. Biogenically synthesized metal nanomaterials, such as gold, silver, iron, copper, zinc, and other metal oxides, possess unique physicochemical properties and morphology [22, 23]. Moreover, biogenic synthesis uses the plant extract containing phytoconstituents such as alkaloids, tannins, terpenoids, phenols, aromatic compounds, and polysaccharides. These compounds can have various pharmacological activities such as antioxidant, anti-inflammatory, anticancer, and antimicrobial activities. These phytoconstituents help stabilize the nanomaterials and reduce the metal ions during the formation of nanomaterials [24–28]. Several microorganisms, such as *Pseudomonas deceptionensis, Pseudomonas aeruginosa, Bacillus licheniformis, Prococcusfuriosus, and Pyrobaculum islandicum*, are most commonly used in the biogenic synthesis for metal and metal oxide nanomaterials [29–32]. Biogenic synthesis is a viable approach for synthesizing stable metal nanoparticles (NPs) having various applications across different sectors, leveraging natural biological processes for sustainable development. The distinctive characteristics of the biogenic synthesized nanomaterials are as shown in Figure 1.1.

Biogenic synthesis primarily occurs through bio-reduction, that is the reduction of metal ions to metal form by biological agents. Generally, the process occurs via transfer of metal ions into the cells and reduction within the cellular environment (intracellularly), and metal ions can also be reduced outside the cells, often using secreted enzymes or metabolites from the microorganisms (extracellularly) [21, 33, 34]. The biogenically synthesized nanomaterials have been demonstrated as potential therapeutic strategies in pharmaceutical products, anticancer, antifungal, cosmetics, clinical diagnosis, and manipulation of targeted medicines [35–37]. In this chapter, we have epitomized the biogenically synthesized nanomaterials, which will provide an understanding of the sustainable methods for synthesizing nanomaterials using plants and microbes for biomedical applications.

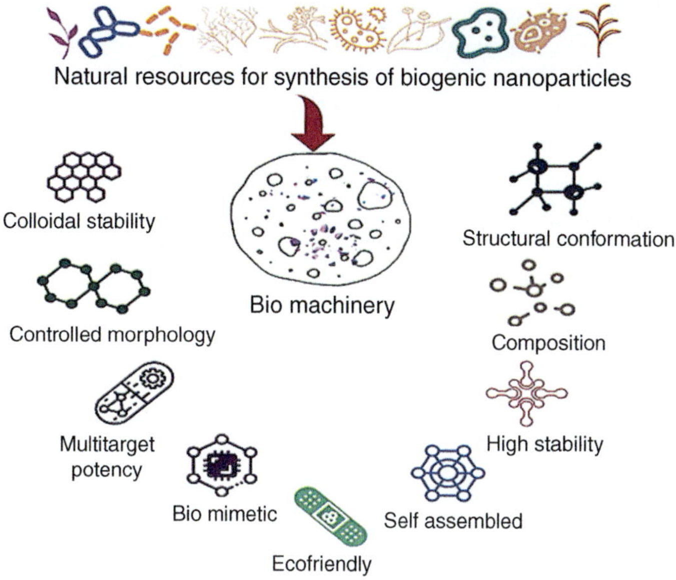

Figure 1.1 Distinctive characteristics of biogenically synthesized nanomaterials. *Source:* [21] / Frontiers Media S.A / CC BY 4.0.

1.2 Approaches for the Synthesis of Nanomaterials

Nanoparticles can be synthesized using two fundamental methods: top-down and bottom-up, as shown in Figure 1.2 for the synthesis of nanomaterials, depending on the desired characteristics of the particles, including size, shape, surface area, and crystallinity. The development of nanoparticles with specific properties and functions is made possible by the distinct techniques and applications of each strategy.

1.2.1 Top-Down Method

It is also known as destructive method, which involves breaking down large portions of material down to its smaller nanosized pieces or atomic blocks. Ball milling, lithography, etching, sputtering, laser ablation, and the thermal decomposition process are examples of top-down techniques [39–41].

1.2.1.1 Ball Milling Method

It involves the reduction of the size of bulk materials into nano-scale particles in a high-energy ball mill using mechanical or physical processes. It is used to

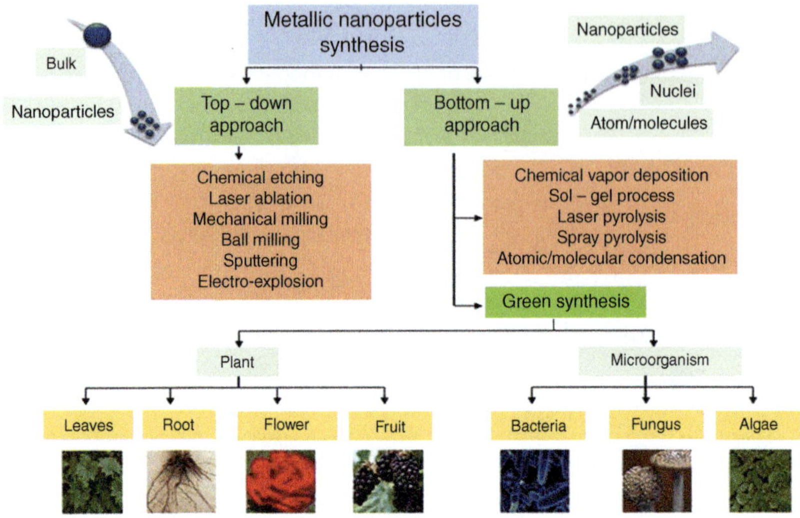

Figure 1.2 General approaches for the synthesis of nanomaterials. *Source:* Singh et al. [38] / Springer Nature / CC BY 4.0.

produce copper-based nano alloys and carbide-strengthened aluminum alloys, wear-resistant spray coatings, aluminum, nickel, magnesium oxide, and various nano composite substances. The ball milling process has several merits, such as being simple and cost-effective and applicable to a wide range of materials. However, it consists of several demerits like high energy consumption, uneven nanoparticle sizes can result from a wide particle size dispersion, possibility of milling media contamination [42–44].

1.2.1.2 Laser Ablation

It involves striking the target material with a strong laser beam to create nanoparticles. The source material vaporizes due to the high intensity of the laser irradiation in the laser ablation process, forming nanoparticles. It can be considered as a green process as it does not require any other chemicals to create noble metal nanoparticles. The average size and distribution of the nanoparticle can be altered by adjusting the fluence, wavelength, and laser salt addition [45].

1.2.1.3 Lithography

To create the nanostructure and shape, it involves printing the required structure or shape on a light-sensitive material and then carefully removing some of the material. A concentrated electron or light beam is used in the practical process of lithography to produce nanomaterials [44].

1.2.2 Bottom-up Approaches

It is also known as the constructive method; it involves the application of chemical or physical forces to synthesize smaller units (atomic or molecular) to form clusters to nanoparticles [40, 45]. The process primarily relies on the self-assembly principle of molecular recognition. Self-assembly is the process of developing more and more traits of one's own type [46].

1.2.2.1 Sol–Gel Method

The process involves the thermal breakdown, hydrolysis, and condensation of metal precursors in solution, forming a stable sol. As hydrolysis or condensation increases viscosity, factors like temperature, pH, and concentration control particle size. Maturation allows solid growth, with solvent elimination, Ostwald ripening, and phase changes occurring over days, leading to nanoparticle formation by separating unstable reagents [47]. It is used mainly to synthesize metal oxides, ceramics, and glasses [46].

1.2.2.2 Pyrolysis

This thermochemical process is widely used in industries for large-scale nanomaterial production. It involves high-temperature decomposition without oxygen, offering a simple, cost-effective, and efficient method with a high yield [44, 46].

1.2.2.3 Chemical Vapor Deposition

It is the method in which a thin film of gaseous exchange is deposited onto the substrate at 300–1200 °C. When the substrate is heated, it comes into contact with a combined gas, leading to a chemical reaction that results in the formation of a thin film product on the substrate surface, which is recovered and reused. The advantages of this technique are that uniform, robust, stiff, and highly pure nanoparticles can be manufactured. The drawback of chemical vapor deposition (CVD) is that it requires special equipment and the gaseous byproducts are highly toxic [48].

1.2.2.4 Biological Synthesis

It involves the synthesis of nanoparticles using plants and microorganisms. It is a conventional method of chemical and physical synthesis for its potential to synthesize the nanoparticles without any toxic chemicals, while in the physical and chemical synthesis process, it produces nanoparticles with a high risk of toxicity to the environment and to the cells [32, 49].

Various plant parts, including leaves, roots, stems, seeds, and fruits, can be used for nanoparticle synthesis. Plant secondary metabolites like alkaloids, terpenoids, and flavonoids play a key role in this process [50]. Plant biomolecules like proteins, carbohydrates, and co-enzymes aid in reducing metal salts to form nanoparticles [51–53].

Microorganisms like bacteria, fungi, algae, and yeast are eco-friendly, cost-effective tools for nanoparticle synthesis. Their reductase enzymes help detoxify heavy metals and reduce metal salts into nanoparticles. This occurs through **intracellular biosynthesis**, where ions enter the cell to form nanoparticles, or **extracellular biosynthesis**, where metal ions are trapped and reduced on the cell surface [54, 55].

1.3 Biogenic Synthesis for Sustainable Nanotechnology

Biological synthesis of nanomaterials using plant extracts offers a sustainable and eco-friendly approach to nanotechnology [56]. Plant extracts act as capping and reducing agents; this method avoids toxic chemicals and hazardous solvents. Nevertheless, research needs to focus more on synthesis conditions and characterize nanoparticles for real-world applications [57]. While physical and chemical methods produce high-purity nanoparticles, they tend to be time-consuming and expensive. Biological methods are faster, cheaper, and environmentally friendly. Phyto-nanotechnology, in the last decade, has found its way into the ever-growing world of nanotechnology for plant science. Green-synthesized nanoparticles have proved to be useful for stress-resilient plants in various abiotic stresses. Plants extract used biosynthesized nanoparticles such as lead, platinum, iron, silver, gold, copper, selenium, and zinc, present in different parts of the plant, viz. leaves, flowers, fruits, seeds, and roots [58]. Phytocompounds, such as alkaloids, flavonoids, and phenolics, stimulate nanoparticle generation in plants; they act as stabilizers and reducers. The main controlling factors of nanoparticle biosynthesis are metal ion concentration, the concentration of plant extract, temperature, pH, and reaction time. Characterization and purification of these nanoparticles should be efficient for research purposes. The mechanisms behind plant-mediated nanoparticle synthesis are still under investigation. It is believed that proteins, amino acids, and secondary metabolites contribute significantly to metal salt reduction and act as stabilizing agents [59].

1.3.1 Microbial-Mediated Synthesis

Nanoparticle (NP) synthesis employing traditional methods like chemical and physical methods requires excessive energy input, is time-consuming, and results in the production of toxic chemicals, leading to environmental pollution [60]. Microbial-mediated synthesized nanoparticles have great advantages due to its eco-friendly, cost-effective, and sustainable nature, which has attracted significant research interest [61, 62]. Synthesizing nanoparticles through biological systems offers great feasibility, eco-friendly nature, time saving, and the ability to control its toxicity traits and size characteristics [29]. Microorganisms such as bacteria, fungi, and yeast can grow in various habitats and possess certain ability to engineer characteristic nanomaterials, including exopolysaccharides, nanocellulose, and nanowires [63]. They can acclimate well in high metal concentrations and can reduce metallic ions via various enzymatic reductions [64].

1.3.1.1 Bacteria-Mediated Synthesis of Nanoparticles

Bacteria-mediated metallic nanoparticles, such as silver (Ag), gold (Au), copper (Cu), selenium (Se), iron (Fe), and other metal oxides, have been synthesized employing bacteria [65]. Bacteria have a unique physiological function to convert heavy metals to nanomaterials and can undergo abiotic stressors. For example, *Pseudomonas stutzeri* and *P. aeruginosa* can tolerate harsh environmental conditions with high metal concentrations [66, 67]. *Escherichia coli*, *Lactobacillus* species, *Bacillus cereus*, *Acinetobacter*, *Pseudomonas* species, *Corynebacterium* species, and *Klebsiella pneumoniae* are widely investigated bacteria for the production of AgNPs [68].

Shahverdi et al. (2007) demonstrated that *E. coli*, *K. pneumoniae*, and *Enterobacter cloacae* may be used for the reduction of Ag ions and produce AgNPs swiftly [69]. Prakash et al. (2011) assessed the extracellular synthesis of spherical AgNPs (size 10–30 nm) using *B. cereus*. Antibacterial activity of synthetic AgNPs was also investigated in *Streptococcus* and *E. coli*, and it was discovered that AgNPs could successfully inhibit bacterial growth [70]. Biosynthesis of AuNPs was developed using several cyanobacteria species [54]. *Plectonema* sp. has been utilized by Lengke et al. [71] and Brayner et al. [72] to fabricate AuNPs. According to Brock and Gustafson [73], the bacteria *Sulfolobus acidocaldarius*, *Thiobacillus thiooxidans*, and *Thiobacillus ferrooxidans* are capable of transforming the ferric ion to the ferrous state when they are cultivated on elemental sulfur-containing energy sources. Additionally, *T. thiooxidans* produced stable ferrous iron and aerobically reduced ferric iron in a low pH medium. Extracellularly biofabricated CuONPs were produced by Hassan et al. [74] using the actinobacteria *Streptomyces* sp. (isolates Oc-5 and Acv-11) that were obtained from the *Oxalis corniculata* L. CuONPs had superior antibacterial, antifungal, and antioxidant activities.

1.3.1.2 Actinobacteria-Mediated Synthesis of Nanoparticles

Gram-positive bacteria, which is a subdomain of the bacterial domain. Additionally, actinobacteria are capable of synthesizing many NPs because of their unique physiological extracellular or intracellular pathways [75]. Study showed that *Rhodococcus* NCIM 2891 enzymes cell wall aid in the intracellular creation of the AgNPs and the biomineralization of Ag ions [76]. In another study, Karthik et al. [77] used the marine bacterium *Streptomyces* sp. LK-3 to reveal the extracellular production of AgNPs. Through an electron transfer reaction, they found that nicotinamide adenine dinucleotide (hydrogen) (NADH)-dependent nitrate reductase facilitated the reduction of Ag ions into stable AgNPs.

1.3.1.3 Fungi-Mediated Synthesis of Nanoparticles

Fungi-mediated biosynthesized NPs have potential cell wall than bacterial cells since they have a greater capacity to withstand metal concentrations and have significant binding potential with metal ions [62]. Since fungi have a greater tendency to accumulate metals than bacteria, utilizing them to produce NPs is more cost-effective and efficient. Furthermore, the fungus-based biosynthesis of NPs makes it easier to handle biomass and process NPs downstream [75]. Mishra et al. [78] studied *Trichoderma viride* and *Hypocrea lixii,* mediated extracellular production of AuNPs, and discussed how temperature and pH affect this process. According to Ramos et al. [79], the fungus *Trichoderma* sp. produced AgNPs through extracellular biosynthesis from plant seeds of *Bertholletia excelsa,* and the bio-fabricated AgNPs demonstrated superior antibacterial action against Gram-negative bacteria.

1.3.1.4 Yeast-Mediated Synthesis of Nanoparticles

Metal ions can be deposited from yeast and can thrive in high-metal environments [80]. Several groups have utilized this characteristic of yeast for producing environmentally friendly NPs [81]. Apte et al. [82] study showed that *Yarrowia lipolytica*-mediated biosynthesized AgNPs and found that melanin was linked to the green synthesis of AgNPs. A strain of *Candida lusitaniae* was used to biosynthesize AgNPs, which ranged in size from 2 to 10 nm and had potent antibacterial action against *K. pneumoniae* and *Staphylococcus aureus* [83]. Elahian et al. [84] successfully biosynthesized silver nanoparticles using the genetically modified yeast strain *Pichia pastoris.*

1.3.1.5 Algae-Mediated Synthesis of Nanoparticles

Algal cells have various secondary metabolites and biologically active phytoconstituents that act as capping agents during the formation of NP [85]. In a study, AgCl NP was bio-fabricated extracellularly by employing the marine algae *Sargassum plagiiophyllum* [86]. Da Silva Ferreira et al. [87] used a dried algae *Chlorella vulgaris* in another study to show the extracellular production of AgClNPs. Fourier transform infrared spectroscopy (FTIR) analysis of synthesized AgClNPs showed that the primary capping agent provides stability, and biosynthesis of AgClNPs was algal proteins. Senapati et al. [88] synthesized using *Tetraselmisko chinensis* to report the intracellular biosynthesized of AuNPs (size 5–35 nm) and Singh et al. [89] biosynthesized using macroalga *Padina gymnospora* to show the extracellular synthesis of AuNPs (size 53–67 nm). *Turbinaria conoides* and

Padina tetrastromatica, two marine brown algae, were used to extracellularly bio-fabricate ZnONPs [90].

1.3.1.6 Virus-Mediated Synthesis of Nanoparticles

Capsid proteins that encase viruses at the nanoscale to synthesized metallic ions with appropriate environment conditions to interact with virus [91]. There are numerous studies which demonstrate the viruses-mediated biogenic synthesis of nanoparticles. For example, M13 bacteriophage is widely employed as a template to synthesize semiconductor nanoparticles of zinc sulfide (ZnS) and cadmium sulfide (CdS) [92]. In a study, red clover necrotic mosaic virus (RCNMV) was used to synthesized nanocarriers that were successfully evaluated for the controlled administration of the medication doxorubicin as an anticancer [93]. Using the tobacco mosaic virus (TMV), Czapar et al. [94] synthesized hollow nanotubes containing a polyanionic interior surface to administer anticancer medication phenanthriplatin precisely. In a recent study, Thangavelu et al. [95] used plant pathogenic squash leaf curl China virus (SLCCNV) to produce Au—Ag composite semiconductor NPs.

1.3.2 Plant-Mediated Synthesis

An emerging area of green nanotechnology is plant-mediated synthesis of metallic nanomaterials, which uses the inherent qualities of plant extracts to create nanomaterials in an environmentally responsible way. Many phytoconstituents, including tannins, vitamins, alkaloids, polysaccharides, flavonoids, terpenoids, and saponins, are found in plants and are crucial to the bio-reduction process that creates and stabilizes nanomaterials. Utilizing fruit, roots, leaves, and bark has the advantage because all plant parts contain these metabolites [36, 96].

1.3.2.1 Nanoparticles Produced from Plant Stem

Plant stem extract has been extensively explored for the biogenic synthesis of metal nanoparticles (Table 1.1). Extracts from stems of plants such as Coleus aromaticus, salvadora persica, and momordica charantia are commonly used for synthesizing silver nanoparticles. Silver nanoparticles were synthesized by using a green and eco-friendly approach through the extract from the stem of Coleus aromaticus [97]. The synthesized nanoparticles exhibited unique properties, including high antimicrobial and catalytic activities, and showed potential application in medicine, environmental science, and materials science. The stem extracts from Leucas lavandulifolia have been used to synthesize spherical selenium nanoparticles [98]. The reduction of selenium ions is facilitated by the presence of water-soluble heterocyclic compounds and polyphenols; surface

Table 1.1 Plant stem-based biogenically synthesized metal nanomaterials.

Family	Scientific name	Types of nanoparticles	Size (nm)	Application	References
Lamiaceae	*Coleus aromaticus*	Silver	42	Antioxidant	[97]
Cucurbitaceae	*Momordica charantia*	Silver	25	Antibacterial activity against both Gram-positive and Gram-negative bacteria, including strains *Escherichia coli*, *Staphylococcus aureus*, and *Pseudomonas aeruginosa*.	[98]
Lamiaceae	*Leucas lavandulifolia* Sm	Selenium	56–75	Antibacterial activity against both Gram-positive (*Staphylococcus aureus*, *Staphylococcus epidermidis*) and Gram-negative bacteria (*Escherichia coli*, *Salmonella typhi*)	[98]
Rubiaceae.	*Mussaenda frondosa*	Zinc oxide	5–25	Antibacterial activity against both Gram-positive (*Staphylococcus aureus*) and Gram-negative bacteria. (*Klebsiella* sp.)	[99]
Gentianaceae	*Swertia chirayita*	Zinc oxide	17.66	Antibacterial activity of ZnO nanoparticles against the Gram-positive (*Staphylococcus aureus*) and Gram-negative (*Escherichia coli*) bacteria.	[18]
Orobanchaceae	*Orobanche aegyptiaca*	Copper	50	Nematicidal properties against Meloidogyne incognita. Antibacterial activity against *Escherichia coli* and *Staphylococcus aureus*.	[100]

plasmon resonance reveals the process through a color shift. Zinc oxide nanoparticles synthesized from Mussaenda frondosa stem extracts have significantly inhibited the α-glucosidase and α-amylase enzymes [99]. Similarly, Swertia chirayita-mediated biosynthesis of zinc oxide nanoparticles have potential photocatalytic and antibacterial activities [18]. Terminalia arjuna stem bark, being highly rich in phytoconstituents such as flavonoids, glycosides, and tannins, has shown immense therapeutic value. Many studies have ascertained its efficacy in treating various diseases, such as cardiovascular diseases. The plant has antioxidant, anti-inflammatory, and lipid-lowering properties, thus contributing to its cardioprotective effects. While more research is necessary to fully elucidate its long-term safety profile, T. Arjuna bark continues to remain a prospective natural remedy for managing cardiovascular disorders and other conditions [100].

1.3.2.2 Nanoparticles Produced from Plant Leaves

Green-synthesized metallic nanoparticles using the plant leaves extract (Table 1.2) are the most common and have great potential as antimicrobial agents [101, 102]. Gold nanoparticles (AuNPs) with particle sizes between 20 and 60 nm, biosynthesized by reducing HAuCl4 using *aloe vera* leaf extract, result in good stability and no agglomeration [103]. The synthesized AgNPs show distinct characteristics, including varying particle sizes and ζ-potentials that suggest stability and potential efficacy in biomedical applications. Antibacterial assays show dose-dependent inhibition against *E. coli, S. aureus*, and *P. aeruginosa*, underlining the nanoparticles' antimicrobial potential. Additionally, significant antioxidant activity was shown through diphenylpicrylhydrazyl (DPPH) and H_2O_2 scavenging assays [104]. The La_2O_3NPs synthesized from *Eucalyptus globulus* leaf extract exhibit significant antioxidant, anti-inflammatory, and antidiabetic activities. They were mainly attributed to their radical scavenging and the inhibition of inflammation and glycolysis. Thus, in terms of their therapeutic applications, the use of La_2O_3NPs may be regarded as beneficial for the treatment of several diseases due to oxidative stress, inflammation, and hyperglycemia. Further research and clinical tests would be required to establish their *in vivo* therapeutic potential and mechanism of action [105]. Biosynthesized silver nanoparticles (AgNPs) by *Melia azedarach* leaf extract were successful. Synthesized AgNPs showed potent antifungal activity against Verticillium dahliae, one of the major pathogens infecting eggplants. This green approach opens up a promising route for sustainable agriculture with alternative, eco-friendly fungicides instead of the conventional chemical fungicides [106].

1.3.2.3 Nanoparticles Produced from Plant Roots

The multidimensional root traits have significantly advanced the field of green synthesis [42]. Studies have demonstrated that the reducing power of several

Table 1.2 Plant leaf-based biogenically synthesized metal nanomaterials.

Family	Scientific name	Types of nanoparticles	Size (nm)	Application effect	References
Asteraceae	Artemisia nilagirica	Silver	70–90	Antibacterial activity against Gram-positive bacteria (*Staphylococcus aureus*) and Gram-negative bacteria: *Escherichia coli, Pseudomonas aeruginosa*	[101]
Crassulaceae.	*Kalanchoe fedtschenkoi*	Silver	42	Shows antimicrobial activity, which inhibits bacterial growth against *Escherichia coli, Staphylococcus aureus*, and *Pseudomonas aeruginosa,* suggesting potential therapeutic applications in wound healing and infection control. The antioxidant activity, demonstrated by DPPH and H_2O_2 radical scavenging assays, reduces oxidation reactions.	[104]
Myrtaceae	*Eucalyptus globulus*	lanthanum oxide	50.37	The antioxidant, anti-inflammatory, and anti-diabetic activities	[105]
Meliaceae	*Melia azedarach*	Silver	23	antifungal agent against Verticillium dahliae in eggplants. Antifungal efficacy against Verticillium dahliae	[106]
Lamiaceae	*Mentha spicata*	Zinc oxide	11–88	Antiviral activity specifically against Tobacco mosaic virus (TMV).	[106]

Table 1.3 Plant root-based biogenically synthesized metal nanomaterials.

Family	Scientific name	Types of nanoparticles	Size (nm)	Application	References
Zingiberaceae	*Zingiber officinale*	Silver, gold	10–20 5–20	Antibacterial activity of AgNPs against foodborne pathogens.	[107]
Ranunculaceae	*Aconitum toxicum* Reichenb	Silver	9–15	Antioxidant	[108]
Licorice	*Glycyrrhiza glabra*	Silver	7–45	Anti-ulcer activity against *Helicobacter pylori.*	[109]
Araliaceae	*Panax ginseng*	Zinc oxide	59.76	Cytotoxicity against L20B tumor cells, demonstrating their potential as a promising anticancer agent.	[108]
Fabaceae.	*Mimosa pudica*	Iron oxide	67	Antimicrobial activity under Gram-positive bacteria: *Bacillus subtilis, Staphylococcus aureus* Gram-negative bacteria: *Escherichia coli.*	[108]
Solanaceae	*Withania somnifera*	Titanium oxide	9.5–12.5	Antibacterial nature against Gram-positive and Gram-negative bacteria. Shows anticancer activity.	[109]

endogenous substances, such as phenylpropanoids, ascorbic acid, oxalic acid, gingerol, zingerone, paradol, and shagaols, controls the biosynthesis of gold and silver nanoparticles from *Zingiber officinale* root extracts (Table 1.3). Similarly, *Withania somnifera* root-mediated biosynthesized titanium oxide nanoparticles have numerous bioactive metabolites such as flavonoids, sitoindosides, amino acids, and anolides [107]. Ethanol extracts of Aconitum toxicum Reichenb roots were utilized for AuNP synthesis, and after three hours at room temperature, AuNPs with sizes ranging from 9 to 15 nm were produced. Biogenic synthesis of silver and zinc nanoparticles using *Glycyrrhiza glabra* and *Panax ginseng* root acts as a reducing and capping agent and has potential anti-gastric ulcer and anticancer activity. The nanoparticles exhibited significant *in vitro* anti-ulcer activity against *Helicobacter pylori* at a concentration of 500 μg/ml, with a minimum inhibitory concentration (MIC) of 250 μg/ml. These findings highlight the potential of plant roots-mediated biogenic synthesis of AgNPs as a sustainable and effective approach for developing novel therapeutic agents for gastrointestinal disorders [108]. Biogenic synthesis of copper nanoparticles (CuNPs) using the root extract of *W. somnifera* has been successfully achieved. Synthesized CuNPs possess considerable antibacterial and antioxidant properties. The average particle size is 6.28 ± 1.13 nm. The study emphasizes the significant potential of biogenic synthesis for the production of nanoparticles with promising biological applications, such as antimicrobial agents and antioxidants [109].

1.3.2.4 Nanoparticles Produced from Flowers and Buds

Flowers contain various phytoconstituents such as polyphenols, flavonoids, carotenoids, and anthocyanins, which exhibit strong pharmacological properties. In addition, buds contain flavanols, glycosides, and polyphenols with oxidizing and reducing activities. Flower extracts (Table 1.4) have been highly utilized as a green and cost-effective method for the bioproduction of nanoparticles [110]. Flower-derived nanoparticles possess unique properties, including antioxidant and antimicrobial activities, which help them withstand environmental stressors. Although the biogenic synthesis of nanoparticles using flower extracts has mainly focused on silver and gold, further exploration is needed. *Syzygium aromaticum* (clove) buds have been widely explored for synthesizing various nanoparticles, such as copper, gold, silver, palladium, and copper oxide. The extracts from these buds contain pharmacologically active compounds like quercetin, eugenol, and gallic acid derivatives and are involved in the synthesis of nanoparticles [111]. Gold nanoparticles biosynthesis using Nyctanthes arbortristis flower extract exhibited spherical shape with a diameter of 19.8 ± 5.0 nm. Similarly, biogenic synthesis of silver nanoparticles using the bud of Couropita guianensis contains phytoconstituents like flavonoids and phenolics, which act as reducing and capping agents in the formation of nanoparticles [112]. Extracts from *Tussilago farfara* contain sesquiterpenoid reducing agents that can create spherical silver and gold nanoparticles. Furthermore, ZnO nanoparticles biosynthesized using *Trifolium pratense* flower extract have potential antimicrobial activity against *S. aureus*, *P. aeruginosa*, and *E. coli*. This eco-friendly method offers a promising alternative for nanoparticle synthesis with applications in medicine and environmental science [113].

1.3.2.5 Nanoparticles Produced from Fruits and Seeds

Fruits and seeds are natural reservoirs of phytoconstituents like polyphenols, flavonoids, and terpenoids act as reducing agents as well as capping agents that can help form and stabilize nanoparticles (Table 1.5). Extracts of fruit and peel

Table 1.4 Plants flower and bud-based biogenically synthesized metal nanomaterials.

Family	Scientific name	Types of nano-particles	Size nm	Application therapeutic effect	Reference
Mytraeace	*Syzygium aromaticum*	Copper	15–20	CuNPs exhibited significant antimicrobial activity, with the highest inhibition observed against *Bacillus* spp. (8 mm zone of inhibition) and *Penicillium* spp. (6 mm zone of inhibition)	[111]
Oleaceae	*Nyctanthes arbortristis*	Gold	19.8–5.0	Antibacterial activity exhibited against *Pseudomonas aeruginosa* and *Staphylococcus aureus*	[112]
Asparagaceae	*Polianthus tuberosa*	Gold	38.76	Antimicrobial activity of PtubAuNPs against *Escherichia coli* and *Staphylococcus aureus* was splendid. Shows cytotoxic response of the PtubAuNPs against MCF-7.	[113]
Asteraceae	*Tussilago farfara*	Silver	13.57 ± 3.26	Antibacterial activity against vancomycin-resistant Enterococci Van-A type *Enterococcus faecium*.	[113]
Lecythidaceae	*Couropita guianensis*	Silver	32	Antibacterial effect on Gram-positive (*Bacillus subtilis*) and Gram-negative (*Escherichia coli*)	[112]
Fabaceae.	*Trifolium pratense*	Zinc oxide	60–70	Antimicrobial activity against clinical and standard strains of *Staphylococcus aureus* and *Pseudomonas aeruginosa* and the standard strain of *Escherichia coli*.	[113]

Table 1.5 Plant fruits and seed-based biogenically synthesized metal nanomaterials.

Family	Scientific name	Types of nanoparticles	Size (nm)	Application	Reference
Phyllanthaceae.	*Emblica officinale* (Indian gooseberry)	Silver and gold	10–25	Antimicrobial activity against *Bacillus subtilis* (Gram-positive), *Escherichia coli* (Gram-negative)	[115]
Myrtaceae	*Syzygium alternifolium* (java plum)	Copper oxide	5–13	Anticancer activity against MDA-MB-231 breast cancer cell lines.	[116]
Lythraceae	*Punica granatum* (pomegranate)	Iron oxide	25–55	Antioxidant activity and excellent photocatalytic activity against the reactive dyes under the blue UV light.	[118]
Fabaceae	*Trigonella foenum graecum* (fenugreek)	Gold	15–25	The catalytic activity was found to be size-dependent, with smaller nanoparticles showing faster activity.	[118]
Rutaceae	*Citrus maxima* (pomelo)	Zinc oxide	10–20	Show good photocatalytic activity for the photodegradation of methylene blue. It shows good antibacterial activity against pathogens and electrochemical sensing properties toward the quantification of dopamine at a trace concentration level.	[114]
Rubiaceae	*Coffea arabica*	Silver	20–30	Antibacterial activity of silver nanoparticles on *Escherichia coli* and *Staphylococcus aureus* demonstrated diminished bacterial growth with the development of well-defined inhibition zones.	[117]

have gained attention in the biogenic synthesis of nanoparticles as stabilizing and reducing agents. Extract of various citrus fruits, i.e., *C. reticulata*, *C. limon*, and *C. sinensis*, was utilized in the synthesis of gold nanospheres, nanoprisms, and nanotriangles. Gold nanoparticles were produced through the bio-reduction of tetrachloroaurate by citrus fruit extracts. *Crataegus douglasii* fruit extracts were used to successfully isolate silver nanospheres, which have potent antibacterial properties against both Gram-negative and Gram-positive bacterial species [114]. In addition, *Emblica officinalis* (Indian Gooseberry) fruits extract contain tannins, amino acids, phenolics, alkaloids, and carbohydrates, which act as reducing and capping agents in the biosynthesis of silver nanoparticles. Biomolecules present in plant extracts, such as polyphenols and sugars, contributed to the stability and reduced toxicity of the nanoparticles. The biosynthesized AgNPs showed antibacterial activity against both Gram-positive and Gram-negative bacteria [115]. *Syzygium alternifolium* extract facilitates the biosynthesis of copper oxide nanoparticles, showing excellent antiviral activity. The seeds of this plant are rich in phytoconstituents like reducing sugars, carbohydrates, and phenolics, which are essential for the synthesis of nanoparticles. Commonly known as mogi or adavinerudu, this fruit is consumed by local tribal communities and used to make squashes, jellies, and local vinegar. It is also employed in various treatments such as stomach ulcers, rheumatic pains, diarrhea, and diabetes [116]. Silver nanoparticles are biosynthesized with the seed extracts of *Sinapis arvensis* and Coffea italica. Zinc oxide and Fe_3O_4 nanoparticles synthesized from the seed extract of *Peganum harmala* and *Punica granatum* have potential Cr absorbent properties and photocatalytic activity [117]. The Fe_2O_3NPs of synthesized *P. granatum* (pomegranate) displayed a semi-spherical shape and size (25–55 nm). Bioactive compounds such as phenolic acids and flavonoids were essential for the formation and stabilization of the nanoparticles, which show excellent photocatalytic activity in degrading reactive blue dye under ultraviolet (UV) light, indicating their potential for environmental remediation. This research underscores the viability of using natural resources for the sustainable production of valuable nanomaterials, using a green and environmental approach involving fenugreek seed extracts to synthesize gold nanoparticles. AuNPs exhibited high crystallinity and a narrow size distribution (15–25 nm). Nanoparticles were evaluated using various analytical techniques, including UV–Vis spectroscopy, transmission electron microscopy (TEM), X-ray diffraction (XRD), and FTIR. The synthesized AuNPs demonstrated potent catalytic activity for the reduction of 4-nitrophenol to 4-aminophenol, exhibiting faster reaction rates. This study highlights the potential of fruits and seed-mediated biosynthesis of nanomaterials with controlled properties and applications in various fields, including catalysis and environmental remediation [118].

1.4 Application of Biogenic Sustainable Nanotechnology

1.4.1 Antimicrobial Activity

AgNPs (10–40 nm) synthesized by the fungus *Schizophyllum radiatum* were evaluated as potent antimicrobial agents and when tested against different microbes, including *B. subtilus*, *B. stearothermophilus*, *Salmonella paratyphi*, *Salmonella typhi*, *S. aureus*, *P. aeruginosa*, *Enterobacter aerogenes*, *E. coli*, and *Klebsiella pneumoniae,* exhibited zone of inhibition (ZOI) ranging 7–19 mm [119]. Zinc sulfide quantum dots (ZnS QDs) were biosynthesized using fungus, Aspergillus sp., and exhibited promising antibacterial activity tested against *S. aureus* (ZOI = 32 ± 0.10 mm), *B. subtilis* (ZOI = 28 ± 0.38 mm), *E. coli* (ZOI = 39 ± 0.5 mm), and *K. pneumoniae* (ZOI = 27.2 ± 0.61 mm). Further, the antibacterial activity of ZnS QDs was affirmed that it relied on bacterial membrane disruption, causing leakage of cytoplasmic contents, leading to ROS-influenced oxidative damage and bacterial cell lysis [120]. Similarly, AgNPs synthesized using *Salvadora persica* aqueous root extracts displayed antibacterial effects with promising inhibition against *E. coli* (MIC = 0.39 μg/ml; MBC = 0.78 μg/ml) and *Staphylococcus epidermidis* (MIC = 0.19 μg/ml; MBC = 0.39 μg/ml) [121]. *Bacillus megaterium* Y-4-mediated palladium NPs showed antibacterial efficacy against *S. aureus* and *E. coli* at a minimal dose of 20 mg/l, exhibiting photothermal bacterial killing efficiency [122]. Selenium NPs (72–100 nm) synthesized using bacterial isolates, viz. *E. coli* and *S. aureus* exhibited strong antibacterial effects against *E. coli*, *S. aureus*, *P. aeruginosa*, and *S. epidermidis* with MIC and analysis minimum bacterial concentration (MBC) values ranging 3–50 μg/ml [123]. Likewise, another AgNPs manufactured using a probiotic strain of *Lactobacillus salivarius* showed promising antibacterial activity against a panel of bacterial strains, including *Chromobacterium violaceum*, *Serratia marcescens*, *P. aeruginosa*, *S. aureus*, and *E. coli*, with MIC values ranging from 50 to 100 μg/ml. Moreover, these AgNPs inhibited 65.3–79.65% bacterial biofilm formation at 1/2 MIC, along with inhibition of quorum-sensing activity of bacteria by reducing the production of quorum-sensing-related pigments, like violacein in *C. violaceum* (80.27), pyocyanin in *P. aeruginosa* (66.95), and prodigiosin in *S. marcescens* (74.23) at 1/2 MIC. Also, AgNPs were effective in inhibiting fungal growth, exhibiting antifungal activity against *Fusarium proliferatum* (ZOI = 20 mm), *Penicillium purpurogenum* (ZOI = 17 mm), *Aspergillus niger* (ZOI = 15 mm), and *Rhizopus stolonifer* (ZOI = 12 mm) [124]. *Streptomyces capillispiralis* Ca-1-mediated copper NPs showed antimicrobial activities against tested microbial strains of *Bacillus dimenuta* (ZOI = 13.3 mm), *B. subtilis* (ZOI = 17.7 mm), *S. aureus* (ZOI = 16.3 mm), *E. coli* (ZOI = 14.3 mm), *P. aeruginosa* (ZOI = 18 mm, *Candida albicans* (ZOI = 18.7 mm), and *Aspergillus brasiliensis* (ZOI = 12 mm) [125]. Biosynthesized gold NPs (5 nm) and AgNPs (30 nm) mediated by the fungus, *C. albicans,* demonstrated antibacterial activity against *S. aureus* and *E. coli,* displaying ZOI of 11 mm,19 mm, 16 mm, and 24 mm, respectively. Also, the MIC_{80} values of AuNPs and AgNPs against *E. coli* were 128 μg/ml and 8 μg/ml, respectively, and for *S. aureus* were 512 μg/ml and 32 μg/ml, respectively [126]. Another AgNPs biosynthesized using bacterium *Bacillus endophyticus* SCU-L (5.1 nm) exhibited potent antimicrobial activity, evidenced by ZOI against *C. albicans* (15 mm), *E. coli* (13 mm), *S. typhi* (11 mm), and *S. aureus* (19 mm) at a minimal dose of 25 μg/ml [127]. *Streptacidiphilus durhamensis* HGG16n isolate-mediated AgNPs showed antimicrobial activity against a panel of microbes, including *P. aeruginosa, S. aureus, Proteus mirabilis, E. coli, K. pneumoniae,* and *B. subtilis* with ZOI ranging 6–10 mm, and MIC values of 6.25–50 μg/ml [128]. *Pichia kudriavzevii* conjugated ZnONPs resulted in potent antibacterial properties tested against bacterial strains of *B. subtilis, S. epidermidis, S. aureus, E. coli, and S. marcescens* with ZOI ranging 6–19 mm [129]. Marine macroalgae *Padina* sp. tethered AgNPs displayed antibacterial efficacy, resulting in ZOI of 15.17 mm and 13.33 mm against *S. aureus* and *P. aeruginosa*, respectively [130].

AgNPs-jujube extract evaluated prominent antibacterial activities displaying MIC values of 1.26 μg/ml, 2.5 μg/ml, and 2.5 μg/ml against *E. coli, K. pneumoniae,* and *S. aureus*, respectively [131]. *Wedelia chinensis*-mediated AgNPs showed antibacterial activities against Gram-negative bacteria *E. coli* (ZOI = 20.5 ± 0.19 mm) and Gram-positive bacteria *Listeria monocytogenes* (ZOI = 14.9 ± 0.84 mm) at a dose of 50 μg/ml [132]. AuNPs and AgNPs prepared using extracts of *Curcuma longa* rhizomes were evaluated for their antibacterial efficacy against *E. coli* and *B. subtilis* and displayed ZOI ranging 10–14 mm [133]. *Oscimum sanctum*-AuNPs showed strong antibacterial activities against different bacterial strains and reported their MIC values of *P. aeruginosa* (17.37 ± 0.67), *V. cholera* (42.09 ± 0.47), *E. coli* (58.52 ± 0.37), and *B. subtilis* (34.65 ± 1.47) [134]. *Zingiber officinale*-Ag doped stannic oxide NPs showed good antimicrobial activities against different microbes, displaying their inhibitory ZOI of 29 ± 0.54 (*S. aureus*), 27 ± 0.05 (*E. coli*), 17 ± 0.05 (*Fusarium oxysporum*), and 15 ± 0.05 mm (*Fusarium graminearum*) [135]. *Rhus chinensis*-AgNPs reported for its antibacterial activity, showing ZOI of 7.83 ± 0.17 mm (*S. aureus*) and 7.0 mm (*K. pneumoniae*) [31].

1.4.2 Larvicidal Activity

Biogenic AgNPs synthesized employing *Agaricus bisporus* isolates exerted larvicidal activity against *Culex* larvae, producing 100% in 5 mg/l, which is a minimal concentration [136]. In a study, CuNPs biosynthesized with bacterial isolate-*S. capillispiralis* Ca-1 have shown mortality against *Culex pipiens* of 96.67% at a concentration of 1.5 mM. Also, these CuNPs showed larvicidal activity against *Musca domistica,* exhibiting mortality of 76.67% at a 10 mM concentration [125]. AgNPs conjugated with the extracts of *Couropita guinensis* flowers and *Lagerstroemia speciosa* fruits exhibited larvicidal effects on *Culex quinquefasciatus* mosquito, displaying an LC_{50} value of 0.742 ppm [137].

1.4.3 Anticancer Activity

Gram-positive and Gram-negative bacteria used for the synthesis of SeNPs were isolates and exhibited anticancer effects against melanoma cells in a dose-dependent manner, exhibiting IC_{50} values ranging from 8 to 20 μg/ml [123]. AgNPs biosynthesized employing *L. salivarius* displaying cytotoxic effects against Michigan Cancer Foundation-7 (MCF-7) breast cancer cells, producing an IC_{50} value of 52.29 μg/ml [124]. Zinc oxide NPs biosynthesized utilizing the endophytic fungus *Alternaria tunuissima* showed potential antiproliferative activity against breast (MCF-7) and liver (HepG2) cancer cells with IC_{50} of 16.87 μg/ml and 18.02 μg/ml, respectively [138]. Biogenic AgNPs synthesized using *jujube* extract result in cytotoxic effects against AGS human gastric adenocarcinoma cell line, causing 82% cell death at a dose of 100 μg/ml [131].

Punica granatum-AgNPs displayed antiproliferative efficacy against human liver cancer cells (HepG2) with an IC_{50} of 70 μg/ml [139]. *Wedelia chinensis*-AgNPs displayed cytotoxic activity against HepG2 cells with an IC_{50} value of 25 μg/ml [132]. *Curcuma longa* mediated AuNPs and AgNPs showed antiproliferative potential against lung cancer cells (A549) and prostate cancer cells (PC3) with IC_{50} values ranging from 31 to 41 μg/ml [133]. Fruit extracts of *Azadirachta indica* were utilized to prepare AgNPs, which displayed inhibitory activities against A549 cell growth with IC_{50} values reported to be 64.46 μg/ml [134]. *Ocimum sanctum* oil was applied to biosynthesize AuNPs that exhibited promising antiproliferative activity against MCF-7 cells with an IC_{50} value of 78.51 ± 2.4 μg/ml [140]. *Brassica juncea*-ZnONPs showed promising anticancer activity against MCF-7 and A549 cells at IC_{50} values of 43.83 μg/ml and 41.48 μg/ml, respectively [141]. *Vinca rosea* mediated ZnONPs and Vanadium-doped ZnONPs showed anticancer activity against MCF-7 with IC_{50} values of 84.7 and 78.7 μg/ml, respectively [142]. *Rhus coriaria* fruit extracts capped ZnONPs were observed to have powerful anti-breast cancer effects, inducing apoptotic cell-death and inhibited metastasis against MCF-7 and MDA-MB-231 cells at IC_{50} values of 35.04–63.71 μg/ml [143].

1.4.4 Drug Delivery

Potato virus X NPs (PVX NPs) were utilized as a nanocarrier to potentially deliver doxorubicin, which continued to exhibit strong anticancer properties tested in

A2780 (IC_{50} = 0.84 μM), MDA-MB-231 (IC_{50} = 0.94 μM), and HeLa (IC_{50} = 5.8 μM). Finally, PVX NPs, when tested in athymic mice bearing MDA-MB-231 xenografts, showed a reduction in tumor growth [144]. *Bifidobacterium longum*-based poly(lactic-co-glycolic acid) (PLGA) NPs conjugated with perfluorohexane were observed to be effective in targeting solid tumors (tested in MDA-MB-231 bearing nude mice model) with improved diagnosis, imaging, treatment utilities, and had better biosafety and biocompatibility characteristics [145]. *Escherichia coli* conjugated mesoporous silica NPs loaded with doxorubicin were revealed to enhance their release and distribution, along with improved tumoral cell mortality [146]. *Garcinia mangostana* mediated polymeric NPs were prepared in Eudragit RL100/RS100 for a targeted intracellular delivery for effective treatment of colon cancers, providing improved endocytic uptake and solubility [147].

1.4.5 Other Biomedical Utilities

AgNPs-*L. salivarius* demonstrated substantial antioxidant activity, efficiently scavenging free radicals of DPPH (IC_{50} = 42.65 μg/ml) and $ABTS^+$ (IC_{50} = 53.77 μg/ml) [124]. ZnONPs-*Alternaria tenuissima* exhibited DPPH radical scavenging potential with an IC_{50} value of 102.13 μg/ml [138]. ZnONPs-*P. kudriavzevii* Yeast Strain showed strong antioxidant activity (DPPH), producing an IC_{50} value of 5.26 ± 0.42 μg/ml [129]. Aqueous extracts of *Wedelia chinensis* mediated AgNPs displayed DPPH radical scavenging potential of 80.2 ± 0.15% at a concentration of 200 μg/ml [132]. *Vinca rosea*-capped ZnONPs and Vanadium-doped ZnONPs exhibited strong antioxidant properties via DPPH and H_2O_2 scavenging assays with IC_{50} values of 17.23 μg/ml [142]. AgNPs-*L. salivarius* demonstrated significant anticoagulant and thrombolytic activities in the management of thrombosis [124].

MgONPs synthesized using floral extracts of *Rosa floribunda* exhibited anti-hyperuricemic activity by inhibiting xanthine oxidase activity with an IC_{50} value of 15.9 μg/ml. Besides, these MgONPs have potential anti-aging activity evaluated by enzymatic activity, including hyaluronidase (IC_{50} = 196.1 μg/ml), tyrosinase (IC_{50} = 158.5 μg/ml), elastase (IC_{50} = 82.5 μg/ml), and collagenase (IC_{50} = 58.7 μg/ml) [148]. Leaf extracts of *P. granatum* embedded AgNPs exhibited potential activities against α-amylase (IC_{50} = 65.2 μg/ml) and α-glucosidase (IC_{50} = 53.8 μg/ml) [139]. Likewise, ZnONPs prepared with aqueous extract of *B. juncea* reported antidiabetic activity on α-amylase assay with an IC_{50} value of 71.76 μg/ml [141]. CuNPs-*S. capillispiralis* Ca-1 exhibited inhibitory activities against plant fungal diseases, resulting in the inhibition of 57.14% (*Alternaria* spp.), 63.81% (*A. niger*), 58.05% (*Pythium* spp.), and 42.61% (*Fusarium* spp.) [125].

1.4.6 Catalytic and Other Utilities

Biogenic ZnS QDs were revealed as alternate applications for good photodegradation of dyes under solar irradiation and sensing potential of heavy metals [120]. *Shewanella loithica* PV-4 mediated ultra-small NPs of Au/Pb/Pt (2–7 nm) exhibited strong catalytic activity on the degradation of methyl orange dye [149]. *Streptomyces griseoruber*-mediated AuNPs showed potential catalytic activity in the photodegradation of methylene blue dyes [150]. Likewise, ZnONPs-*A. tenuissima* revealed effective degradation activity of methylene blue dyes [138]. ZnO-*B. licheniformis* nanoflowers (620 nm) displayed enhanced photocatalytic activity against methylene blue dyes with 83% degradation in a span of 60 minutes [151]. ZnO-*A. niger* NPs were observed to degrade 89% Bismarck brown dyes at a concentration of 100 μl, and the same can be utilized in wastewater treatment [152]. The application of mold *Sporotrichum thermophile* BJTLRMDU7 aided in the synthesis of AgNPs and reduced *p*-nitrophenol into *p*-aminophenol in a green synthesis method due to light and $NaBH_4$ [153].

AgNPs conjugated with *jujube* extracts showed that eriochrome black T (EBT) and rhodamine b (RhB) as cationic and anionic contaminants under UV–visible light irradiation, demonstrating promising catalytic properties. AgNPs decomposed 90.9% (RhB) and 84.7% (EBT) after 80 minutes of exposure to UV light [131]. *Moringa oleifera*-CuONPs showed good photocatalytic activity in the decomposition of harmful dyes like malachite green (84.7%) and titan yellow (79.03%) [154]. AuNPs biosynthesized utilizing floral extracts of *M. oleifera* exhibit a promising catalytic activity, resulting in the rapid decomposition of *p*-nitrophenol and *p*-nitroaniline [155]. In another report, *A. niger* mediated ZnONPs were effective in dye degrading capabilities and can be combined with cotton fabrics for industrial applications as it also showed antibacterial effects when tested against *S. aureus* and *E. coli* [152]. Methanol blue and reactive black 5 were among the dyes that were decolored by ZnONPs-Pseudochrobactrum sp. C5. Moreover, it helped in the reduction of wastewater effluents like chemical oxygen demand, total dissolved solids, pH, and electrical conductivity [156].

1.5 Conclusion and Future Aspects

Biogenic synthesis of nanomaterials has seen significant advancements, providing more economical and environmentally friendly substitutes for conventional chemical techniques. The naturally extracted biosynthesized nanomaterials leverage their inherent biochemical pathways for reduction and stabilization. These biological systems can efficiently and selectively produce nanomaterials with optimized sizes and shapes, making them highly desirable for various applications in medicine, electronics, and environmental remediation. Biogenic synthesis reduces the toxic chemicals and energy-intensive processes, further highlighting its potential as a sustainable approach to nanomaterials production. The surface charge and morphology of the nanoparticles determine their biostability. More investigation is required to determine the exact molecular mechanisms that trigger intracellular plant signaling responses after applying nanoparticles in less-than-ideal circumstances. However, the limited reports of *in vivo* and kinetic study profiles are minimal, therefore further study is necessary. Additionally, the safety and stability of nanomaterials need to be assessed as per regulatory guidelines. Biogenic synthesis using bacteria, fungi, yeast, and algae must be closely considered when employing a biological process that requires hazardous chemicals and high temperatures. Optimization and advanced characterization techniques need to be employed to gain insight into the structures and properties of nanomaterials. Similarly, scaling up techniques for the production of nanomaterials for commercial application remains a significant challenge.

Research has shown that nanoparticles synthesized through biological methods exhibit significant advantages. Biogenic nanoparticles have demonstrated antimicrobial and anticancer properties, which have a potential therapeutic application. Regardless of the benefits, several hurdles remain in the biogenic synthesis of nanoparticles. Issues such as the complexity of phytochemical interactions, variability in synthesis outcomes, scalability of production processes, and ensuring product stability need to be addressed, and ongoing research is crucial to overcome existing challenges and unlock its full potential across various industries.

References

1 Maynard AD, Dudley SM. Navigating advanced technology transitions: using lessons from nanotechnology. Nat Nanotechnol [Internet]. 2023;18(10):1118–1120. Available from: https://www.nature.com/articles/s41565-023-01481-5.

2 Nyandoro VO, Masioge HK, Malago ZL. Biogenic synthesis of metal nanoparticles: promoting green nanotechnology and sustainable development goals. Clean Techn Environ Policy [Internet]. 2024:0123456789. Available from: https://doi.org/10.1007/s10098-024-02966-0.

3 Malik S, Muhammad K, Waheed Y. Nanotechnology: a revolution in modern industry. Molecules [Internet]. 2023 Jan 9;28(2):661. Available from: https://www.mdpi.com/1420-3049/28/2/661.

4 Sim S, Wong N. Nanotechnology and its use in imaging and drug delivery (Review). Biomed Rep. [Internet]. 2021;14(5):42. Available from: http://www.spandidos-publications.com/10.3892/br.2021.1418.

5 Luque-Jacobo CM, Cespedes-Loayza AL, Echegaray-Ugarte TS, Cruz-Loayza JL, Cruz I, de Carvalho JC, et al. Biogenic synthesis of copper nanoparticles: a systematic review of their features and main applications. Molecules [Internet]. 2023;28(12):4838. Available from: https://www.mdpi.com/1420-3049/28/12/4838.

6 Huston M, DeBella M, DiBella M, Gupta A. Green synthesis of nanomaterials. Nanomaterials [Internet]. 2021;11(8):2130. Available from: https://www.mdpi.com/2079-4991/11/8/2130.

7 Paramasivam G, Palem VV, Sundaram T, Sundaram V, Kishore SC, Bellucci S. Nanomaterials: synthesis and applications in Theranostics. Nanomaterials [Internet]. 2021;11(12):3228. Available from: https://www.mdpi.com/2079-4991/11/12/3228.

8 Baptista A, Silva F, Porteiro J, Míguez J, Pinto G. Sputtering physical vapour deposition (PVD) coatings: a critical review on process improvement and market trend demands. Coatings. 2018;8(11):402.

9 Alqarni LS, Alghamdi MD, Alshahrani AA, Nassar AM. Green nanotechnology: recent research on bioresource-based nanoparticle synthesis and applications. Singh AK, editor. J Chem [Internet]. 2022 Aug 9;2022:1–31. Available from: https://www.hindawi.com/journals/jchem/2022/4030999.

10 Abuzeid HM, Julien CM, Zhu L, Hashem AM. Green synthesis of nanoparticles and their energy storage, Environmental, and Biomedical Applications. Crystals. 2023;13(11):1576.

11 Akhter MS, Rahman MA, Ripon RK, Mubarak M, Akter M, Mahbub S, et al. A systematic review on green synthesis of silver nanoparticles using plants extract and their bio-medical applications. Heliyon [Internet]. 2024 Jun;10(11):e29766. Available from: https://doi.org/10.1016/j.heliyon.2024.e29766.

12 Ashique S, Upadhyay A, Hussain A, Bag S, Chaterjee D, Rihan M, et al. Green biogenic silver nanoparticles, therapeutic uses, recent advances, risk assessment, challenges, and future perspectives. J Drug Deliv Sci Technol [Internet]. 2022 Nov;77:103876. Available from: https://linkinghub.elsevier.com/retrieve/pii/S1773224722007870.

13 Ali ARI, Salam B. A review on nanofluid: preparation, stability, thermophysical properties, heat transfer characteristics and application. SN Appl Sci [Internet]. 2020 Oct 5;2(10):1636. Available from: https://link.springer.com/10.1007/s42452-020-03427-1.

14 Joshi DR, Adhikari N. An overview on common organic solvents and their toxicity. J Pharm Res Int [Internet]. 2019 Jun;29:1–18. Available from: https://journaljpri.com/index.php/JPRI/article/view/1135.

15 Tobiszewski M, Namieśnik J, Pena-Pereira F. Environmental risk-based ranking of solvents using the combination of a multimedia model and multi-criteria decision analysis. Green Chem. 2017;19:1034–1042.

16 Teoh WY, Amal R, Mädler L. Flame spray pyrolysis: an enabling technology for nanoparticles design and fabrication. Nanoscale [Internet]. 2010;2(8):1324. Available from: https://xlink.rsc.org/?DOI=c0nr00017e.

17 Saratale RG, Karuppusamy I, Saratale GD, Pugazhendhi A, Kumar G, Park Y, et al. A comprehensive review on green nanomaterials using biological systems: recent perception and their future applications. Colloids Surf B: Biointerfaces. 2018;170:20–35.

18 Chakraborty S, Singh A, Roychoudhury A. Biogenic nanoparticles and generation of abiotic stress-resilient plants: a new approach for sustainable agriculture. Plant Stress [Internet]. 2022 Dec;6:100117. Available from: https://linkinghub.elsevier.com/retrieve/pii/S2667064X22000628.

19 Jounaki K, Soltani KM, Vahidi H, Barabadi H. Bioengineering of nanomaterials using biological resources: biofabrication mechanisms, characterizations, and biomedical applications. In: Bioengineered Nanomaterials for Wound Healing and Infection Control [Internet]. 2023. Elsevier, p. 239–286. Available from: https://linkinghub.elsevier.com/retrieve/pii/B9780323953764000058.

20 Shah A, Khalil AT, Ahmad K, Iqbal J, Shah H, Shinwari ZK, et al. Biogenic nanoparticles: synthesis, mechanism, characterization and applications. In: Biogenic Nanoparticles for Cancer Theranostics [Internet]. 2021. Elsevier, p. 27–42. Available from: https://linkinghub.elsevier.com/retrieve/pii/B9780128214671000100.

21 Kulkarni D, Sherkar R, Shirsathe C, Sonwane R, Varpe N, Shelke S, et al. Biofabrication of nanoparticles: sources, synthesis, and biomedical applications. Front Bioeng Biotechnol [Internet]. 2023 May 2;(11):1159193. Available from: https://www.frontiersin.org/articles/10.3389/fbioe.2023.1159193/full.

22 Kumar H, Bhardwaj K, Kuča K, Kalia A, Nepovimova E, Verma R, et al. Flower-based green synthesis of metallic nanoparticles: applications beyond Fragrance. nanomaterials [Internet]. 2020 Apr 16;10(4):766. Available from: https://www.mdpi.com/2079-4991/10/4/766.

23 Kumar H, Bhardwaj K, Sharma R, Nepovimova E, Kuča K, Dhanjal DS, et al. Fruit and vegetable peels: utilization of high value horticultural waste in novel industrial applications. Molecules. 2020;25(12):2812.

24 Priya AS, Afzal O, Khalid M, Faruque Ahmad M, Upadhyay A, et al. Biogenic nanoparticles from waste fruit peels: synthesis, applications, challenges and future perspectives. Int J Pharm [Internet]. 2023 Aug;643:123223. Available from: https://linkinghub.elsevier.com/retrieve/pii/S0378517323006439.

25 Saratale RG, Saratale GD, Shin HS, Jacob JM, Pugazhendhi A, Bhaisare M, et al. New insights on the green synthesis of metallic nanoparticles using plant and waste biomaterials: current knowledge, their agricultural and environmental applications. Environ Sci Pollut Res. 2018;25:10164–10183.

26 Annu SM, Singh S, Prajapati S, Verma DK, Shin DK. From green chemistry to biomedicine: the sustainable symphony of cobalt oxide nanoparticles. RSC Adv [Internet]. 2024;14(45):32733–32758. Available from: https://xlink.rsc.org/?DOI=D4RA05872K.

27 Campaña AL, Saragliadis A, Mikheenko P, Linke D. Insights into the bacterial synthesis of metal nanoparticles. Front Nanotechnol [Internet]. 2023 Aug 10;5. Available from: https://www.frontiersin.org/articles/10.3389/fnano.2023.1216921/full.

28 Arora A, Lashani E, Turner RJ. Bacterial synthesis of metal nanoparticles as antimicrobials. Microb Biotechnol [Internet]. 2024 Aug 16;17(8):1–15. Available from: https://sfamjournals.onlinelibrary.wiley.com/doi/10.1111/1751-7915.14549.

29 Salem SS, Fouda A. Green synthesis of metallic nanoparticles and their prospective biotechnological applications: an overview. Biol Trace Elem Res [Internet]. 2021;199:344–370. Available from: https://doi.org/10.1007/s12011-020-02138-3.

30 Jang HD, Kim SK, Chang H, Choi JH, Cho BG, Jo EH, et al. Three-dimensional crumpled graphene-based platinum-gold alloy nanoparticle composites as superior electrocatalysts for direct methanol fuel cells. Carbon N Y. 2015.

31 Bhusal M, Pathak I, Bhadel A, Shrestha DK, Sharma KR. Synthesis of silver nanoparticles assisted by aqueous root and leaf extracts of Rhus chinensis Mill and its antibacterial activity. Heliyon [Internet]. 2024;10(13):e33603. Available from: https://doi.org/10.1016/j.heliyon.2024.e33603.

32 Pandit C, Roy A, Ghotekar S, Khusro A, Islam MN, Bin ET, et al. Biological agents for synthesis of nanoparticles and their applications. J King Saud Univ Sci. 2022;34(3):101869.

33 Khandel P, Yadaw RK, Soni DK, Kanwar L, Shahi SK. Biogenesis of metal nanoparticles and their pharmacological applications: present status and application prospects. J Nanostructure Chem [Internet]. 2018 Sep 11;8(3):217–254. Available from: http://link.springer.com/10.1007/s40097-018-0267-4.

34 Patil S, Chandrasekaran R. Biogenic nanoparticles: a comprehensive perspective in synthesis, characterization, application and its challenges. J Genet Eng Biotechnol. 2020;18:67.

35 Bhat SA, Kumar V, Dhanjal DS, Gandhi Y, Mishra SK, Singh S, et al. Biogenic nanoparticles: pioneering a new era in breast cancer therapeutics—a comprehensive review. Discov Nano [Internet]. 2024 Aug 3;19(1):121. Available from: https://doi.org/10.1186/s11671-024-04072-y.

36 Alsaiari NS, Alzahrani FM, Amari A, Osman H, Harharah HN, Elboughdiri N, et al. Plant and microbial approaches as green methods for the synthesis of nanomaterials: synthesis, applications, and future perspectives. Molecules [Internet]. 2023 Jan 3;28, 1:463. Available from: https://www.mdpi.com/1420-3049/28/1/463.

37 Danish MSS, Estrella-Pajulas LL, Alemaida IM, Grilli ML, Mikhaylov A, Senjyu T. Green synthesis of silver oxide nanoparticles for photocatalytic environmental remediation and biomedical applications. Metals (Basel) [Internet]. 2022 Apr 29;12(5):769. Available from: https://www.mdpi.com/2075-4701/12/5/769.

38 Singh J, Dutta T, Kim KH, Rawat M, Samddar P, Kumar P. 'Green' synthesis of metals and their oxide nanoparticles: applications for environmental remediation. J Nanobiotechnol [Internet]. 2018 Dec 30;16(1):84. Available from: https://jnanobiotechnology.biomedcentral.com/articles/10.1186/s12951-018-0408-4.

39 Ijaz I, Gilani E, Nazir A, Bukhari A. Detail review on chemical, physical and green synthesis, classification, characterizations and applications of nanoparticles. Green Chem Lett Rev. 2020;13:223–245.

40 Khan Y, Sadia H, Ali Shah SZ, Khan MN, Shah AA, Ullah N, et al. Classification, synthetic, and characterization approaches to nanoparticles, and their applications in various fields of nanotechnology: a review. Catalysts [Internet]. 2022 Nov 8;12(11):1386. Available from: https://www.mdpi.com/2073-4344/12/11/1386.

41 Srinivasan LV, Rana SS. A critical review of various synthesis methods of nanoparticles and their applications in biomedical, regenerative medicine, food packaging, and environment. Discov Appl Sci [Internet]. 2024 Jul 6;6(7):371. Available from: https://doi.org/10.1007/s42452-024-06040-8.

42 Sunny NE, Mathew SS, Chandel N, Saravanan P, Rajeshkannan R, Rajasimman M, et al. Green synthesis of titanium dioxide nanoparticles using plant biomass and their applications- a review. Chemosphere. 2022;300(April):134612.

43 Samuel MS, Ravikumar M, John A, Selvarajan E, Patel H, Chander PS, et al. A review on green synthesis of nanoparticles and their diverse biomedical and environmental applications. Catalysts. 2022;12(5):469.

44 Mekuye B, Abera B. Nanomaterials: an overview of synthesis, classification, characterization, and applications. Nano Sel [Internet]. 2023 Aug 10;4(8):486–501. Available from: https://onlinelibrary.wiley.com/doi/10.1002/nano.202300038.

45 Baig N, Kammakakam I, Falath W. Nanomaterials: a review of synthesis methods, properties, recent progress, and challenges. Mater Adv [Internet]. 2021;2(6):1821–1871. Available from: http://xlink.rsc.org/?DOI=D0MA00807A.

46 Patil N, Bhaskar R, Vyavhare V, Dhadge R, Khaire V, Patil Y. Overview on methods of synthesis of nanoparticles. Int J Curr Pharm Res [Internet]. 2021 Mar;15:11–16. Available from: https://innovareacademics.in/journals/index.php/ijcpr/article/view/41556.

47 Li J, Wu Q, Wu J. Synthesis of nanoparticles via solvothermal and hydrothermal methods. In: Handbook of Nanoparticles [Internet]. 2016. Springer International Publishing, Cham, p. 295–328. Available from: https://link.springer.com/10.1007/978-3-319-15338-4_17.

48 Stoian M, Maurer T, Lamri S, Fechete I. Techniques of preparation of thin films: catalytic combustion. Catalysts [Internet]. 2021 Dec 15;11(12):1530. Available from: https://www.mdpi.com/2073-4344/11/12/1530.

49 Karunakaran G, Sudha KG, Ali S, Cho EB. Biosynthesis of nanoparticles from various biological sources and its biomedical applications. Molecules. 2023;28(11):4527.

50 Makarov VV, Love AJ, Sinitsyna OV, Makarova SS, Yaminsky IV, Taliansky ME, et al. "Green" nanotechnologies: synthesis of metal nanoparticles using plants. Acta Nat. 2014;6:35–44.

51 Yadi M, Mostafavi E, Saleh B, Davaran S, Aliyeva I, Khalilov R, et al. Current developments in green synthesis of metallic nanoparticles using plant extracts: a review. Artif Cells Nanomed Biotechnol. 2018;46:336–343.

52 Antunes Filho S, dos Santos MS, dos Santos OAL, Backx BP, Soran ML, Opriş O, et al. Biosynthesis of nanoparticles using plant extracts and essential oils. Molecules. 2023;28(7):3060.

53 Khan F, Shariq M, Asif M, Siddiqui MA, Malan P, Ahmad F. Green nanotechnology: plant-mediated nanoparticle synthesis and application. Nanomaterials. 2022;12(4):673.

54 Koul B, Poonia AK, Yadav D, Jin JO. Microbe-mediated biosynthesis of nanoparticles: applications and future prospects. Biomolecules. 2021;11(6):886.

55 Raj NB, Swamy MK, Purushotham B, Sukrutha SK. Applications of microbe-based nanoparticles in agriculture: present state and future challenges. Microbial Nanobiotechnology. 2021:343–382.

56 Hano C, Abbasi BH. Plant-based green synthesis of nanoparticles: production, characterization and applications. Biomolecules [Internet]. 2021 Dec 25;12(1):31. Available from: https://www.mdpi.com/2218-273X/12/131.

57 Sharma D, Kanchi S, Bisetty K. Biogenic synthesis of nanoparticles: a review. Arab J Chem. 2019;12(8):3576–3600.

58 Jadoun S, Arif R, Jangid NK, Meena RK. Green synthesis of nanoparticles using plant extracts: a review. Environ Chem Lett. 2021;19(1):355–374.

59 Kuppusamy P, Yusoff MM, Maniam GP, Govindan N. Biosynthesis of metallic nanoparticles using plant derivatives and their new avenues in pharmacological applications – an updated report. Saudi Pharm J. 2016;24(4):473–484.

60 Nath D, Banerjee P. Green nanotechnology – a new hope for medical biology. Environ Toxicol Pharmacol [Internet]. 2013 Nov;36(3):997–1014. Available from: https://linkinghub.elsevier.com/retrieve/pii/S1382668913001932.

61 Fariq A, Khan T, Yasmin A. Microbial synthesis of nanoparticles and their potential applications in biomedicine. J Appl Biomed [Internet]. 2017 Nov 1;15(4):241–248. Available from: http://jab.zsf.jcu.cz/doi/10.1016/j.jab.2017.03.004.html.

62 Singh P, Kim YJ, Zhang D, Yang DC. Biological synthesis of nanoparticles from plants and microorganisms. Trends Biotechnol [Internet]. 2016 Jul;34(7):588–599. Available from: https://linkinghub.elsevier.com/retrieve/pii/S0167779916000408.

63 Golmohammadi H, Morales-Narváez E, Naghdi T, Merkoçi A. Nanocellulose in sensing and biosensing. Chem Mater. 2017;29:5426–5446.

64 Li X, Xu H, Chen ZS, Chen G. Biosynthesis of nanoparticles by microorganisms and their applications. J Nanomater. 2011;2011:1–16.

65 Grasso G, Zane D, Dragone R. Microbial nanotechnology: challenges and prospects for green biocatalytic synthesis of nanoscale materials for sensoristic and biomedical applications. Nanomaterials. 2020;10:11.

66 Bridges K, Kidson A, Lowbury EJL, Wilkins MD. Gentamicin- and silver-resistant pseudomonas in a burns unit. Br Med J. 1979;1:446–449.

67 Haefeli C, Franklin C, Hardy K. Plasmid-determined silver resistance in pseudomonas stutzeri isolated from a silver mine. J Bacteriol. 1984;158:389–392.

68 Iravani S. Bacteria in nanoparticle synthesis: current status and future prospects. Int Sch Res Not. 2014;2014:1–18.

69 Shahverdi AR, Minaeian S, Shahverdi HR, Jamalifar H, Nohi AA. Rapid synthesis of silver nanoparticles using culture supernatants of Enterobacteria: a novel biological approach. Process Biochem. 2007;42:919–923.

70 Prakash A, Sharma S, Ahmad N, Ghosh A, Sinha P. Synthesis of Agnps by bacillus cereus bacteria and their antimicrobial potential. J Biomater Nanobiotechnol [Internet]. 2011;2(2):155–161. Available from: http://www.scirp.org/journal/doi.aspx?DOI=10.4236/jbnb.2011.22020.

71 Lengke MF, Fleet ME, Southam G. Morphology of gold nanoparticles synthesized by filamentous cyanobacteria from gold(I)-thiosulfate and gold(III)-chloride complexes. Langmuir. 2006;22:2780–2787.

72 Brayner R, Barberousse H, Hemadi M, Djedjat C, Yéprémian C, Coradin T, et al. Cyanobacteria as bioreactors for the synthesis of Au, Ag, Pd, and

Pt nanoparticles via an enzyme-mediated route. J Nanosci Nanotechnol. 2007;7:2696–2708.

73 Brock TD, Gustafson J. Ferric iron reduction by sulfur- and iron-oxidizing bacteria. Appl Environ Microbiol. 1976;32:567–571.

74 Hassan SED, Fouda A, Radwan AA, Salem SS, Barghoth MG, Awad MA, et al. Endophytic actinomycetes Streptomyces spp mediated biosynthesis of copper oxide nanoparticles as a promising tool for biotechnological applications. JBIC J Biol Inorg Chem [Internet]. 2019 May 26;24(3):377–393. Available from: http://link.springer.com/10.1007/s00775-019-01654-5.

75 Gahlawat G, Choudhury AR. A review on the biosynthesis of metal and metal salt nanoparticles by microbes. RSC Adv. 2019;9:12944–12967.

76 Raba J, Ortega F, Fernández Baldo M, Fernández J, Serrano M, Sanz M, et al. Study of antitumor activity in breast cell lines using silver nanoparticles produced by yeast. Int J Nanomedicine [Internet]. 2015 Mar;2021. Available from: http://www.dovepress.com/study-of-antitumor-activity-in-breast-cell-lines-using-silver-nanopart-peer-reviewed-article-IJN.

77 Karthik L, Kumar G, Kirthi AV, Rahuman AA, Bhaskara Rao KV. Streptomyces sp. LK3 mediated synthesis of silver nanoparticles and its biomedical application. Bioprocess Biosyst Eng [Internet]. 2014 Feb 15;37(2):261–267. Available from: http://link.springer.com/10.1007/s00449-013-0994-3.

78 Mishra A, Kumari M, Pandey S, Chaudhry V, Gupta KC, Nautiyal CS. Biocatalytic and antimicrobial activities of gold nanoparticles synthesized by Trichoderma sp. Bioresour Technol. 2014;166:235–242.

79 Ramos MM, dos S. Morais E, da S. Sena I, Lima AL, de Oliveira FR, de Freitas CM, et al. Silver nanoparticle from whole cells of the fungi Trichoderma spp. isolated from Brazilian Amazon. Biotechnol Lett. 2020;42:833–843.

80 Shah M, Fawcett D, Sharma S, Tripathy SK, Poinern GEJ. Green synthesis of metallic nanoparticles via biological entities. Materials. 2015;8:7278–7308.

81 Waghmare SR, Mulla MN, Marathe SR, Sonawane KD. Ecofriendly production of silver nanoparticles using *Candida utilis* and its mechanistic action against pathogenic microorganisms. 3 Biotech. 2015;5:33–38.

82 Apte M, Sambre D, Gaikawad S, Joshi S, Bankar A, Kumar AR, et al. Psychrotrophic yeast Yarrowia lipolytica NCYC 789 mediates the synthesis of antimicrobial silver nanoparticles via cell-associated melanin. AMB Express. 2013;3(1):32.

83 Eugenio M, Müller N, Frasés S, Almeida-Paes R, Lima LMTR, Lemgruber L, et al. Yeast-derived biosynthesis of silver/silver chloride nanoparticles and their antiproliferative activity against bacteria. RSC Adv. 2016;6:9893–9904.

84 Elahian F, Reiisi S, Shahidi A, Mirzaei SA. High-throughput bioaccumulation, biotransformation, and production of silver and selenium nanoparticles using genetically engineered Pichia pastoris. Nanomed Nanotechnol Biol Med

[Internet]. 2017 Apr;13(3):853–861. Available from: https://linkinghub.elsevier.com/retrieve/pii/S1549963416301794.

85 Fawcett D, Verduin JJ, Shah M, Sharma SB, Poinern GEJ. A review of current research into the Biogenic synthesis of metal and metal oxide nanoparticles via marine algae and seagrasses. J Nanosci. 2017;2017:1–15.

86 Stalin Dhas T, Ganesh Kumar V, Karthick V, Jini Angel K, Govindaraju K. Facile synthesis of silver chloride nanoparticles using marine alga and its antibacterial efficacy. Spectrochim Acta - Part A Mol Biomol Spectrosc. 2014;120:416–420.

87 da Silva FV, ConzFerreira ME, Lima LMTR, Frasés S, de Souza W, Sant'Anna C. Green production of microalgae-based silver chloride nanoparticles with antimicrobial activity against pathogenic bacteria. Enzym Microb Technol [Internet]. 2017 Feb;97:114–121. Available from: https://linkinghub.elsevier.com/retrieve/pii/S014102291630223X.

88 Senapati S, Syed A, Moeez S, Kumar A, Ahmad A. Intracellular synthesis of gold nanoparticles using alga Tetraselmis kochinensis. Mater Lett [Internet]. 2012 Jul;79:116–118. Available from: https://linkinghub.elsevier.com/retrieve/pii/S0167577X12004879.

89 Singh M, Kalaivani R, Manikandan S, Sangeetha N, Kumaraguru AK. Facile green synthesis of variable metallic gold nanoparticle using Padina gymnospora, a brown marine macroalga. Appl Nanosci [Internet]. 2013 Apr 15;3(2):145–151. Available from: http://link.springer.com/10.1007/s13204-012-0115-7.

90 Rajeshkumar S, Kannan C, Annadurai G. Green synthesis of silver nanoparticles using marine brown Algae turbinaria conoides and its antibacterial activity. Int J Pharm Bio Sci. 2012;3(4):502–510.

91 Kobayashi M, Tomita S, Sawada K, Shiba K, Yanagi H, Yamashita I, et al. Chiral meta-molecules consisting of gold nanoparticles and genetically engineered tobacco mosaic virus, 24856. Opt Express [Internet]. 2012 Oct 22;20(22). Available from: https://opg.optica.org/oe/abstract.cfm?uri=oe-20-22-24856.

92 Mao C, Flynn CE, Hayhurst A, Sweeney R, Qi J, Georgiou G, et al. Viral assembly of oriented quantum dot nanowires. Proc Natl Acad Sci USA. 2003;100:6946–6951.

93 Cao J, Guenther RH, Sit TL, Opperman CH, Lommel SA, Willoughby JA. Loading and release mechanism of red clover necrotic mosaic virus derived plant viral nanoparticles for drug delivery of doxorubicin. Small. 2014;10:5126–5136.

94 Czapar AE, Zheng YR, Riddell IA, Shukla S, Awuah SG, Lippard SJ, et al. Tobacco mosaic virus delivery of Phenanthriplatin for cancer therapy. ACS Nano. 2016;10:4119–4126.

95 Thangavelu RM, Ganapathy R, Ramasamy P, Krishnan K. Fabrication of virus metal hybrid nanomaterials: an ideal reference for bio semiconductor. Arab J Chem. 2020;13:2750–2765.

96 Kapoor RT, Salvadori MR, Rafatullah M, Siddiqui MR, Khan MA, Alshareef SA. Exploration of microbial factories for synthesis of nanoparticles – a sustainable approach for bioremediation of environmental contaminants. Front Microbiol [Internet]. 2021 Jun 4:12. Available from: https://www.frontiersin.org/articles/10.3389/fmicb.2021.658294/full.

97 Bao Y, He J, Song K, Guo J, Zhou X, Liu S. Plant-extract-mediated synthesis of metal nanoparticles. J Chem. 2021;2021:1–14.

98 Kirupagaran R, Saritha A, Bhuvaneswari S. Leucas lavandulifolia Sm. and their application. J Nanosci Technol. 2016;2(5):224–226.

99 Jayappa MD, Ramaiah CK, Kumar MAP, Suresh D, Prabhu A, Devasya RP, et al. Green synthesis of zinc oxide nanoparticles from the leaf, stem and in vitro grown callus of Mussaenda frondosa L.: characterization and their applications. Appl Nanosci. 2020;10(8):3057–3074.

100 Gaikwad D, Jadhav N. A review on biogenic properties of stem bark of Terminalia Arjuna: an update. Asian J Pharm Clin Res. 2018;11(8): 35–39.

101 Vijayakumar M, Priya K, Nancy FT, Noorlidah A, Ahmed ABA. Biosynthesis, characterisation and anti-bacterial effect of plant-mediated silver nanoparticles using Artemisia nilagirica. Ind Crop Prod. 2013;41(1):235–240.

102 Magdum AB, Waghmode RS, Shinde KV, Mane MP, Kamble MV, Kamble RS, et al. Biogenic synthesis of silver nanoparticles from leaves extract of Decaschistia trilobata an endemic shrub and its application as antioxidant, antibacterial, anti-inflammatory and dye reduction. Catal Commun. 2024;(187, November 2023):106865.

103 Tao J, Fu Z, Dong C, Wang X, Yang X. Green synthesis and characterization of monodisperse gold nanoparticles using aloe vera leaf extract. Xiyou Jinshu Cailiao Yu Gongcheng/Rare Met Mater Eng. 2019;48(11):3470–3475.

104 Mejía-Méndez JL, Sánchez-Ante G, Cerro-López M, Minutti-Calva Y, Navarro-López DE, Lozada-Ramírez JD, et al. Green synthesis of silver nanoparticles with extracts from Kalanchoe fedtschenkoi: characterization and bioactivities. Biomolecules. 2024;14(7):782.

105 Maheshwaran G, Selva Muneeswari R, Nivedhitha Bharathi A, Krishna Kumar M, Sudhahar S. Eco-friendly synthesis of lanthanum oxide nanoparticles by Eucalyptus globulus leaf extracts for effective biomedical applications. Mater Lett. 2021;283(1):128799.

106 Mehmood A, Murtaza G, Bhatti TM, Kausar R. Phyto-mediated synthesis of silver nanoparticles from Melia azedarach L. leaf extract: characterization and antibacterial activity. Arab J Chem. 2017;10(November):S3048–S3053.

107 Antonio-Pérez A, Durán-Armenta LF, Pérez-Loredo MG, Torres-Huerta AL. Biosynthesis of copper nanoparticles with medicinal plants extracts: from extraction methods to applications. Micromachines. 2023;14(10):1–32.

108 Sreelakshmy V, Mk D. Green synthesis of silver nanoparticles from Glycyrrhiza glabra root extract for the treatment of gastric ulcer. J Dev Drugs. 2016;5(2).

109 Ahmed S, Annu CSA, Ikram S. A review on biogenic synthesis of ZnO nanoparticles using plant extracts and microbes: a prospect towards green chemistry. J Photochem Photobiol B Biol. 2017;166:272–284.

110 Mathew S, Victório CP, Sidhi MSJ, Baby BT. Biosynthesis of silver nanoparticle using flowers of Calotropis gigantea (L.) W.T. Aiton and activity against pathogenic bacteria. Arab J Chem. 2020;13(12):9139–9144.

111 Rajesh KM, Ajitha B, Reddy YAK, Suneetha Y, Reddy PS. Assisted green synthesis of copper nanoparticles using Syzygium aromaticum bud extract: physical, optical and antimicrobial properties. Optik (Stuttg). 2018;154: 593–600.

112 Das RK, Gogoi N, Bora U. Green synthesis of gold nanoparticles using Nyctanthes arbortristis flower extract. Bioprocess Biosyst Eng. 2011;34(5): 615–619.

113 Alghuthaymi MA, Rajkuberan C, Santhiya T, Krejcar O, Kuča K, Periakaruppan R, et al. Green synthesis of gold nanoparticles using polianthes tuberosa l. Floral extract. Plants [Internet]. 2021;10(11):1–12. Available from: https://doi.org/10.3390/plants10112370.

114 Naikoo GA, Mustaqeem M, Hassan IU, Awan T, Arshad F, Salim H, et al. Bioinspired and green synthesis of nanoparticles from plant extracts with antiviral and antimicrobial properties: a critical review. J Saudi Chem Soc. 2021;25(9):101304.

115 Ramesh PS, Kokila T, Geetha D. Plant mediated green synthesis and antibacterial activity of silver nanoparticles using Emblica officinalis fruit extract. Spectrochim Acta - Part A Mol Biomol Spectrosc. 2015;142:339–343.

116 Yugandhar P, Vasavi T, Jayavardhana Rao Y, Uma Maheswari Devi P, Narasimha G, Savithramma N. Cost effective, green synthesis of copper oxide nanoparticles using fruit extract of Syzygium alternifolium (Wt.) Walp., characterization and evaluation of antiviral activity. J Clust Sci. 2018;29(4): 743–755.

117 Lacerda JS, Martinez HEP, Pedrosa AW, Clemente JM, Santos RHS, Oliveira GL, et al. Importance of zinc for Arabica coffee and its effects on the chemical composition of raw grain and beverage quality. Crop Sci. 2018;58(3):1360–1370.

118 Savekar PL, Nadaf SJ, Killedar SG, Kumbar VM, Hoskeri JH, Bhagwat DA, et al. Citric acid cross-linked pomegranate peel extract-loaded pH-responsive β-cyclodextrin/carboxymethyl tapioca starch hydrogel film for diabetic wound healing. Int J Biol Macromol. 2024;274(P1):133366.

119 Metuku RP, Pabba S, Burra S, Hima Bindu NSVSSSL, Gudikandula K, Singara Charya MA. Biosynthesis of silver nanoparticles from Schizophyllum radiatum HE 863742.1: their characterization and antimicrobial activity. 3 Biotech. 2014;4(3):227–234.

120 Jacob JM, Rajan R, Tom TC, Kumar VS, Kurup GG, Shanmuganathan R, et al. Biogenic design of ZnS quantum dots – insights into their in-vitro cytotoxicity, photocatalysis and biosensing properties. Ceram Int. 2019;45(18):24193–24201.

121 Arshad H, Sami MA, Sadaf S, Hassan U. *Salvadora persica* mediated synthesis of silver nanoparticles and their antimicrobial efficacy. Sci Rep. 2021;11:5996.

122 Chen Y, Chen Z, Yang D, Zhu L, Liang Z, Pang Y, et al. Novel microbial palladium nanoparticles with a high photothermal effect for antibacterial applications. ACS Omega [Internet]. 2023 Jan 10;8(1):1534–1541. Available from: https://pubs.acs.org/doi/10.1021/acsomega.2c07037.

123 Medina-Cruz D, Truong LB, Sotelo E, Martínez L, González MU, Huttel Y, et al. Bacterial-mediated selenium nanoparticles as highly selective antimicrobial agents with anticancer properties. RSC Sustain. 2023;1:1436–1448.

124 Abdelgadir A, Adnan M, Patel M, Saxena J, Alam MJ, Alshahrani MM, et al. Probiotic Lactobacillus salivarius mediated synthesis of silver nanoparticles (AgNPs-LS): a sustainable approach and multifaceted biomedical application. Heliyon [Internet]. 2024;10(18):e37987. Available from: https://doi.org/10.1016/j.heliyon.2024.e37987.

125 Hassan SED, Salem SS, Fouda A, Awad MA, El-Gamal MS, Abdo AM. New approach for antimicrobial activity and bio-control of various pathogens by biosynthesized copper nanoparticles using endophytic actinomycetes. J Radiat Res Appl Sci. 2018;11:262–270.

126 Ahmad T, Wani IA, Manzoor N, Ahmed J, Asiri AM. Biosynthesis, structural characterization and antimicrobial activity of gold and silver nanoparticles. Colloids Surf B: Biointerfaces [Internet]. 2013 Jul;107:227–234. Available from: https://linkinghub.elsevier.com/retrieve/pii/S0927776513001136.

127 Gan L, Zhang S, Zhang Y, He S, Tian Y. Biosynthesis, characterization and antimicrobial activity of silver nanoparticles by a halotolerant bacillus endophyticus SCU-L. Prep Biochem Biotechnol [Internet]. 2018 Aug 9;48(7):582–588. Available from: https://www.tandfonline.com/doi/full/10.1080/10826068.2018.1476880.

128 Buszewski B, Railean-Plugaru V, Pomastowski P, Rafińska K, Szultka-Mlynska M, Golinska P, et al. Antimicrobial activity of biosilver nanoparticles produced by a novel Streptacidiphilus durhamensis strain. J Microbiol Immunol Infect. 2018;51:45–54.

129 Moghaddam AB, Moniri M, Azizi S, Rahim RA, Bin AA, Saad WZ, et al. Biosynthesis of ZnO nanoparticles by a new *Pichia kudriavzevii* yeast strain and evaluation of their antimicrobial and antioxidant activities. Molecules. 2017;22(6):872.

130 Bhuyar P, Rahim MHA, Sundararaju S, Ramaraj R, Maniam GP, Govindan N. Synthesis of silver nanoparticles using marine macroalgae *Padina* sp. and its antibacterial activity towards pathogenic bacteria. Beni-Suef Univ J Basic Appl Sci. 2020;9:3.

131 Naghizadeh A, Mizwari ZM, Ghoreishi SM, Lashgari S, Mortazavi-Derazkola S, Rezaie B. Biogenic and eco-benign synthesis of silver nanoparticles using jujube core extract and its performance in catalytic and pharmaceutical applications: removal of industrial contaminants and in-vitro antibacterial and anticancer activities. Environ Technol Innov. 2021;23:101560.

132 Paul Das M, Rebecca Livingstone J, Veluswamy P, Das J. Exploration of Wedelia chinensis leaf-assisted silver nanoparticles for antioxidant, antibacterial and in vitro cytotoxic applications. J Food Drug Anal. 2018;26:917–925.

133 Sharma M, Monika TP, Saini RV, Kumar R, Torino E. Unveiling antimicrobial and anticancerous behavior of AuNPs and AgNPs moderated by rhizome extracts of *Curcuma longa* from diverse altitudes of Himalaya. Sci Rep. 2020;10:10934.

134 Alharbi NS, Alsubhi NS. Green synthesis and anticancer activity of silver nanoparticles prepared using fruit extract of Azadirachta indica. J Radiat Res Appl Sci. 2022;15:335–345.

135 Tiki YL, Tolesa LD, Tiwikrama AH, Chala TF. Ginger (Zingiber officinale)-mediated green synthesis of silver-doped tin oxide nanoparticles and evaluation of its antimicrobial activity. ACS. Omega. 2024;9:11443–11452.

136 Dhanasekaran D, Thangaraj R. Evaluation of larvicidal activity of biogenic nanoparticles against filariasis causing Culex mosquito vector. Asian Pacific J Trop Dis. 2013;3:174–179.

137 Babu V, Arokiyaraj S, Sakthi Sri SP, George M, Ragavan RM, Dharmalingam D, et al. Antibacterial, antioxidant, Larvicidal and anticancer activities of silver nanoparticles synthesized using extracts from fruits of lagerstroemia speciose and flowers of Couroupita guianensis. Molecules [Internet]. 2022 Nov 12;27(22):7792. Available from: https://www.mdpi.com/1420-3049/27/22/7792.

138 Abdelhakim HK, El-Sayed ER, Rashidi FB. Biosynthesis of zinc oxide nanoparticles with antimicrobial, anticancer, antioxidant and photocatalytic activities by the endophytic Alternaria tenuissima. J Appl Microbiol. 2020;128:1634–1646.

139 Saratale RG, Shin HS, Kumar G, Benelli G, Kim DS, Saratale GD. Exploiting antidiabetic activity of silver nanoparticles synthesized using *punica granatum* leaves and anticancer potential against human liver cancer cells (HepG2). Artif Cells Nanomed Biotechnol. 2018;46:211–222.

140 Mohanta YK, Biswas K, Mishra AK, Patra B, Mishra B, Panda J, et al. Amelioration of gold nanoparticles mediated through Ocimum oil extracts induces reactive oxygen species and mitochondrial instability against MCF-7 breast carcinoma. RSC Adv. 2024;14(38):27816–27830.

141 Jeyaraj S, Mary SA. Bio-fabrication of zinc oxide nanoparticles from Brassica juncea (L.) Czern extract: biological activity and DNA binding capability. Results Chem [Internet]. 2024;11(August):101747. Available from: https://doi.org/10.1016/j.rechem.2024.101747.

142 Chandrasekaran S, Anbazhagan V. Green synthesis of ZnO and V-doped ZnO nanoparticles using Vinca rosea plant leaf for biomedical applications. Appl Biochem Biotechnol. 2024;196:50–67.

143 Mongy Y, Shalaby T. Green synthesis of zinc oxide nanoparticles using Rhus coriaria extract and their anticancer activity against triple-negative breast cancer cells. Sci Rep [Internet]. 2024;14(1):1–15. Available from: https://doi.org/10.1038/s41598-024-63258-7.

144 Le DHT, Lee KL, Shukla S, Commandeur U, Steinmetz NF. Potato virus X, a filamentous plant viral nanoparticle for doxorubicin delivery in cancer therapy. Nanoscale. 2017;9:2348–2357.

145 Luo Y, Xu D, Gao X, Xiong J, Jiang B, Zhang Y, et al. Nanoparticles conjugated with bacteria targeting tumors for precision imaging and therapy. Biochem Biophys Res Commun. 2019;514:1147–1153.

146 Moreno VM, Álvarez E, Izquierdo-Barba I, Baeza A, Serrano-López J, Vallet-Regí M. Bacteria as nanoparticles carrier for enhancing penetration in a Tumoral matrix model. Adv Mater Interfaces. 2020;7(11):1901942.

147 Aisha AFA, Abdulmajid AMS, Ismail Z, Alrokayan SA, Abu-Salah KM. Development of polymeric nanoparticles of *Garcinia mangostana* Xanthones in Eudragit RL100/RS100 for anti-colon cancer drug delivery. J Nanomater. 2015;7:1–12.

148 Younis IY, El-Hawary SS, Eldahshan OA, Abdel-Aziz MM, Ali ZY. Green synthesis of magnesium nanoparticles mediated from Rosa floribunda charisma extract and its antioxidant, antiaging and antibiofilm activities. Sci Rep [Internet]. 2021 Aug 19;11(1):16868. Available from: https://www.nature.com/articles/s41598-021-96377-6.

149 Ahmed E, Kalathil S, Shi L, Alharbi O, Wang P. Synthesis of ultra-small platinum, palladium and gold nanoparticles by Shewanella loihica PV-4 electrochemically active biofilms and their enhanced catalytic activities. J Saudi Chem Soc. 2018;22:919–929.

150 Ranjitha VR, Rai VR. Actinomycetes mediated synthesis of gold nanoparticles from the culture supernatant of *Streptomyces griseoruber* with special reference to catalytic activity. 3 Biotech. 2017;7:299.

151 Tripathi RM, Bhadwal AS, Gupta RK, Singh P, Shrivastav A, Shrivastav BR. ZnO nanoflowers: novel biogenic synthesis and enhanced photocatalytic activity. J Photochem Photobiol B Biol. 2014;141:288–295.

152 Kalpana VN, Kataru BAS, Sravani N, Vigneshwari T, Panneerselvam A, Devi RV. Biosynthesis of zinc oxide nanoparticles using culture filtrates of aspergillus Niger: antimicrobial textiles and dye degradation studies. OpenNano. 2018;3:48–55.

153 Shankar A, Kumar V, Kaushik NK, Kumar A, Malik V, Singh D, et al. Sporotrichum thermophile culture extract-mediated greener synthesis of silver nanoparticles: eco-friendly functional group transformation and anti-bacterial study. Curr Res Green Sustain Chem [Internet]. 2020 Jun;3:100029. Available from: https://linkinghub.elsevier.com/retrieve/pii/S2666086520300321.

154 Alhamdi HW, Majani SS, Dilliraj S, Seal A, Nagesh N, Shivamallu C, et al. Exploring the anticancer potency and photocatalytic efficiency of bio-derived CuO nanoparticles using Moringa oleifera leaf extract. Results Chem. 2024;7:101430.

155 Anand K, Gengan RM, Phulukdaree A, Chuturgoon A. Agroforestry waste Moringa oleifera petals mediated green synthesis of gold nanoparticles and their anti-cancer and catalytic activity. J Ind Eng Chem [Internet]. 2015 Jan;21:1105–1111. Available from: https://linkinghub.elsevier.com/retrieve/pii/S1226086X14002706.

156 Siddique K, Shahid M, Shahzad T, Mahmood F, Nadeem H, M S u R, et al. Comparative efficacy of biogenic zinc oxide nanoparticles synthesized by Pseudochrobactrum sp. C5 and chemically synthesized zinc oxide nanoparticles for catalytic degradation of dyes and wastewater treatment. Environ Sci Pollut Res. 2021;28:28307–28318.

2

Biogenic Synthesis of Nanoparticles by Plant Extract and Green Nanotherapeutics

L. Ronibala Singha[1], Punamjyoti Das[2], Abirami Raja[3], Pallabi Borah[4], Rachna Sharma[1], Rajlakshmi Gogoi[1], and Malay K. Das[1]

[1] *Drug Delivery Research Laboratory, Department of Pharmaceutical Sciences, Dibrugarh University, Dibrugarh, Assam, India*
[2] *School of Pharmaceutical Sciences, Girijananda Chowdhury University-Tezpur Campus, Tezpur, India*
[3] *C L Baid Metha College of Pharmacy, OMR Thoraipakkam, Chennai, Tamilnadu, India*
[4] *Centre for Biotechnology and Bioinformatics, Dibrugarh University, Dibrugarh, India*

2.1 Introduction

The multitude of research available in the literature appraising the application of nanomaterials is evidence of its indispensable purpose in different fields, particularly in the field of medicine and pharmaceuticals. Several synthesis strategies that have been adopted to fabricate nanocarriers encompass the conventional physical and chemical methods (Figure 2.1) wherein extremely high temperature, instruments with complicated mechanisms, high energy consumption, toxic chemicals, etc., are involved [1]. These factors have raised concerns among researchers and environmentalists regarding the negative effects on the environment, thereby forming a need to look for options that are not only safe and effective but also very gentle to our environment. Here, the biogenic nanoparticles come to the rescue, as these are nanoparticles that are synthesized from biological sources such as fungi, viruses, algae, bacteria, and plants, including marine biologics (Figure 2.2) [2]. The biogenic mode of nanoparticles synthesis utilizes the principle of green chemistry, which involves processes where sustainability toward the environment is given priority. Green or biogenic synthesis of nanoparticles is considered superior to that of conventional methods as these processes avoid the usage of toxic chemicals and solvents and are economical, biocompatible, energy saving,

Biogenic Synthesis of Nanoparticles and Sustainable Nanotherapeutics,
First Edition. Edited by Malay K. Das.

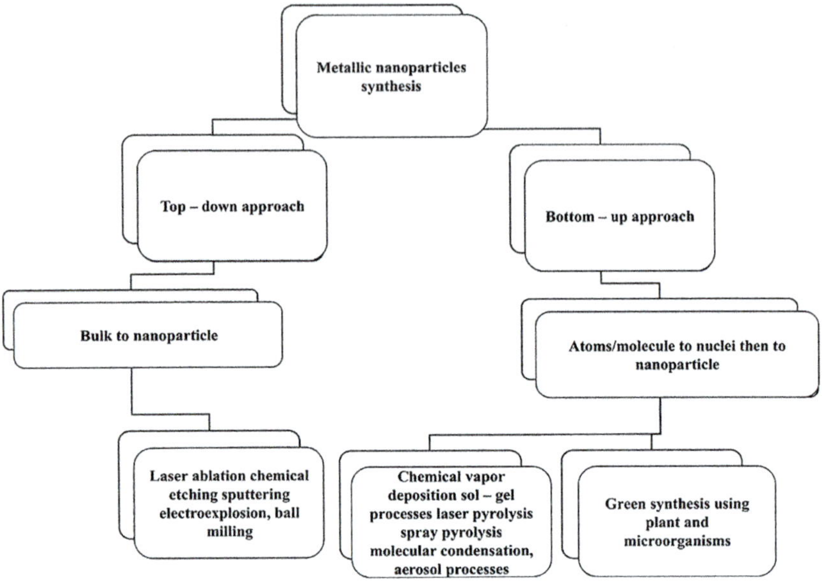

Figure 2.1 Conventional methods of nanoparticle synthesis.

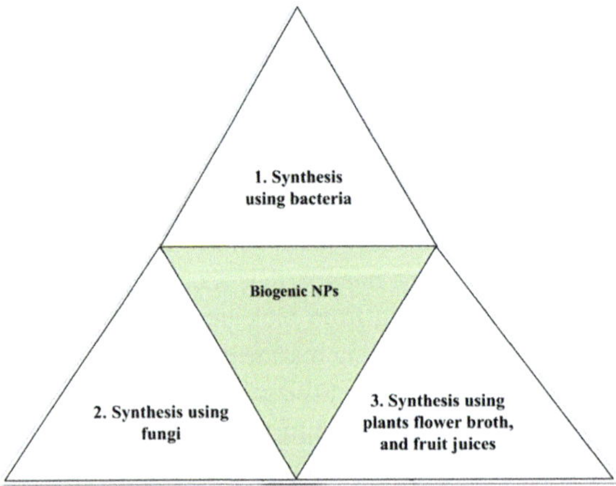

Figure 2.2 Different sources of biogenic nanoparticle synthesis.

thus providing easy scale-up [3]. Among the diverse range of nanoparticles, metal, metalloid, and metal oxide nanoparticles have gained undivided attention for biogenic synthesis as they are blessed with unique catalytic, electrical, and optical properties, as they are extremely small in size (1–100 nm) and have a larger surface area-to-volume ratio [4]. The metal-based and metal oxide nanoparticles, which have found profound biomedical and pharmaceutical applications, include silver nanoparticles (AgNPs), gold nanoparticles (AuNPs), selenium nanoparticles (SeNPs), zinc oxide nanoparticles (ZnONPs), copper oxide nanoparticles (CuONPs), iron oxide nanoparticles, aluminum oxide nanoparticles (Al_2O_3NPs), and manganese dioxide nanoparticles (MnO_2NPs) [5]. Regardless of the source for biogenic synthesis, whether it be plants or microorganisms, the basic reaction that occurs is redox, where the metabolites of these sources reduce the metal ions, using different enzymes/proteins present in them to form a "complex broth" of stable nanomaterials [6]. These specific enzymes and certain proteins act as reducing agents or capping agents and also function as chemical surfactants in conventional synthesis. These bio-capping agents not only impart stability to the obtained biogenic nanoparticles but are also biocompatible, safe, and cytocompatible [7]. The capping agents are also useful in surface functionalization of nanoparticles, thus providing the potential to tune their size, morphology, and shape and also preventing aggregation [8]. The phytocompounds of plants and plant parts function as reducing or capping agents in itself, for biogenic nanomaterial synthesis, thereby avoiding the necessity for additional chemical surfactants, whose main drawback is the difficulty in achieving complete removal and biodegradation, which further leads to accumulation in long-term usage. The plant extracts from different parts like leaves, roots, flowers, fruits, bark, seeds, and latex have the ability to convert the metals into a more stable form of nanoparticles within their cells. Also, the bioactive phytomolecules like alkaloids, terpenoids, phenols, flavonoids, tannins, saponins, and polysaccharides have the potential to reduce metals into stable biogenic nanoparticles [9]. Easy availability of plant materials, inexpensive, broad range of raw materials from a single plant, and easy scale-up to mass production are some of the advantages provided by plant sources in terms of biogenic nanomaterial synthesis as compared to other biological sources [10].

In this chapter, the weightage has been given to the applicability of various metal, metalloid, and metal oxide nanoparticles that are biogenetically synthesized from different plant extracts, their basic chemistry, and preparation. A few challenges that occur during the process have also been briefly covered.

2.2 Basic Mechanism Involved in Biogenic Synthesis of Nanoparticles from Plant Source

Plants are capable of storing metal as a depot which can be purposed to create nanoparticles inside their cells. Plants being nonpathogenic also serve as a major advantage by providing critical chemical and physical attributes to the biogenic nanoparticles production. Not only does plant-based biogenic synthesis result in well-characterized nanoparticle production, but it also provides unique uses for them [11]. There is ample amount of evidence suggesting the plant's ability to absorb, hyperaccumulate, and degrade inorganic metal and their oxide ions available in its vicinity. Depending on this, their size, form, and absorbance at particular wavelengths are determined. Different functional groups of plant extracts, such as OH, CXH_2, NH_2, CN, and COOH, react electrostatically with metallic ion (M^+), followed by reduction of M^+ to M^0 [12].

Plant-mediated biogenic nanoparticles can be synthesized in three ways [13]:

1) Intracellular (inside the cellular structure of plants)
2) Extracellular (employing phyto-extracts in the extracellular region)
3) Individually (employing isolated photoactive).

However, in all of the three above listed ways, the bio-reduction of plant-based biogenic metal nanoparticles occurs at three phases [14]:

1) Activation phase – reduction, followed by nucleation of metal ions.
2) Growth phase – consolidation of surrounding nanoparticles to form a larger entity and an increase in thermodynamic stability of metal NPs.
3) Termination phase – formation of the final shape of nanoparticles.

2.3 Biogenic Synthesis of Nanoparticles by Plant Source

2.3.1 Biogenic Synthesis of Gold Nanoparticles by Plant Extract

Gold nanoparticles (AuNPs) are small particles of gold, typically starting from 1 to 100 nm, exhibiting unique electronic, optical, and chemical properties [15]. Their exceptional attributes, such as surface plasmon resonance, biocompatibility, and high surface area, deem them ideal for their applicability in catalysis, drug delivery, biosensing, and imaging. For instance, AuNPs synthesized using citrate reduction are widely used in cancer mitigation due to their ability to target tumor cells while reducing damage to healthy tissue [16]. The biogenic synthesis of gold nanoparticles involves the reduction of gold ions (Au^{3+}), typically from chloroauric acid ($HAuCl_4$) to elemental gold (Au^0). This eco-friendly process utilizes bioactive compounds found in plant extracts, such as polyphenols, terpenoids, and flavonoids, as reducing agents to convert ions and stabilizing agents to prevent nanoparticle agglomeration [17]

$$HAuCl_4 + \text{Phytochemicals} \rightarrow Au^0 + \text{By-products}$$

Gold nanoparticles are formed through the reduction of gold ions (Au^{3+}) from precursors like chloroauric acid ($HAuCl_4$) into elemental gold (Au^0). The process involves three main steps:

1) Reduction: reducing agents (e.g. sodium citrate, plant phytochemicals) donate electrons to Au^{3+}, converting it into Au^0.
2) Nucleation: reduced gold atoms aggregate to form nuclei.
3) Growth and Stabilization: gold atoms deposit on nuclei, forming nanoparticles. Stabilizing agents (e.g. polyphenols) prevent agglomeration, ensuring stability.

The preparation of biogenic AuNPs involves the collection of plant parts initially after they are extracted using appropriate solvents and extraction methods, followed by isolation of bioactive compounds, where required. Then, a gold solution is prepared by preparing a solution of chloroauric acid ($HAuCl_4$) at a specific concentration. This is followed by mixing the plant extract with the gold solution at controlled pH, temperature, and stirring speed. The final indication suggesting the formation of biogenic AuNPs is usually when the shift in color occurs to ruby red from that of pale yellow. Purification by centrifugation to separate the formed nanoparticles is the last step, which is followed by drying and characterization of nanoparticles by different analytical techniques such as Fourier-transform infrared spectroscopy (FTIR), X-ray diffraction (XRD), Ultraviolet - Visible (UV)–Vis spectroscopy, transmission electron microscopy (TEM), etc. [18].

Gold nanoparticles (AuNPs) demonstrate a diverse range of mechanisms tailored to their specific applications, making them highly versatile in various scientific and medical fields. For example, a study reported by Kepsutlu et al. investigated the cellular interaction ability of AuNPs are internalized by cells through endocytosis, where they localize to specific organelles, enabling precise therapeutic interventions or diagnostic imaging [19]. Another instance is where a study reported by Kowsalya et al. demonstrated the reactive oxygen species (ROS) production of AuNPs, where they generated oxidative stress to induce ROS generation, which disrupts cellular components like lipids, proteins, and deoxyribonucleic acid (DNA). This oxidative imbalance leads to cellular damage and programmed cell death (apoptosis), a mechanism particularly effective in targeting cancer cells due to their heightened susceptibility to ROS [20]. AuNPs also exhibited photothermal properties, enabling them to absorb light energy and conversion of the same into heat when exposed to laser irradiation. This localized heating effect effectively destroys cancerous tissues by inducing thermal ablation, causing irreversible damage to cancer cells, while sparing surrounding healthy tissues [21]. A study by Wang et al. reported biosynthesis of biogenic AuNPs from the source, *Scutellaria barbata,* and investigated its antineoplastic potential on pancreatic cancer cell lines [22]. Dorosti and Jamshidi utilized the *Dracocephalum kotschyi* leaf extract to synthesize AuNPs to evaluate their anti-cholinesterase and anticancer activity, where the resultant AuNPs exhibited a significant inhibitory effect against BChE and AChE, dose-dependent cytotoxicity of HeLa and K562 cell lines [23]. Badeggi et al. investigated the antidiabetic property of biogenic AuNPS synthesized from the procyanidins present in *Leucosidea sericea* by enzymatic activity [24]. Ateeq et al. reported the hepatoprotection and urease inhibition potential of biosynthesized garlic extract-based AuNPs and found it to exhibit better potential in the prevention of CCl4-induced acute hepatic injury than Silymarin, the positive control group [25].

2.3.2 Biogenic Synthesis of Silver Nanoparticles by Plant Extract

Silver nanoparticles (AgNPs) stand out as one of the most promising innovations in the nanotechnology sector among various metallic nanoparticles. Current research in nanotechnology focuses on developing reliable methods for synthesizing AgNPs [26]. These nanoparticles have diverse applications due to their unique magnetic, optical, and electrical properties. They can be utilized in antiviral, antibacterial, and antifungal products, as well as in composite fibers, biosensor materials, cosmetics, food industry applications, and electrical parts. AgNPs are also abundantly found in everyday products that come into direct contact with humans, such as detergents, shampoos, toothpaste, soaps, and multiple cosmetic-based products [27–29]. In the biomedical domain, AgNPs serve as antibacterial [30], antifungal [31], anti-inflammatory [32], antiviral [33], and antidiabetic [34] agents. Reported studies in the literature have explored their prospects in cancer diagnosis and management, as well as their use as bioactive carriers through passive or active mechanisms [35]. The antibacterial potential of silver has been recorded and utilized since ancient times, and it is currently used to circumvent bacterial growth in various applications, including burn management, catheters, and dental work. It is well-established that silver ions and silver-based nanomaterials exhibit very dominant biocidal effects and are found to be strongly toxic toward microorganisms [36].

AgNPs are formed when silver ions (Ag^+) reduce into elemental silver (Ag^0), followed by nucleation and growth. The process involves

- **Reduction:** a reducing agent (sodium citrate or borohydride) donates electrons to silver ions, forming silver atoms (Ag^0) [37].

$$Ag^+ + e^- \rightarrow Ag^0$$

- **Nucleation:** silver atoms cluster together to form tiny nuclei, which serve as seeds for nanoparticle growth [38].
- **Growth and Stabilization:** additional silver atoms deposit onto the nuclei, growing them into nanoparticles. Stabilizing agents (e.g. citrate or polyvinylpyrrolidone [PVP]) prevent aggregation and control size and shape [39].

The basic preparation method of biogenic AgNPs is similar to that of gold nanoparticles, except that the indication of the termination phase is a change in color to reddish brown [40].

A study by He et al. synthesized quasi-spherical-shaped and well-dispersed AgNPs, bearing a size of 12.6 nm from the seed extract of *Alpinia katsumadai* using the sonication process in deionized water. The prepared biogenic AgNPs were investigated for cytotoxicity, antioxidative, and antibacterial properties [41]. Another study by Torres-Martínez et al. biosynthesized AgNPs from the aqueous extract of leaves belonging to *Mentha spicata* and investigated for *in vitro* anticancer activity, where a decrease in cell viability and formation of ROS on Chang and SiHa cells were observed, thus producing *in vitro* anticancer and cytotoxicity activities [42]. Sharma et al., synthesized and optimized biogenic AgNPs from aqueous extract of *Ocimum gratissimum* leaves and tested their antibacterial activity against *M. luteus, Bacillus subtilis, Klebsiella pneumoniae, Escherichia coli,* and *Staphylococcus aureus*. The determination of antibacterial activity was done using a standard disc diffusion assay against Gram-positive and Gram-negative bacteria, where the sensitivity of Gram-negative bacteria was found to be more toward the obtained biogenic AgNPs [43].

2.3.3 Biogenic Synthesis of Selenium Nanoparticles by Plant Extract

Selenium, a vital trace element commonly found in dietary sources in the form of selenoproteins (selenomethionine and selenocysteine), is abundantly found in plant sources like brazil nuts, almonds, sunflower seeds, grains, cereals, baked beans, asparagus, cruciferous vegetables, garlic, onions, bananas, and mushrooms. Selenium has qualities of both metallic and non-metallic, and therefore it is considered as a metalloid. After neutron activation, six stable isotopes and seven unstable isotopes of selenium occur naturally. The isotopes 75Se and 81Se have been reported for the purpose of measuring selenium quantitatively, facilitated by neutron activation analysis. The isotope 75Se has been found to have great application for biological experimentation as it has a long half-life of 120 days, relatively. Selenium has been revealed to exist in an amorphous state or in either of the three crystalline states (alpha-monoclinic, beta-monoclinic, and hexagonal form). Amorphous selenium, at temperatures beyond 230°C, is a free-flowing liquid wherein the reduction in temperature causes viscosity to increase. Certain biologically available agents have the ability to reduce selenium compounds to form their nanomaterial counterparts [44].

Selenium supplements are used to generally manage all ramifications caused by its deficiency. The commonly used forms are selenium-enriched yeast, selenium methionine, sodium selenate, sodium selenite, and protein or amino acid chelated form. Adults recommended dietary allowance is 55 mcg, and during pregnancy and lactation 60–70 mcg. Selenium is marketed as a combination form like multivitamin/mineral supplements, such as a combination of selenium with ingredients like vitamin E and or stand-alone selenium supplements [45].

Phytochemicals, such as carbohydrates, amino acids, flavonoids, alkaloids, proteins, saponins, and cardiac glycosides, and saponins present in plant extracts are responsible for the reduction of selenate. Sharma et al. reported a preparation method of very small (3–18 nm) sized and uniformly shaped Se nanoballs using dried extract of *Vitis vinifera*, which played an indispensable role in fabricating lignin-capped SeNPs. Selenous acid (H_2SeO_3) was the initial raw material used, which was then followed by the addition of this fruit extract. The hypothetical mechanism was the selenous acid, while reacting with the constituents of the fruit, selenium is separated after reduction, and while heating with the reducing agent lignin, the lignin-capped selenium nanoparticles were precipitated. The fabrication of biogenic SeNPs was found to have been completed when the colorless solution of selenous acid was converted to a brick red color. Then the formed biogenic SeNPs were characterized by atomic absorption spectroscopy (AAS), dynamic light scattering (DLS), and TEM techniques [46]. A study by Li S et al. reported the biosynthesis of SeNPs from *Capsicum annuum* L. extract by reacting it with selenium acid (H_2SeO_3) aqueous solution by reducing selenium ions (SeO_3^{2-}) at room temperature, with the protein (molecular weight of 30 kDa) from *C. annuum* L. This protein along with reducing SeO_3^{2-} ions to Se^0, also dictates nucleation, followed by growth of Se^0, and ultimately leads to the formation of amorphous selenium/protein composites, which is indicated by a final brick red color formation. The findings of this investigation open doors for preparing inorganic/organic biocomposites [47]. Cittrarasu et al. synthesized biogenic SeNPs by aqueous extract of tuber of *Ceropegia bulbosa* and characterized them by UV–Vis spectroscopy, FTIR, differential scanning calorimetry (DSC), DLS, zeta potential analysis, XRD, field emission scanning electron microscope with energy dispersive spectroscopy (FE-SEM-EDS) mapping, and HR-TEM. The anticancer efficacy of the prepared biogenic SeNPs against MDA-MB-231 has been investigated and found to downregulate the cancer cells significantly when compared to normal HBL100 cell lines. These biosynthesized SeNPs were also reported to have inhibitory effects on the multiplication of some clinical pathogens like *E. coli, B. subtilis*, and effective larvicidal activity against *Aedes albopictus* [48].

2.3.4 Biogenic Synthesis of Zinc Oxide Nanoparticles by Plant Extract

Zinc is one of the essential micronutrients taken up by plant roots and leaves from the soil to assist in their growth and certain other functions like the synthesis of proteins, activation of necessary enzymes responsible for metabolism of carbohydrates, regulation of plant hormones like auxin, and formation of pollen. The zinc is taken up by plant parts in Zn^{2+} ions or as complexes with organic acid chelates [49].

The first step in the synthesis of biogenic zinc oxide nanoparticles, ZnONPs by plant extract, is the same synthesis of other previously discussed nanoparticles, i.e. the collection of plant materials, cleaning, and processing, followed by extraction using appropriate solvents. After which, the resultant extract of plant biomass is made to react with zinc precursor solutions such as zinc acetate, zinc acetate dehydrate, zinc chloride, zinc sulfate, and zinc nitrate [50]. It has been observed that when nanoparticles are fabricated using zinc acetate as precursor, spherical-shaped biogenic ZnONPs are formed, whereas ZnONPs are formed with zinc nitrate is used as metal precursor. The optimum concentration of the zinc metal precursor is decided based on the pattern of the effects it displays on the size of NPs, which is that with increasing concentration of zinc precursor, the average diameter of the resulting ZnONPs increases to an extent after which it drops. Apart from varying types and concentrations of zinc precursor, different temperatures and pH also affect the nature of biogenic nanoparticles obtained [51]. After pouring the plant extract into the metal precursor solution, a precipitate forms, resulting from the reaction, after which it is calcined to finally obtain biogenic ZnONPs (white mass). The obtained white mass can be crystalline or amorphous or exist in a metastable crystallographic phase, depending on the effect of surface tension [52].

Matinise et al. prepared ZnONPs from the extract of *Moringa oleifera*, wherein the authors suggested that the main constituent of the extract, L-ascorbic acid, was converted to L-dehydro ascorbic acid with the aid of free radicals. Following this, the Zn-ascorbic acid complex resulted due to the electrostatic interaction between anions of L-hydro ascorbic acid and Zn^{2+} ions, which ultimately led to the formation of biogenic plant extract-based ZnONPs with the help of a calcination step at high temperature [53]. Another study by Sharma et al. reported the fabrication of ZnONPs of varying morphologies using milk latex from *Carica papaya* as a reducing agent. The formation of the latex-based nanoflowers was characterized with FTIR, TEM, high resolution transmission electron microscopy (HRTEM), and SEM. The fabricated nanoflowers were investigated for photocatalytic activity by degradation of Alizarin Red-S, courtesy of surface defects, small crystallite size, wide band gap, and ability to reduce electron–hole recombination.

Apart from this, the antibacterial activity of these nanoflowers was also exhibited against *Pseudomonas aeruginosa* and *S. aureus* [54].

2.3.5 Biogenic Synthesis of Copper Oxide Nanoparticles by Plant Extract

Recently metal oxide nanoparticles, like copper oxide nanoparticles (CuONPs), are gaining significant attention in therapeutic research for their strong antimicrobial, bactericidal, catalytic, and anti-carcinogenic properties. Copper oxide has found its application across various other fields like near-infrared filters, development of supercapacitors, semiconductors, and sensors, due to their electrical, optical, and magnetic properties [55]. Copper oxide is a member of the metal oxide family with various oxidation states: Cu0, CuI, CuII, and CuIII, and it exists mainly in two stable oxidation states: Cupric oxide (CuO) and Cuprous oxide (Cu_2O). Cuprous oxide is in the +1-oxidation state and exhibits a brownish-red color. It has a cubic crystal structure where each oxygen atom is surrounded by a tetrahedron of copper atoms. Cupric oxide is in the +2-oxidation state and is a black solid powder. CuO exists in a monoclinic crystal structure with copper ions bonded in square planar geometry to four equivalent oxygen atoms (Figure 2.3) [56]. This structure contributes to its anisotropic properties. Cu_2O forms due to partial oxidation of Cu metal, whereas CuO is formed as the final product when Cu undergoes complete oxidation [57]. CuO is more stable in comparison to Cu_2O. They have a 1.2–2.6 eV band gap energy, show better thermal stability, and are naturally occurring p-type semiconductors due to the presence of defects like copper vacancy [58].

For the synthesis of copper oxide NPs, the plant extract is mixed with metal salt (copper salt), and the reaction takes 1–3 hours at room temperature.

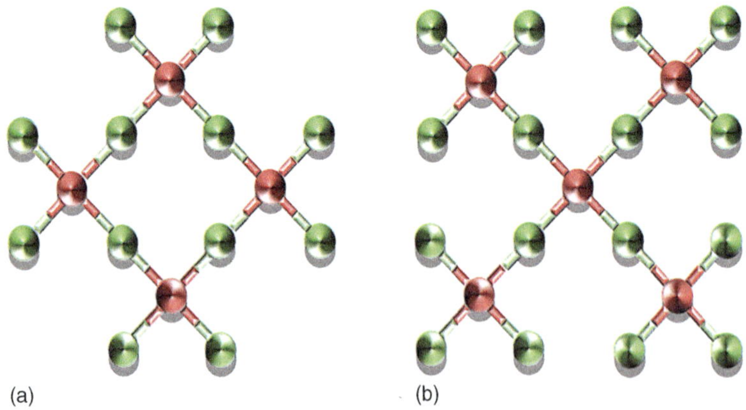

Figure 2.3 Structure of (a) CuO and (b) Cu_2O.

The phytoactives present in the plant extracts like flavonoids, terpenoids, phenolics, tannins, and alkaloids act as both reducing agents and stabilizing agents [59]. Jabeen et al. synthesized Aloe vera (Aloe barbadensis Miller) leaf extract for green synthesis of CuONPs. Here, copper (II) nitrate trihydrate acted as a precursor for the initiation of the reaction with the leaf extract under stirring, which is followed by the addition of sodium hydroxide solution. After processing, the formed CuONPs were found to be tapioca-pearl shaped as evident from the findings of SEM and XRD analysis. *in vitro* antibacterial effects against *P. aeruginosa, Listeria monocytogenes, K. pneumoniae,* and *Salmonella* typhi were established. The photocatalytic activity against the methylene blue dye was also demonstrated [60]. Another study by Asemani et al. reported fabrication of walnut (*Juglans regia* L.) leaf extract for biogenic synthesis of CuONPs. Copper nitrate was used as a precursor to react with walnut leaf extract to obtain biogenic CuONPs, which were evaluated for antioxidant and antibacterial activities against *E. coli, Salmonella, S. aureus,* and *Proteus* [61]. Further, the leaves of *Aegle marmelos* plant were used for the green synthesis of copper oxide nanoparticles. The prepared CuO NPs showed significant antibacterial and antifungal activities and photocatalytic degradation of organic dyes in the presence of sunlight [62]. The CuONPs synthesized with the leaf extract of *Abutilon indicum* have shown potential bactericidal activity against both Gram-negative (*E. coli* and *S. typhi*) and Gram-positive (*B. subtilis* and *S. aureus*) bacteria. It has also shown *in vitro* anticancer activity against human breast MDA-MB-231 and lung A549 cancer cell lines [63]. CuONPs have a range of scientific applications across multiple fields. They are highly effective against a wide range of pathogenic microbes, enabling their uses as antimicrobial agents as well as in food preservation and agricultural fields. As these CuONPs show biocidal and antiviral potential, they are also used as antimicrobial-based textile coatings and wound dressings. They are also known to show antibacterial and anticancer activities. They have also been used as gas sensors, batteries, catalysts, high-temperature superconductors, and for solar energy conversion tools [64].

2.3.6 Biogenic Synthesis of Manganese Oxide Nanoparticles by Plant Extract

Manganese (Mn) is one of the most abundantly available transition elements of the Earth, and it exists in many oxidation states (+2, +3, +4, +6, and +7). MnO, MnO_2, Mn_2O_3, Mn_3O_4, and Mn_5O_8 are some of the structures of manganese oxide, which can be useful in many fields. When these manganese oxides are synthesized by green technology using biological resources (plant source specifically), sustainable biogenic manganese oxide NPs are formed [65]. Jayandran et al. reported the development of MnONPs wherein the extract of lemon fruits acted as a reducing agent to reduce metal ions. Manganese acetate solution was used as a precursor solution to initiate the reaction with pure lemon fruit extract at a maintained pH and temperature. The confirmation of the metal ion reduction was denoted by the color change of the solution from pale green to yellow. This step was followed by the addition of fresh curcumin extract from turmeric (BSR-01 variety) to stabilize the manganese nanoparticles, which was confirmed when the color transitions from yellow to yellowish brown, which finally turned to reddish brown color after achieving complete stabilization. The eclipsed morphological structure of the formed NPs was confirmed through TEM and SEM. The fabricated manganese nanoparticles were also investigated for antibacterial and antifungal activities [66]. In a study by Lu et al., the biogenic synthesis of manganese dioxide nanoparticles from *Viola betonicifolia* leaves extract was used, wherein it was suggested that the secondary metabolites present in the leaves extract acted as capping and reducing agents. In this study, manganese acetate was also used as a precursor for reaction with leaf extract. After calcination, the obtained green manganese dioxide nanoparticles were characterized using suitable analytical techniques. The leaves extract-based MnO_2NPs were found to exhibit potent antioxidant, antifungal, antibacterial, and biofilm inhibition activities against the microbial specimens under test [67]. Hoseinpour et al. investigated the synthesis of MnO_2NPs from *Yucca gloriosa* leaf extract, which served as a reducing agent, with manganese acetate as precursor, and curcumin as a stabilizer. This was followed by the evaluation of fabricated biogenic NPs using UV–Vis, XRD, FTIR, and FESEM. The synthesized biogenic MnO_2NPs were investigated for photocatalytic activity against Acid Orange dye, where its degradation was tested in the visible region of light [68].

2.3.7 Biogenic Synthesis of Iron Oxide Nanoparticles by Plant Extract

Iron is one of the crucial elements and micronutrients, and lack of iron in soil gives rise to many agricultural challenges like problems in chlorophyll formation; iron is essential for many biochemical and physiological processes of plants. Iron is a major component of the plant enzymes involved in respiration, nitrogen fixation, and photosynthesis, and iron is also involved in the synthesis of ribonucleic acid (RNA), DNA, and proteins [69]. Iron oxide nanoparticles have a great deal of applications due to their characteristics like less susceptibility to oxidation, superparamagnetism, longer blood half-lives, firmness in liquid form, and flexible surface chemistry. Due to these characteristics, iron oxide nanoparticles have found their purpose in the domain of magnetic resonance imaging (MRI), tissue engineering, magnetic particle imaging, stem cell tracking, magnetic cell sorting, immunoassays, bioengineering, biosensing, cosmeceuticals, antimicrobial, antifungal activities, etc. Iron oxide NPs have qualities of both metallic and hydrous iron oxide due to their demonstration of core and shell morphology. Iron NPs also act as nanozymes due to their inherent enzyme-mimicking potential similar to catalase, sulfite oxidase, peroxidases, and superoxide dismutase. Iron oxide NPs also have a great potential for interacting with various biological compounds like nucleic acids, fatty acids, lipids, metabolites, and peptides. For formulating biogenic NPs, the precursor salts like $FeSO_4$, $FeCl_3 \cdot 6H_2O$ $(Fe\,NO_3)_3 \cdot 9H_2O$, $FeSO_4 \cdot 5H_2O$, $FeCl_2 \cdot 4H_2O$ are used in solution form at different molar concentrations. These metal ion precursor solution is generally made to react with plant extract in order to fabricate biogenic iron oxide NPs, after which the mixture is incubated until a change in color is observed, depending on the metal precursor [70]. Linima et al. biologically fabricated iron oxide nanoparticles from aqueous leaves extract of *Ricinus communis* and $FeCl_3{\cdot}6H_2O$ as metal precursor, and the formation of iron oxide nanoparticles was confirmed when the color changed from light yellow to black. The flavonoids and phenolic constituents present in the *R. communis* plant were found to act as reducing agents during the synthesis. The formed NPs were tested for antimicrobial activity against *Aspergillus flavus* and *S. aureus*, and *S. typhi* [71]. Batool et al. fabricated *Phoenix dactylifera* extract iron oxide NPs where iron sulfate heptahydrate was used as metal precursor, and conversion in color from yellow to black confirmed the formation of biogenic iron oxide NPs. Then the nanoparticles were evaluated for antibacterial activity against *B. subtilis, K. pneumoniae, P. dactylifera,* and *E. coli* [72]. Another study by Shabbir et al. reported the biological reduction of $FeSO_4$ solution to FeONPs by *Madhuca Indica* extract as a bio reductant, and the reaction end was indicated with the conversion in color from yellow to dark brown [73].

2.4 Challenges Associated with Plant Extract-Based Biogenic Nanoparticles

Plants as natural resources are mostly dominantly preferred sources to biologically synthesize nanoparticles owing to their presence of a wide range of phytoconstituents, cost-effectiveness, ethnobotanical value, abundant availability, etc. However, they also present a few challenges that call for the attention of researchers, so that efficient nanoparticles can be derived through biologically mediated synthesis. Some of the key challenges associated with green synthesis of biogenic nanoparticles are listed below:

- Insufficient comprehensive understanding to create green nanoparticles using plant entities.
- It is important to guarantee nanoparticle uniformity. It is seen that many variations in size, shape, and structure are produced by plant-mediated nanoparticle synthesis.

- The rational approach to creating green nanoparticles with distinct sizes and shapes should be used.
- It is necessary to clarify the exact function of plant components in nanoparticles as these compounds have both stabilizing and reducing properties.
- The primary issue that needs to be resolved is the conversion of salt to ions. The goal of plant-based synthesis is to convert as much salt into ions as possible.
- To produce NPs from the laboratory to the industrial level, technology transfer procedures must be used tactfully.
- For environmental applications, it is also essential to comprehend the ecotoxicological viewpoint of metallic nanoparticles. It is also necessary to conduct research on the aquatic ecosystem, different niche habitats, and nontarget creatures as well.
- Because different biological substances are used for synthesis, the obtained nanoparticles must be made sure to be homogeneous in terms of size, shape, and effectiveness.
- A benign approach that prioritizes ease of synthesis, resource utilization, particle formation (monodispersity, homogeneity, and repeatability), waste management, and toxicity perspective should have been used in the industrial production of NPs [74].

2.5 Clinical Translation of Green Nanotherapeutics

Green synthetic nanomedicine has adopted rapid evolution and significant achievements in fundamental discoveries and medication development for possible clinical translation, all under the aegis of biomedical investigation. Across the past several decades, a number of innovative ideas for medicine delivery, illness prevention, intervention, and monitoring have been put forth employing manufactured nanoparticles or nanomaterials [75]. In clinical trials, more than a hundred other nanoparticle candidates are being clinically studied as diagnostics, vaccines, and therapeutics. However, the success of these candidates can only be partially warranted by their clinical translation. In many scientific studies, strategies for nanotherapeutics development, materials synthesis, and analysis often do not consider translational potential or clinically relevant solutions. It is practically difficult to transform an intriguing idea into a scalable, economically feasible, clinically applicable, and marketable medical product because of limitations associated with preclinical evidence of principle, chemistry, manufacturing, clinical efficacy, safety, regulatory compliance, and issues of post-approval [76]. Researchers studying green-synthesized nanomedicine have thus lately identified

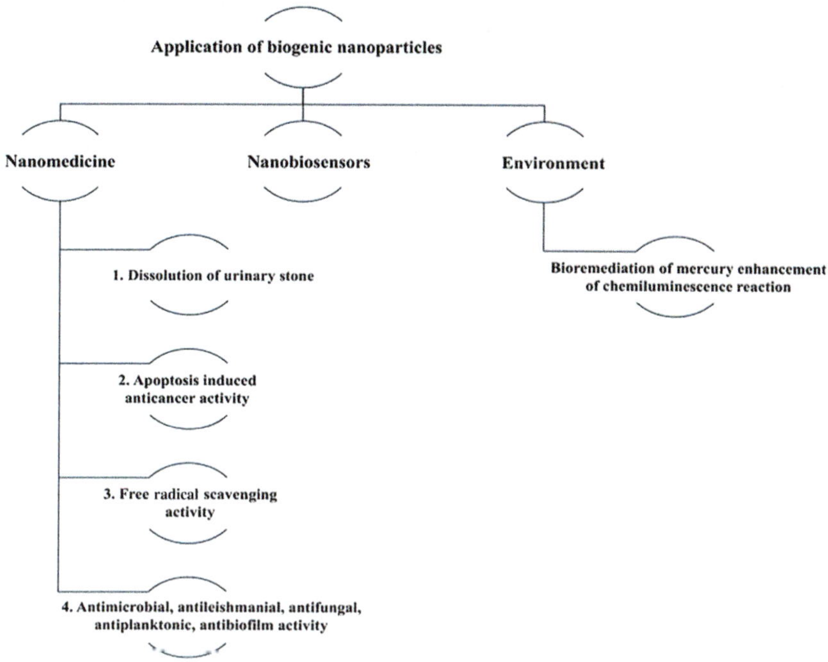

Figure 2.4 Applications of biogenic nanoparticles.

a number of developmental obstacles that need to be addressed, including those related to formulation, toxicity, physiology, immunology, and translation. More significantly, a successful clinical translation requires close coordination with clinical, pharmaceutical, and regulatory groups in order to evaluate the drug's essential quality attributes, such as therapeutic index, dosimetry, pharmacokinetics, and cost-effectiveness, including the complexity of a disease, too [77]. General applications of biogenic nanoparticles are represented in Figure 2.4.

Biogenic nanoparticles synthesized from plants have great potential for a number of treatments because of their special qualities, which include low toxicity, safety, biocompatibility, and the inclusion of bioactive constituents with numerous antidisease effects [78]. Here, in Table 2.1, different plant-based biogenic nanoparticles, which are being investigated for major disease treatments, are listed.

Although the green biogenically synthesized nanoparticles have proven their purpose at initial stages, namely pre-clinical settings, they have a long road ahead in order to arrive at their full clinical potential, which requires direct association with patients for their disease mitigation and management.

Table 2.1 Various plant-mediated biogenic nanoparticles and their therapeutic activity.

Biogenic Nanoparticles	Plant material	Plant part	Therapeutic use	Results	Reference
Silver Nanoparticle (AgNPs)	*Curcuma longa*	Turmeric powder extract	Antibacterial activity	Nanoparticles having a size of 5–35 nm exhibited a total of 4 log CFU/ml of reduction of bacterial cells compared to the control.	[79]
	Azadirachta indica	Leaves	Cervical cancer	NPs having a size of 2–8 nm, hexagonal or triangular in shape, and IC_{50} values of ≤4.25 g/ml.	[80]
	Sapindus mukorossi	Fruit pericarp extract	Antibacterial activity	Spherical AgNPs formed were found to display antibacterial activity against *Pseudomonas aeruginosa*, with zone of inhibition (ZOI) 12.0 mm, and against *Staphylococcus aureus* with ZOI 27.3 mm.	[81]
	Eysenhardtia polystachya	Bark extract	Antidiabetic	Spherical NPs having a size of 10–12 nm exhibited good antidiabetic activity by promoting insulin secretion, pancreatic β-cell survival, and enhancing hyperlipidemia and hyperglycemia.	[82]
	Capparis zeylanica L.	Leaf extract	Antimicrobial and Antiproliferation	Results indicate spherical AgNPs having 23 nm average size exhibit admirable antimicrobial activity against *Staphylococcus epidermidis, Enterococcus faecalis, Salmonella paratyphi*, and *Shigella dysenteriae* (ZOI = 30 mm, 26 mm, 23 mm, and 20 mm, respectively) and also have excellent antiproliferative properties against the A549 lung cancer cell line.	[83]

Table 2.1 (Continued)

Biogenic Nanoparticles	Plant material	Plant part	Therapeutic use	Results	Reference
	Ipomoea aquatica	Leaf extract	Antimicrobial	Spherical-shaped AgNPs having a size of 5–30 nm exhibited excellent antibacterial activity with ZOI for *Escherichia coli* (9.25 ± 1.15 mm), *Staphylococcus* sp. (12.67 ± 1.04 mm), and *Salmonella* (9.10 ± 0.50 mm).	[84]
	Teucrium polium L.	Stems and flowers	Antifungal	Spherical-shaped AgNPs with a mean particle size range of 10–100 nm displayed dose-dependent fungicidal action and caused a reduction in colony diameter and mean growth inhibition of *Fusarium oxysporum*.	[85]
	Andrographis paniculata, *Phyllanthus niruri*, and *Tinospora cordifolia*	Stems and leaves	Antiviral activity against chikungunya virus.	AgNPs from all three plants were found to be spherical in shape with an average size of 80 nm. AgNPs from *Andrographis paniculata* were found to have better cytotoxic activity with a non-toxic dose of 31.25 μg/ml, compared to that of AgNPs from *Tinospora cordifolia* (250 μg/ml) and *Phyllanthus niruri* (125 μg/ml).	[86]
	Pinus roxburghii	Fresh needles	Lung cancer and Prostate cancer	AgNPs formed were spherical with an average size of 80 nm, and they exhibited significant cytotoxicity toward both lung and prostatic cell carcinomas with IC_{50} values of 11.28 ± 1.28 μg/ml and 56.27 ± 1.17 μg/ml, respectively.	[87]

(Continued)

Table 2.1 (Continued)

Biogenic Nanoparticles	Plant material	Plant part	Therapeutic use	Results	Reference
Gold Nanoparticles (AuNPs)	*Parkia biglobosa*	Leaf and stem	Antimicrobial	*Parkia* leaf -AuNPs exhibited spherical, triangular, and rod structures with sizes 2.5–25 nm, while *Parkia* stem AuNPs were found to be nanorod, truncated, pentagonal, spherical shaped, having sizes of 1–35 nm. Both AuNPs displayed antibacterial activity against *Escherichia coli, Pseudomonas syringae, Staphylococcus aureus*, and *Bacillus anthracis*.	[88]
	Benincasa hispida	Fresh peel	Cervical cancer	AuNPs formed were spherical in shape and had a mean particle size of 22.18 ± 2 nm, which shows a potent cytotoxic effect with an IC_{50} value of 2.25 μg/ml on HeLa cells.	[89]
	Abies spectabilis	Aerial parts	Bladder cancer	Results revealed that spherical, oval-shaped particles having a size of 20–200 nm were obtained, which displayed significant cytotoxicity against bladder T24 cancer cells with an IC_{50} at 15 μg/ml.	[90]
	Hibiscus sabdariffa	Whole part	Amyloid disease	The formed AuNPs displayed an inhibitory effect on the aggregation of α-lactalbumin by decreasing the aggregation and fibril formation of reduced α-lactalbumin.	[91]

Table 2.1 (Continued)

Biogenic Nanoparticles	Plant material	Plant part	Therapeutic use	Results	Reference
	Platycodon grandiflorum	Leaf	Antibacterial	Spherical-shaped AuNPs with a size of 15 nm were found to exhibit antibacterial activity against *Escherichia coli* with a ZOI of 16 mm.	[92]
	Zingiber officinale	Rhizome	Antioxidant, Antimicrobial, Anticancer	Spherical-shaped AuNPs, having a size range of 5–53 nm, were found to have antioxidant activity (87.6 ± 0.5%) and cytotoxic effects on hepatic cancer cell lines with an IC_{50} value of 131.9 ± 9.34 μg/ml. The AuNPs were also found to display antimicrobial activity against *Bacillus subtilis*, *Austromerope brasiliensis*, *Pseudomonas aeruginosa*, *Escherichia coli*, *Staphylococcus aureus*, and *Candida albicans*.	[93]
	Nigella arvensis	Leaf	Anticancer, Antioxidant, Antibacterial	AuNPs of spherical structure having a size of 3–37 nm, displayed cytotoxicity activity against MCF-7 and H1299 cancer cell lines with IC_{50} values of 25 and 10 μg/ml, respectively. The NPs also displayed antibacterial activity against *Staphylococcus aureus*, *Bacillus subtilis*, and *Escherichia coli* with ZOI of 10, 11, and 11 mm, respectively.	[94]
	Tabebuia argentiea	Flower	Antioxidant	Spherical NPs sized 56 nm exhibited a maximum reducing power of 71.46% at 500 μg/ml concentration.	[95]

(Continued)

Table 2.1 (Continued)

Biogenic Nanoparticles	Plant material	Plant part	Therapeutic use	Results	Reference
	Trachy-spermum ammi	Seed	Hepatic cancer	Spherical-shaped AuNPs with a size range of 3–23 nm exhibited potential inhibition of HepG2 cells' growth in a concentration and time-dependent manner, with an IC_{50} at 92.453 μg/ml, and significant depletion of Glutathione levels in the cell line.	[96]
Zinc oxide Nanoparticles (ZnO-NPs)	*Cassia fistula* and *Melia azadarach*	Leaf	Antibacterial	Results show NPs' shapes were spherical and their size range from 3 to 68 nm. The extract-mediated ZnONPs showed strong antimicrobial activity. C. fistula shows a ZOI of 21–44 mm and 14–32 mm for *Escherichia coli* and *Staphylococcus aureus*, respectively and *Melia azadarach* shows ZOI of 20–40 mm and 21–38 mm for *Escherichia coli* and *Staphylococcus aureus*, respectively.	[97]
	Artemisia pallens	Leaf along with the stem	Antimicrobial	Resulted NPs were hexagonal in shape, having sizes of 50 nm and 100 nm, and they exhibit significant antimicrobial activity with minimum inhibitory concentration of 62.5 μg/ml, 31.25 μg/ml, 62.5 μg/ml, for *Escherichia coli*, *Staphylococcus aureus*, and *Bacillus subtilis*, respectively.	[98]

Table 2.1 (Continued)

Biogenic Nanoparticles	Plant material	Plant part	Therapeutic use	Results	Reference
	Beta vulgaris	Root	Antioxidant	Results revealed the formation of hexagonal NPs having a size of 52–76 nm, which exhibited good antioxidant activity through inhibition of free radicals (DPPH) with an IC_{50} value of 4400 μg/ml.	[99]
	Myristica fragrans	Fruit	Antidiabetic	Elliptical or spherical-shaped ZnONPs having a size of 41.23 nm displayed significant inhibitory action against the enzymes protein kinase (12.23 ± 0.42), α-glucosidase (65.21 ± 0.49), and α-amylase (73.23 ± 0.42).	[100]
	Raphanus sativus var. Longipinnatus	Leaves	Lung cancer	Results show that ZnONPs are spherical, hexagonal-shaped, with an average size of 209 nm, and they also exhibit a significant cytotoxic effect against the A549 cell line with an IC_{50} of 40 mg/ml.	[101]
	Tinospora Cordifolia	Stem	Colorectal Cancer	Spherical-shaped ZnONPs having a mean particle size of 29.5 nm displayed potent cytotoxic activity against Caco-2 cell lines (IC_{50} = 36.675 ± 0.916 μg/ml) and HCT-116 cell lines (IC_{50} = 31.419 ± 0.682 μg/ml).	[102]
	Loranthus cordifolius	Leaves	Antibacterial	Quasi-spherical-shaped ZnONPs with a mean particle size of 14.47 nm displayed good inhibitory action against Gyrase-B activity with an IC_{50} of 0.78 ± 0.2 M.	[103]

(Continued)

Table 2.1 (Continued)

Biogenic Nanoparticles	Plant material	Plant part	Therapeutic use	Results	Reference
	Cocos nucifera	Leaves	Antimicrobial, Antioxidant	Hexagonal wurtzite-structured NPs with a mean size of 16.6 nm displayed antioxidant activity with IC_{50} at 764 μg/ml, antifungal action against *Aspergillus niger* with ZOI 10 mm, and antibacterial action against *Staphylococcus aureus* with ZOI 10 mm.	[104]
Magnesium oxide Nanoparticle (MgO-NPs)	*Carica papaya, Azadirachta indica, Mangifera indica*	Leaves	Antibacterial	Cubic-shaped Magnesium oxide nanoparticles of *Carica papaya* and *Azadirachta* indica, having a size of 10.25–27.08 nm, were found to exhibit bacteriostatic activity at 200 and 100 μg/ml, respectively.	[105]
	Abrus precatorius	Bark	Antioxidant, Anticancer	MgO nanoparticles have an agglomerate shape and size ranging from 100 to 200 nm. At a concentration of 120 μg/ml, 65.93% free radical scavenging capacity was exhibited, and it showed significant cytotoxicity against the melanoma cell line (IC_{50} = 20 μg/ml)	[106]
	Melia azedarach	Seed	Larvicidal, Antioxidant	Spherically shaped NPs of size having size 44–73 nm exhibited larvicidal activity in a dose-dependent manner, with an IC_{50} value of 82.55 μg/ml and an IC_{90} value of 150.95 μg/ml.	[107]

Table 2.1 (Continued)

Biogenic Nanoparticles	Plant material	Plant part	Therapeutic use	Results	Reference
	Moringa oleifera	Leaf	Antibacterial	The synthesized MgO-NPs were spherical and size range from 20 to 50 nm, and they showed potent antibacterial activity, having a ZOI against *Staphylococcus aureus* (6.3 mm) and *Escherichia coli* (6 mm).	[108]
	Crataegus aronia L.	fruit	Colon Cancer	The MgONPs of size 20–40 nm exhibited cytotoxic activity against colon cancer cell lines (IC_{50} = 98 μg/ml)	[109]
Copper Nanoparticles	*Alstonia scholaris*	Leaves	Antimicrobial	NPs of irregular shapes (spherical, triangular, cylindrical, and polygonal) were formed. The metabolic extracts have shown significant zones of inhibition against *Staphylococcus aureus* (21.5 mm) and *Escherichia coli* (23 mm), respectively.	[110]
	Rumex nepalensis Spreng.	Whole plant	Antibacterial, Antioxidant	Spherical-shaped NPs with sizes ranging from 21 to 97 nm were formed. It also revealed to possess antimicrobial activity against *Pseudomonas* aeruginosa *and Escherichia coli*, with a ZOI of 15 mm and 14 mm, respectively.	[111]
	Annona muricata	Fruits	Breast cancer	Synthesized NPs were spherical, with sizes ranging from 16 to 31 nm, and they precipitated anti-proliferative effects by triggering cell death through apoptosis with IC_{50} concentration for AMJ13 (17.04 μg/ml) and MCF7 cells (18.92 μg/ml).	[112]

(Continued)

Table 2.1 (Continued)

Biogenic Nanoparticles	Plant material	Plant part	Therapeutic use	Results	Reference
	Pistacia khinjuk	Leaf	Prostate cancer	Spherical structures CuNPs of size 10–70 nm, were formed. The NPs exhibited an IC_{50} of 215 μg/ml in DPPH scavenging activity. It also showed efficacy in inhibiting prostate cancer growth, specifically against NCI-H660, DU 145, LNCaP clone FGC, and LNCaP clone FGC-Luc2 prostate cancer cells.	[113]
	Dillenia indica	Bark	Breast cancer	CuNPs synthesized were found to be spherical in shape and crystalline in nature at a size range of 20.87–45.73 nm. CuNPs exhibited significant free radical scavenging activity at an IC_{50} of 37.2 μg/ml, as well as cytotoxic activity against MCF-7 and A549 cancer cell lines.	[114]
Selenium Nanoparticles	*Brassica oleracea*	Leaves		Results showed spherical NPs with an average particle size of 15.2 ± 1.9 nm, and also exhibited effective antimicrobial activity against *Streptococcus mutans, Staphylococcus aureus, E. faecalis, Lacto-bacillus,* and *C. albicans* (ZOI ranges from 10 to 15 mm)	[115]
	Amphipterygium glaucum	Leaves	Antifungal	NPs obtained were of size 8 nm and had antifungal activity against *Colletotrichum gloeosporioides* and *Fusarium oxysporum* at concentrations of 0.25 mg/ml.	[116]

Table 2.1 (Continued)

Biogenic Nanoparticles	Plant material	Plant part	Therapeutic use	Results	Reference
	Moringa oleifera	Leaves	Antioxidant and Antidiabetic	SeNPs were found to be spherical in shape, with a size of 20–250 nm, and at 1000 μg/ml concentration exhibited DPPH scavenging activity of 84%, with 454.1 μg/ml (IC_{50}). The formed SeNPs also showed dose-dependent antidiabetic properties against α-amylase and α-glucosidase enzymes.	[117]
	Rosmarinus officinalis	Leaves	Antibacterial	SeNPs (sized 20–40 nm) in this study exhibited antibacterial activity against *Escherichia coli, Pseudomonas* aeruginosa, *S mutans,* and *Mycobacterium tuberculosis*, with MICs of 128, 64, 32, and 256 μg/ml, respectively.	[118]
Iron Oxide Nanoparticles	*Psidium guajava*	Leaves	Antimicrobial	The formed NPs were found to have a cubic and slightly spherical shape of size 12.64 nm and were found to exhibit antibacterial activity against *Staphylococcus aureus* and *Shigella sonnei* with 13 mm ZOI for both.	[119]
	Curcuma amada	Rhizome	Antioxidant, Anticancer, Antibacterial	NPs, at a concentration of 140 μg/ml, showed scavenging activity of 76.09% and potent cytotoxicity against the breast cancer cell line with IC_{50} (125 μg/ml), and also possessed bactericidal activity with a higher ZOI than the extract of turmeric rhizome.	[120]

2.6 Conclusion and Future Perspectives

The necessity of developing biogenic ways of nanoparticles preparation arises from the setbacks associated with chemical and physical methods for their utilization of toxic reagents, solvents, high-energy-consuming instruments, and the requirement of specific technical personnel. The biogenic synthesis of nanoparticles from biomass sources like fungi, bacteria, algae, and plant extracts raises major safety and toxicity concerns. Moreover, these biomaterials also act as reducing agents, capping agents, and stabilizing agents in certain cases, which negates the additional involvement of extrinsic harmful agents during the preparation process. In this chapter, the main highlight has been given to the metallic, metal oxide nanoparticles derived from plant extract. The plant sources, compared to other biomass, are easily available, inexpensive, and abundant, and the main issue of culturing processes to be dealt with in case of microbes is totally avoided. A single plant provides multiple sources as its leaves, fruits, flowers, bark, seeds, bulb, etc. are extracted using suitable methods and solvents to fabricate these biogenic nanoparticles, as evidenced from the literature. For the synthesis of biogenic nanoparticles, the main component apart from the plant extract is the metal precursor in salt solution form. We have observed that a change in color of the solution is an indicative factor in deciding the cessation of the preparation process, which can be evaluated using different analytical techniques. These biogenic nanoparticles have exhibited their potential in many pathological conditions that have been discussed in the clinical translation section.

However, we have a long way to go to fully assimilate the biodegradation pathways and understand their mechanism, the effect of the size at the cellular level of these biogenic metal and metal oxide nanoparticles in the future. The most important concern is the effects associated with variation in the species of these biological sources.

References

1 Soni V, Raizada P, Singh P, Cuong HN, Rangabhashiyam S, Saini A, et al. Sustainable and green trends in using plant extracts for the synthesis of biogenic metal nanoparticles toward environmental and pharmaceutical advances: a review. Environ Res. 2021;202:111622.

2 El-Seedi HR, El-Shabasy RM, Khalifa SAM, Saeed A, Shah A, Shah R, et al. Metal nanoparticles fabricated by green chemistry using natural extracts: biosynthesis, mechanisms, and applications. RSC Adv. 2019; 9(42):24539–24559.

3 Ahmed SF, Mofijur M, Rafa N, Chowdhury AT, Chowdhury S, Nahrin M, et al. Green approaches in synthesising nanomaterials for environmental nanobioremediation: technological advancements, applications, benefits and challenges. Environ Res. 2022;204(Pt A):111967.

4 Altammar KA. A review on nanoparticles: characteristics, synthesis, applications, and challenges. Front Microbiol. 2023;14:155622.

5 Shah DD, Chorawala MR, Mansuri MKA, Parekh PS, Singh S, Prajapati BG. Biogenic metallic nanoparticles: from green synthesis to clinical translation. Naunyn Schmiedeberg's Arch Pharmacol. 2024;397(11):8603–8631.

6 Srivastava SK, Ogino C, Kondo A. Nanoparticle synthesis by biogenic approach. In: Basiuk VA, Basiuk EV (eds). Green Processes for Nanotechnology: From Inorganic to Bioinspired Nanomaterials. 2015. Springer International Publishing, Cham, p. 237–257.

7 Sharma D, Kanchi S, Bisetty K. Biogenic synthesis of nanoparticles: a review. Arab J Chem. 2019;12:3576–3600.

8 Gittins DI, Bethell D, Schiffrin DJ, Nichols RJ. A nanometre-scale electronic switch consisting of a metal cluster and redox-addressable groups. Nature. 2000;408(6808):67–69.

9 Vijayaraghavan K, Ashokkumar T. Plant-mediated biosynthesis of metallic nanoparticles: a review of literature, factors affecting synthesis, characterization techniques and applications. J Environ Chem Eng. 2017;5(5):4866–4883.

10 Singh J, Dutta T, Kim KH, Rawat M, Samddar P, Kumar P. Green synthesis of metals and their oxide nanoparticles: applications for environmental remediation. J Nanobiotech. 2018;16(1):84.

11 Jeyaraj M, Gurunathan S, Qasim M, Kang MH, Kim JH. A comprehensive review on the synthesis, characterization, and biomedical application of platinum nanoparticles. Nanomaterials. 2019;9(12):1719.

12 Tangahu BV, Sheikh Abdullah SR, Basri H, Idris M, Anuar N, Mukhlisin M. A review on heavy metals (As, Pb, and Hg) uptake by plants through phytoremediation. Int J Chem Eng. 2011;2011:939161.

13 Das RK, Pachapur VL, Lonappan L, Naghdi M, Pulicharla R, Maiti S, et al. Biological synthesis of metallic nanoparticles: plants, animals and microbial aspects. Nanotechnol Environ Eng. 2017;2(1):18.

14 Md Ishak NAI, Kamarudin SK, Timmiati SN. Green synthesis of metal and metal oxide nanoparticles via plant extracts: an overview. Mater Res Express. 2019;6(11):112004.

15 Alex S, Tiwari A. Functionalized gold nanoparticles: synthesis, properties and applications-a review. J Nanosci Nanotechnol. 2015;15(3):1869–1894.

16 Sani A, Cao C, Cui D. Toxicity of gold nanoparticles (AuNPs): a review. Biochem Biophys Rep. 2021;26:00991.

17 Hesan MA, Kormin F, Ahmed Moujdin I, Gzara L, Alsaiari O, A, Abdel Salam M. A new tropical wild fern for the green synthesis of gold nanoparticles: comprehensive characterization and application in water treatment. J Indian Chem Soc. 2024;101(12):101427.

18 Bharadwaj KK, Rabha B, Pati S, Sarkar T, Choudhury BK, Barman A, et al. Green synthesis of gold nanoparticles using plant extracts as beneficial prospect for cancer Theranostics. Molecules. 2021;26(21):6389.

19 Kepsutlu B, Wycisk V, Achazi K, Kapishnikov S, Pérez-Berná AJ, Guttmann P, et al. Cells undergo major changes in the quantity of cytoplasmic organelles after uptake of gold nanoparticles with biologically relevant surface coatings. ACS Nano. 2020;14(2):2248–2264.

20 Kowsalya E, MosaChristas K, Jaquline CRI, Balashanmugam P, Devasena T. Gold nanoparticles induced apoptosis via oxidative stress and mitochondrial dysfunctions in MCF-7 breast cancer cells. Appl Organomet Chem. 2021;35(1):e6071.

21 Al-Barram LF. Laser enhancement of cancer cell destruction by photothermal therapy conjugated glutathione (GSH)-coated small-sized gold nanoparticles. Lasers Med Sci. 2021;36(2):325–337.

22 Wang L, Xu J, Yan Y, Liu H, Karunakaran T, Li F. Green synthesis of gold nanoparticles from *Scutellaria barbata* and its anticancer activity in pancreatic cancer cell (PANC-1). Artif Cells Nanomed Biotechnol. 2019;47(1):1617–1627.

23 Dorosti N, Jamshidi F. Plant-mediated gold nanoparticles by *Dracocephalum kotschyi* as anticholinesterase agent: synthesis, characterization, and evaluation of anticancer and antibacterial activity. J Appl Biomed. 2016;14(3):235–245.

24 Badeggi UM, Ismail E, Adeloye AO, Botha S, Badmus JA, Marnewick JL, et al. Green synthesis of gold nanoparticles capped with Procyanidins from *Leucosidea sericea* as potential antidiabetic and antioxidant agents. Biomolecules. 2020;10(3):452.

25 Ateeq M, Shah MR, Ali H, Kabir N, Khan A, Nadeem S. Hepatoprotective and urease inhibitory activities of garlic conjugated gold nanoparticles. New J Chem. 2015;39:5003–5007.

26 Beyene HD, Werkneh AA, Bezabh HK, Ambaye TG. Synthesis paradigm and applications of silver nanoparticles (AgNPs), a review. Sustain Mater Technol. 2017;13:18–23.

27 Senapati S, Ahmad A, Khan MI, Sastry M, Kumar R. Extracellular biosynthesis of bimetallic Au–Ag alloy nanoparticles. Small. 2005;1(5):517–520.

28 Klaus-Joerger T, Joerger R, Olsson E, Granqvist CG. Bacteria as workers in the living factory: metal-accumulating bacteria and their potential for materials science. Trends Biotechnol. 2001;19(1):15–20.

29 Banerjee P, Satapathy M, Mukhopahayay A, Das P. Leaf extract mediated green synthesis of silver nanoparticles from widely available Indian plants: synthesis, characterization, antimicrobial property and toxicity analysis. Bioresour Bioprocess. 2014;1:1–10.

30 Sondi I, Salopek-Sondi B. Silver nanoparticles as antimicrobial agent: a case study on *E. coli* as a model for Gram-negative bacteria. J Colloid Interface Sci. 2004;275(1):177–182.

31 Kuppusamy P, Yusoff MM, Maniam GP, Govindan N. Biosynthesis of metallic nanoparticles using plant derivatives and their new avenues in pharmacological applications – An updated report. Saudi Pharm J. 2016;24(4):473–484.

32 Gurunathan S, Kalishwaralal K, Vaidyanathan R, Venkataraman D, Pandian SRK, Muniyandi J, et al. Biosynthesis, purification and characterization of silver nanoparticles using *Escherichia coli*. Colloids Surf B: Biointerfaces. 2009;74(1):328–335.

33 Suriyakalaa U, Antony JJ, Suganya S, Siva D, Sukirtha R, Kamalakkannan S, et al. Hepatocurative activity of biosynthesized silver nanoparticles fabricated using *Andrographis paniculata*. Colloids Surf B: Biointerfaces. 2013;102:189–194.

34 Olaleye MT, Akinmoladun AC, Ogunboye AA, Akindahunsi AA. Antioxidant activity and hepatoprotective property of leaf extracts of *Boerhaavia diffusa* Linn against acetaminophen-induced liver damage in rats. Food Chem Toxicol. 2010;48(8–9):2200–2205.

35 Akhtar MS, Panwar J, Yun YS. Biogenic synthesis of metallic nanoparticles by plant extracts. ACS Sustain Chem Eng. 2013;1(6):591–602.

36 Sanghi R, Verma P. Biomimetic synthesis and characterisation of protein capped silver nanoparticles. Bioresour Technol. 2009;100(1):501–504.

37 Ahmad A, Haneef M, Ahmad N, Kamal A, Jaswani S, Khan F. Biological synthesis of silver nanoparticles and their medical applications. World Acad Sci J. 2024;6(3):22.

38 El-Batal A, Amin M, Shehata M, Hallol M. Synthesis of silver nanoparticles by *Bacillus stearothermophilus* using gamma radiation and their antimicrobial activity. World Appl Sci J. 2013;22(1):1–16.

39 Szczyglewska P, Feliczak-Guzik A, Nowak I. Nanotechnology–general aspects: a chemical reduction approach to the synthesis of nanoparticles. Molecules. 2023;28(13):4932.

40 Vishwanath R, Negi B. Conventional and green methods of synthesis of silver nanoparticles and their antimicrobial properties. Curr Res Green Sustain Chem. 2021;4:100205.

41 He Y, Wei F, Ma Z, Zhang H, Yang Q, Yao B, et al. Green synthesis of silver nanoparticles using seed extract of *Alpinia katsumadai*, and their antioxidant, cytotoxicity, and antibacterial activities. RSC Adv. 2017;7:39842–39851.

42 Torres-Martínez Y, Arredondo-Espinoza E, Puente C, González-Santiago O, Pineda-Aguilar N, Balderas-Rentería I, et al. Synthesis of silver nanoparticles using a *Mentha spicata* extract and evaluation of its anticancer and cytotoxic activity. PeerJ. 2019;7:e8142.

43 Sharma K, Guleria S, Razdan V. Green synthesis of silver nanoparticles using *Ocimum gratissimum* leaf extract: characterization, antimicrobial activity and toxicity analysis. J Plant Biochem Biotechnol. 2020;29:213–224.

44 Zambonino MC, Quizhpe EM, Mouheb L, Rahman A, Agathos SN, Dahoumane SA. Biogenic selenium nanoparticles in biomedical sciences: properties, current trends, novel opportunities and emerging challenges in theranostic nanomedicine. Nanomaterials (Basel). 2023;13(3):424.

45 Kieliszek M, Błażejak S. Current knowledge on the importance of selenium in food for living organisms: a review. Molecules. 2016;21(5):609.

46 Sharma G, Sharma AR, Bhavesh R, Park J, Ganbold B, Nam JS, Lee SS. Biomolecule-mediated synthesis of selenium nanoparticles using dried *Vitis vinifera* (raisin) extract. Molecules. 2014;19(3):2761–2770.

47 Li S, Shen Y, Xie A, Yu X, Zhang X, Yang L, Li C. Rapid, room temperature synthesis of amorphous selenium/protein composites using *Capsicum annuum* L extract. Nanotechnology. 2007;18:405101.

48 Cittrarasu V, Kaliannan D, Dharman K, Maluventhen V, Easwaran M, Liu WC, et al. Green synthesis of selenium nanoparticles mediated from *Ceropegia bulbosa* Roxb extract and its cytotoxicity, antimicrobial, mosquitocidal and photocatalytic activities. Sci Rep. 2021;11(1):1032.

49 Rani S, Kumar P, Dahiya P, Dang AS, Suneja P. Biogenic synthesis of zinc nanoparticles, their applications, and toxicity prospects. Front Microbiol. 2022;13:824427.

50 Fakhari S, Jamzad M, Kabiri FH. Green synthesis of zinc oxide nanoparticles: a comparison. Green Chem Lett Rev. 2019;12:19–24.

51 Al-Kordy HMH, Sabry SA, Mabrouk MEM. Statistical optimization of experimental parameters for extracellular synthesis of zinc oxide nanoparticles by a novel haloalaliphilic Alkalibacillus sp.W7. Sci Rep. 2021;11, 1:10924.

52 Ngom B, Mpahane T, Manikandan E, Maaza M. ZnO nano-discs by lyophilization process: size effects on their intrinsic luminescence. J Alloys Compd. 2016;656:758–763.

53 Matinise N, Kaviyarasu K, Mongwaketsi N, Khamlich S, Kotsedi L, Mayedwa N, Maaza M. Green synthesis of novel zinc iron oxide ($ZnFe_2O_4$) nanocomposite via *Moringa oleifera* natural extract for electrochemical applications. Appl Surf Sci. 2016;446:66–73.

54 Sharma SC. ZnO nano-flowers from *Carica papaya* milk: degradation of Alizarin Red-S dye and antibacterial activity against *Pseudomonas aeruginosa* and *Staphylococcus aureus*. Opt Int J Light Electron Opt. 2016;127:6498–6512.

55 Grigore M, Biscu E, Holban A, Gestal M, Grumezescu A. Methods of synthesis, properties and biomedical applications of CuO nanoparticles. Pharmaceuticals. 2016;9(4):75.

56 Naz S, Gul A, Zia M, Javed R. Synthesis, biomedical applications, and toxicity of CuO nanoparticles. Appl Microbiol Biotechnol. 2023;107(4):1039–1061.

57 Leitner J, Sedmidubský D, Lojka M, Jankovský O. The effect of nanosizing on the oxidation of partially oxidized copper nanoparticles. Materials. 2020;13(12):2878.

58 Xu L, Zheng G, Pei S, Wang J. Investigation of optical bandgap variation and photoluminescence behavior in nanocrystalline CuO thin films. Optik. 2018;158:382–390.

59 Waris A, Din M, Ali A, Ali M, Afridi S, Baset A, et al. A comprehensive review of green synthesis of copper oxide nanoparticles and their diverse biomedical applications. Inorg Chem Commun. 2021;123:108369.

60 Jabeen S, Siddiqui VU, Bala S, Mishra N, Mishra A, Lawrence R, et al. Biogenic synthesis of copper oxide nanoparticles from Aloe vera: antibacterial activity, molecular docking, and photocatalytic dye degradation. ACS Omega. 2024;9(28):30190–30204.

61 Asemani M, Anarjan N. Green synthesis of copper oxide nanoparticles using *Juglans regia* leaf extract and assessment of their physico-chemical and biological properties. Green Process Synth. 2019;8(1):557–567.

62 Ali SG, Haseen U, Jalal M, Khan RA, Alsalme A, Ahmad H, Khan HM. Green synthesis of copper oxide nanoparticles from the leaves of *Aegle marmelos* and their antimicrobial activity and photocatalytic activities. Molecules. 2023;28(22):7499.

63 Sathiyavimal S, Durán-Lara EF, Vasantharaj S, Saravanan M, Sabour A, Alshiekheid M, et al. Green synthesis of copper oxide nanoparticles using *Abutilon indicum* leaves extract and their evaluation of antibacterial, anticancer in human A549 lung and MDA-MB-231 breast cancer cells. Food Chem Toxicol. 2022;168:113330.

64 Küünal S, Rowel P, Rowel E. Plant extract mediated synthesis of nanoparticles. Emerging Applications of Nanoparticles and Architecture Nanostructures. 2018:411–446.

65 Patil SD, Morbale ST. Biogenic synthesis of manganese nanoparticles: a review. Int J Biol Pharm Allied Sci. 2022;11(11):5578–5586.

66 Jayandran M, Muhamed Haneefa M, Balasubramanian V. Green synthesis and characterization of manganese nanoparticles using natural plant extracts and its evaluation of antimicrobial activity. J App Pharm Sci. 2015;5(12):105–110.

67 Lu H, Zhang X, Khan SA, Li W, Wan L. Biogenic synthesis of MnO_2 nanoparticles with leaf extract of *Viola betonicifolia* for enhanced antioxidant, antimicrobial, cytotoxic, and biocompatible applications. Front Microbiol. 2021;12:761084.

68 Hoseinpour V, Souri M, Ghaemi N. Green synthesis, characterisation, and photocatalytic activity of manganese dioxide nanoparticles. Sci Appl. 2015;8:1–17.

69 Ahmad S, Ahmad N, Islam MS, Ahmad MA, Ercisli S, Ullah R, et al. Rice seeds biofortification using biogenic ıron oxide nanoparticles synthesized by using *Glycyrrhiza glabra*: a study on growth and yield ımprovement. Sci Rep. 2024;14(1):12368.

70 Priya N, Kaur K, Sidhu AK. Green synthesis: an eco-friendly route for the synthesis of iron oxide nanoparticles. Front Nanotechnol. 2021;3: 655062.

71 Linima VK, Ragunathan R, Johney J. Biogenic synthesis of RICINUS COMMUNIS mediated iron and silver nanoparticles and its antibacterial and antifungal activity. Heliyon. 2023;9(5):e15743.

72 Batool F, Iqbal MS, Khan SU, Khan J, Ahmed B, Qadir MI. Biologically synthesized iron nanoparticles (FeNPs) from *Phoenix dactylifera* have anti-bacterial activities. Sci Rep. 2021;11(1):22132.

73 Shabbir MA, Naveed M, Rehman SU, Ain NU, Aziz T, Alharbi M, et al. Synthesis of iron oxide nanoparticles from *Madhuca indica* plant extract and assessment of their cytotoxic, antioxidant, anti-inflammatory, and anti-diabetic properties via different Nanoinformatics approaches. ACS Omega. 2023;8(37):33358–33366.

74 Patil S, Chandrasekaran R. Biogenic nanoparticles: a comprehensive perspective in synthesis, characterization, application and its challenges. J Genet Eng Biotechnol. 2020;18(1):67.

75 Ali MA, Ahmed T, Wu W, Hossain A, Hafeez R, Masum MMI, et al. Advancements in plant and microbe-based synthesis of metallic nanoparticles and their antimicrobial activity against plant pathogens. Nanomaterials. 2020;10:1146.

76 Anselmo AC, Mitragotri S. Nanoparticles in the clinic: an update post COVID-19 vaccines. Bioeng Transl Med. 2021;6(3):e10246.

77 Ying S, Guan Z, Ofoegbu PC, Clubb P, Rico C, He F, Hong J. Green synthesis of nanoparticles: current developments and limitations. Environ Technol Innov. 2022;26:102336.

78 Singh H, Desimone MF, Pandya S, Jasani S, George N, Adnan M, et al. Revisiting the green synthesis of nanoparticles: uncovering influences of plant extracts as reducing agents for enhanced synthesis efficiency and its biomedical applications. Int J Nanomedicine. 2023;18:4727–4750.

79 Alsammarraie FK, Wang W, Zhou P, Mustapha A, Lin M. Green synthesis of silver nanoparticles using turmeric extracts and investigation of their antibacterial activities. Colloids Surf B: Biointerfaces. 2018;171:398–405.

80 Mishra A, Mehdi SJ, Irshad M, Ali A, Sardar M, Rizvi MMA. Effect of biologically synthesized silver nanoparticles on human cancer cells. Sci Adv Mater. 2012;4:1200–1206.

81 Huong VTL, Nguyen NT. Green synthesis, characterization and antibacterial activity of silver nanoparticles using *Sapindus mukorossi* fruit pericarp extract. Mater Today Proc. 2021;42:88–93.

82 Campoy AHG, Gutierrez RMP, Manriquez-Alvirde G, Ramirez AM. Protection of silver nanoparticles using *Eysenhardtia Polystachya* in peroxide-induced pancreatic β-cell damage and their antidiabetic properties in zebrafish. Int J Nanomedicine. 2018;13:2601–2612.

83 Nilavukkarasi M, Vijayakumar S, Prathip KS. Biological synthesis and characterization of silver nanoparticles with *Capparis zeylanica* L. leaf extract for potent antimicrobial and anti proliferation efficiency. Mater Sci Energy Technol. 2020;3:371–376.

84 Khan MR, Hoque SM, Hossain KFB, Siddique MAB, Uddin MK, Rahman MM. Green synthesis of silver nanoparticles using *ipomoea Aquatica* leaf extract and its cytotoxicity and antibacterial activity assay. Green Chem Lett Rev. 2020;13:39–51.

85 Ghojavand S, Madani M, Karimi J. Green synthesis, characterization and antifungal activity of silver nanoparticles using stems and flowers of Felty germander. J Inorg Organomet Polym Mater. 2020;30:2987–2997.

86 Sharma V, Kaushik S, Pandit P, Dhull D, Yadav JP, Kaushik S. Green synthesis of silver nanoparticles from medicinal plants and evaluation of their antiviral potential against chikungunya virus. Appl Microbiol Biotechnol. 2019;103:881–891.

87 Kumari R, Saini AK, Kumar A, Saini RV. Apoptosis induction in lung and prostate cancer cells through silver nanoparticles synthesized from *Pinus roxburghii* bioactive fraction. J Biol Inorg Chem. 2020;25:23–37.

88 Davids JS, Ackah M, Okoampah E, Fometu SS, Guohua W, Jianping Z. Biocontrol of bacteria associated with pine wilt nematode, Bursaphelenchus xylophilus by using plant mediated gold nanoparticles. Int J Agric Biol. 2021;26:517–526.

89 Al Saqr A, Khafagy ES, Alalaiwe A, Aldawsari MF, Alshahrani SM, Anwer MK, et al. Synthesis of gold nanoparticles by using green machinery: characterization and in vitro toxicity. Nanomaterials. 2021;11:808.

90 Wu T, Duan X, Hu C, Wu C, Chen X, Huang J, et al. Synthesis and characterization of gold nanoparticles from *Abies spectabilis* extract and its anticancer activity on bladder cancer T24 cells. Artif Cell Nanomed Biotechnol. 2019;47:512–523.

91 Talebpour F, Ghahghaei A. Effect of green synthesis of gold nanoparticles (AuNPs) from *Hibiscus sabdariffa* on the aggregation of α-lactalbumin. Int J Pept Res Ther. 2020;26(2):2297–2306.

92 Anbu P, Gopinath SCB, Jayanthi S. Synthesis of gold nanoparticles using *Platycodon grandiflorum* extract and its antipathogenic activity under optimal conditions. Nanomater Nanotechnol. 2020;10:1–9.

93 Fouda A, Eid AM, Guibal E, Hamza MF, Hassan SED, Alkhalifah DHM, El-Hossary D. Green synthesis of gold nanoparticles by aqueous extract of *Zingiber officinale*: characterization and insight into antimicrobial, antioxidant, and in vitro cytotoxic activities. Appl Sci. 2022;12:e12879.

94 Chahardoli A, Karimi N, Sadeghi F, Fattahi A. Green approach for synthesis of gold nanoparticles from *Nigella arvensis* leaf extract and evaluation of their antibacterial, antioxidant, cytotoxicity and catalytic activities. Artif Cell Nanomed Biotechnol. 2018;46:579–588.

95 Kavitha H, Chandrasekhar N, Govindappa M, Ramachandra YL, Chandrappa CP. Synthesis of gold nanoparticles by the flower extracts of *Tabebuia argentiea* and their antioxidant activity. Int J Pharm Bio Sci. 2018;8:379–383.

96 Perveen K, Husain FM, Qais FA, Khan A, Razak S, Afsar T, et al. A microwave-assisted rapid green synthesis of gold nanoparticles using seed extract of *Trachyspermum ammi*: ROS mediated biofilm inhibition and anticancer activity. Biomolecules. 2021;11:197.

97 Naseer M, Aslam U, Khalid B, Chen B. Green route to synthesize zinc oxide nanoparticles using leaf extracts of *Cassia fistula* and *Melia azadarach* and their antibacterial potential. Sci Rep. 2020;10(1):9055.

98 Gomathi R, Suhana H. Green synthesis, characterization and antimicrobial activity of zinc oxide nanoparticles using *Artemisia pallens* plant extract. Inorg Nano-Metal Chem. 2021;51:1663–1672.

99 Kumar MAP, Suresh D, Nagabhushana H, Sharma SC. Beta vulgaris aided green synthesis of ZnO nanoparticles and their luminescence, photocatalytic and antioxidant properties. Eur Phys J Plus. 2015;130:1–7.

100 Faisal S, Jan H, Shah SA, Shah S, Khan A, Akbar MT, et al. Green synthesis of zinc oxide (ZnO) nanoparticles using aqueous fruit extracts of Myristica fragrans: their characterizations and biological and environmental applications. ACS Omega. 2021;6:9709–9722.

101 Umamaheswari A, Prabu SL, John SA, Puratchikody A. Green synthesis of zinc oxide nanoparticles using leaf extracts of *Raphanus sativus var. Longipinnatus* and evaluation of their anticancer property in A549 cell lines. Biotechnol Rep. 2021;29:e00595.

102 Berehu HM, Patnaik S. Biogenic zinc oxide nanoparticles synthesized from Tinospora Cordifolia induce oxidative stress, mitochondrial damage and apoptosis in colorectal cancer. Nanotheranostics. 2024;8(3):312–329.

103 Mazhar MW, Ishtiaq M, Maqbool M, Arshad A, Alshehri MA, Alhelaify SS, et al. Green synthesis of anethole-loaded zinc oxide nanoparticles enhances antibacterial strategies against pathogenic bacteria. Sci Rep. 2024;14: 24671.

104 Rahman F, Patwary MAM, Bashar MS, Haque MA, Akter B, Rashid R, et al. Green synthesis of zinc oxide nanoparticles using Cocos nucifera leaf extract: characterization, antimicrobial, antioxidant and photocatalytic activity. R Soc Open Sci. 2022;9:220858.

105 Rotti RB, Sunitha DV, Manjunath R, Roy A, Mayegowda SB, Gnanaprakash AP, et al. Grren synthesis of MgO nanoparticles and its antibacterial properties. Front Chem. 2023;11:1143614.

106 Ali S, Sudha KG, Thirumalaivasan N, Ahamed M, Pandiaraj S, Rajeswari VD, et al. Green synthesis of magnesium oxide nanoparticles by using Abrus precatorius bark extract and their photocatalytic, antioxidant, antibacterial, and cytotoxicity activities. Bioengineering. 2023;10:1–20.

107 Dinga E, Ekennia A, Ogbonna CU, Udu DA, Mthiyane MN, Marume U, Onwudiw DC. Phyto-mediated synthesis of MgO nanoparticles using Melia azedarach seed extract: larvicidal and antioxidant activities. Sci Afr. 2022;17:e01366.

108 Fatiqin A, Amrulloh H, Simanjuntak W. Green synthesis of MgO nanoparticles using moringa oleifera leaf aqueous extract for antibacterial activity. Bull Chem Soc Ethiop. 2021;35:161–170.

109 Oran SA, Ghidan AY, Yousef IF, Alkhatib MS. Effect of green synthesized magnesium oxide nanoparticles using Crataegus aronia L. on human colon cancer-CaCo-2 cell line. Trop J Nat Prod Res. 2024;8(9):8385–8392.

110 Labaran AN, Zango ZU, Tailor G, Alsadig A, Usman F, Mukhtar MT, et al. Biosynthesis of copper nanoparticles using Alstonia scholaris leaves and its antimicrobial studies. Sci Rep. 2024;14:5589.

111 Abhimanyu P, Arvind M, Kishor N. Biosynthesis of CuO nanoparticles using plant extract as a precursor: characterization, antibacterial, and antioxidant activity. Nano Biomed Eng. 2023;15(4):369–377.

112 Mahmood RI, Kadhim AA, Ibraheem S, Albukhaty S, Mohammed Salih HS, Abbas RH, et al. Biosynthesis of copper oxide nanoparticles mediated Annona muricata as cytotoxic and apoptosis inducer factor in breast cancer cell lines. Sci Rep. 2022;12:16165.

113 Wang Z, Li M, Zhang Q, Xu L. Green formulation of copper nanoparticles by Pistacia khinjuk leaf aqueous extract: introducing a novel chemotherapeutic drug for the treatment of prostate cancer. Open Chem. 2024;22:20240034.

114 Mohanta L, Jena BS. Dillenia indica bark extract mediated bio-fabrication of copper nanoparticles: characterisation, antioxidant and anticancer activity in vitro. J Exp Nanosci. 2023;18:2249241. doi:10.1080/17458080.2023.2249241.

115 Dhanraj G, Rajeshkumar S. Anticarcinogenic effect of selenium nanoparticles synthesized using Brassica oleracea. J Nanomater. 2021:8115585.

116 Lazcano-Ramirez HG, Garza-Garcia JJO, Hernandez-Diaz JA, Leon-Morales JM, Macias-Sandoval AS, Garcia-Morales S. Antifungal activity of selenium nanoparticles obtained by plant-mediated synthesis. Antibiotics. 2023;12:115.

117 Tarmizi AA, Nik Ramli NN, Adam SH, Abdul Mutalib M, Mokhtar MH, Tang SGH. Phytofabrication of selenium nanoparticles with Moringa oleifera (MO-SeNPs) and exploring its antioxidant and antidiabetic potential. Molecules. 2023;28:5322.

118 Ghaderi RS, Adibian F, Sabouri Z, Davoodi J, Kazemi M, Amel Jamehdar S, et al. Green synthesis of selenium nanoparticle by Abelmoschus esculentus extract and assessment of its antibacterial activity. Mater Technol. 2022;37:1289–1297.

119 Adhikari A, Chhetri K, Acharya D, Pant B, Adhikari A. Green synthesis of iron oxide nanoparticles using Psidium guajava L. leaves extract for degradation of organic dyes and anti-microbial applications. Catalysts. 2022;12(10):1188.

120 Shah SS, Turakhia BP, Purohit N, Kapadiya KM, Sahoo CR, Prajapati A. Facile green synthesis of iron oxide nanoparticles and their impact on cytotoxicity, antioxidative properties and bactericidal activity. Iran Biomed J. 2024;28:71–81.

3

Biogenic Synthesis of Nanoparticles by Plant Latex and Green Nanotherapeutics

Bapan Banik[1], *Nikita Dey*[2], *Barnali Gogoi*[3], *and Malay K. Das*[4]

[1]*Department of Pharmaceutical Science, Rabindranath Tagore University, Hojai, Assam, India*
[2]*Department of Pharmaceutical Chemistry, NEF College of Pharmaceutical Education and Research, Nagaon, Assam, India*
[3]*Department of Pharmacognosy, Assam Pharmacy Institute, Titabor, Assam, India*
[4]*Drug Delivery Research Laboratory, Department of Pharmaceutical Sciences, Dibrugarh University, Dibrugarh, Assam, India*

3.1 Introduction

The last 10 years have made a remarkable contribution to the field of control-released nanoparticle fabrication. Modification of shape, particle size, and crystalline nature has potential uses, like bio-medical, biosensors, bacterial biotoxin removal catalysts, and affordable cost electrode [1, 2]. Because of their vast therapeutic application, nanosheets, nanotubes, and nanowires have widespread applications [1, 3, 4]. The unique chemistry of nano-particles makes them an intermediary between molecular structures and bulk materials, which makes them a good candidate for applications including trace-substance detection, bio-medical devices, catalysis, and electrochemistry [5].

Nanoparticles with diverse morphology and size have been manufactured with different approaches but a fundamental comprehension of enhanced manufacturing processes is necessary, which can be utilized at the industrial and business level in order to have better-built, more durable, cleaner, safer, and more innovative products, such as household appliances, technologies for communication, pharmaceuticals, transportation, industries, and agriculture. Hence, the primary emphasis is on creating NPs using environmentally benign methods. So, these provide remedies to emerging difficulties pertaining to the environmental issues [6, 7]. On the basis of this, the present review was centered with special focus on plant latex and its application as nanotherapeutics (Figure 3.1).

Biogenic Synthesis of Nanoparticles and Sustainable Nanotherapeutics,
First Edition. Edited by Malay K. Das.

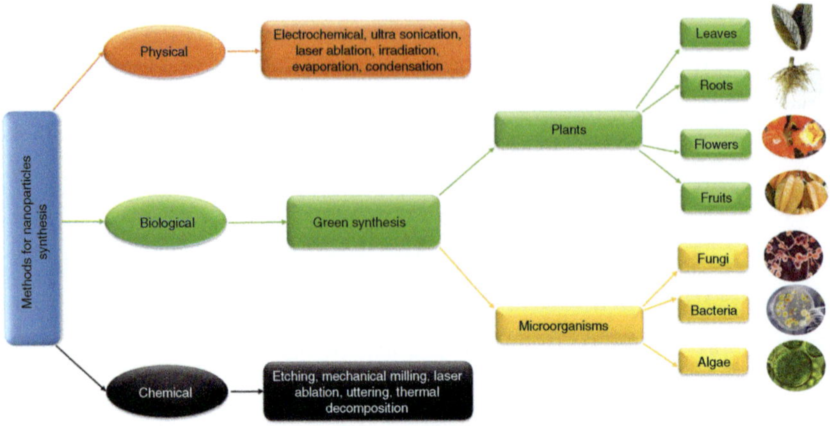

Figure 3.1 Various synthetic methodologies for nanoparticle formation. *Source:* (Roots) Vasin / Adobe Stock Photos, (Fungi) Dr_Microbe / Adobe Stock Photos, (Bacteria) AndreasReh / Getty Images, (Algae) Мальцев Евгений/ Adobe Stock Photos.

3.2 Plants in Nanoparticle Synthesis

Nanomaterial production utilizing plant extracts as a precursor is gaining popularity due to its minimal complexity, ecologically friendly nature, and non-toxicity. Several chemical entities found in plant extracts, including flavonoids, terpenoids, alcohols, polyphenols, carbohydrates, and sugars, operate as reducing and stabilizing agents during the synthesis process. Green synthesis has proven to be a viable alternative to traditional nanomaterials synthesis processes [8]. There are several ways to go green, such as employing plant extracts and creating nanoparticles using bacteria, viruses, fungi, and algae [9]. Regarding environmental preservation, each of these eco-friendly techniques offers unique benefits and drawbacks. When it comes to creating safe, biocompatible, and environmentally friendly nanoparticles, the first green route method – plant extracts – is thought to be the most dependable method [6, 9, 10]. Additionally, the method uses different plant parts – fruits, flowers, latex, leaves, and roots – with different biomolecules, including carbohydrates and amino acids [9, 10].

Plants can remove metal ions from the site of ion penetration, according to research. There have been reports of metals being deposited as nanoparticles [7, 11]. Plant extracts are utilized in a few recently established *in vitro* methods for the metal ion bioreduction to produce nanoparticles [11–13]. By altering the medium pH and reaction temperature, the size and shape of the nanoparticles can be adjusted in this process. Using extracts from a wide range of plant species in conjunction with different metal salts and acids, including copper, gold, silver, platinum, iron, and many more, the *in vitro* method has been shown to work [13–15]. Secondary metabolites, including sugars, terpenoids, polyphenols, alkaloids, phenolic acids, and proteins, have been demonstrated by Fourier transform infrared spectroscopy (FTIR) to play a crucial role in reducing metal ions into nanoparticles and promoting their stability afterward. The size of nanostructures is determined by how this secondary metabolite interacts with metal ions [12, 16]. For example, when cultivated using silver nitrate as a base substance, *Brassica juncea* (mustard greens) and *Medicago sativa* (alfalfa) have been shown to gather 50 nm silver nanoparticles. However, when developed on substrates that contained salts of the corresponding metals, *M. sativa* was discovered to have gold icosahedra that were 4 nm in size, while *Iris pseudacorus* was found to have semi-spherical copper particles that were 2 nm in size [7, 12]. The extracts of *Pelargonium graveolens* are reported to stabilize and reduce gold ions into decahedral icosahedral nanoparticles with a size of 20–40 nm [17].

3.3 Biogenic Production of Nanoparticles from Plant Components

Biogenic synthesis, also known as green synthesis, is an environmentally friendly approach that utilizes biological entities like plants, microbes, or enzymes to produce metallic nanoparticles. Among these, synthesis mediated by plants is especially promising owing to the abundance of phytochemicals present in different plant parts, which serve as natural reducing, capping, and stabilizing agents. These biomolecules assist in transforming metal ions into stable nanoparticles under mild conditions without the use of harmful chemicals. Plant parts such as the stem, latex, callus root, seed, peel, leaves, fruit, and flower are used in biological processes to make metallic nanoparticles of various sizes and types [7, 18, 19]. Nickel, titanium oxide, copper, zinc, silver, platinum, gold, and magnetite are some of the metal-based nanoparticles that are widely generated from plant parts [18] (Table 3.1).

3.3.1 Stem as a Source for Biogenic Synthesis of Metallic Nanoparticles

Stems of certain plants are valuable in green nanoparticle synthesis because of their reservoir of bioactive constituents such as polyphenols, saponins, and glycosides. These compounds facilitate the metal ion reduction and provide stability to the synthesized nanoparticles. These compounds facilitate the metal ion reduction and provide stability to the synthesized nanoparticles. Utilizing plant stems as a biogenic source is advantageous due to the ease of extract preparation and the presence of potent reducing agents. For instance, the stem extract of

Table 3.1 Plant parts are applied for the synthesis of metallic nanoparticles and their pharmacological activity.

Extract of compound	Plant parts	Types of MNPs	Pharmacological application	Reference
Almond extract	Seed	AuNPs-Quercetin AuNPs-Camptothecin	Anti-inflammatory, antineoplastic, anti-angiogenic	[20]
Embelia ribes	Seed	AuNPs AgNPs	Antioxidant, antimicrobial, anticancer	[21]
Taraxacum officinale	Leaf	AgNPs	Antioxidant, antimicrobial, anticancer	[22]
Ocimum sanctum	Leaf	AuNPs	Antioxidant, reducing ability	[23]
Lantana Camara	Leaf	AgNPs	Antibacterial, catalytic	[24]
Melia azedarach	Leaf	AgNPs	Antioxidant, antibacterial, wound healing effect, antidiabetic, cytotoxicity	[25]
Punica granatum	Leaf	AgNPs	Antidiabetic and anticancer	[26]
Perilla frutescens	Leaf	AgNPs	Antioxidant, antibacterial, and anticancer	[27]
Mukia maderaspatna	Fresh Leaf	AuNPs AgNPs	Anticancer	[28]
Mangifera indica(mangiferin) Xanthone derivative	Leaves	AuNPs	Harmless to healthy humans breast cell line	[29]
Artemisia nilagiria	Leaf	AgNPs	Antimicrobial	[30]
Andrographis paniculata	Leaf	AgNPs	Hepatoprotective activity	[31]
Acalypha indica	Leaf	AgNPs, AuNPs	Antibacterial	[32]
Alfaroa *mexicana*	Leaf	AgNPs	Antimicrobial	[33]
Tinospora cordifolia	Leaf	AgNPs	Antilarvicidal	[34]
Menthe piperita	Leaf	AgNPs AuNPs	Antibacterial	[35]

Table 3.1 (Continued)

Extract of compound	Plant parts	Types of MNPs	Pharmacological application	Reference
Lippia citriodora	Leaf	AgNPs	Antimicrobial	[36]
Euphorbia prostrata	Leaf	AgNPs	Antiplasmodial	[37]
Cinnamon zeylanicum	Leaf	AgNPs	Antibacterial	[38]
Backhousia citriodora	Leaf	AuNPs	Antioxidant, anticancer	[39]
Olea europaea	Leaf	CuO-NPs	Anticancer, nontoxic to healthy cells	[40]
Ficus religiosa	Leaf	CuO-NPs	Anticancer	[41]
Cissus quadrangularis and Piper betle	Leaf	CuO-NPs	Antibacterial	[42]
Olive (Olea europaea)	Leaf	ZnO-NPs	Antioxidant	[43]
Falcaria vulgaris	Leaf	CuO-NPs	Anticancer, antioxidant, Antifungal, antibacterial, cutaneous wound recovery	[44]
Cotyledon orbiculata	Fresh leaf	AgNPs	Anti-inflammatory	[45]
Withania somnifera	Leaf	AgNPs	Antimicrobial	
Nigella arvensis	Leaf	AuNPs	Antibacterial, antioxidant, anticancer, catalytic	[46]
Couroupita guianensis Aubl	Fruit	AuNPs	Antioxidant	[47]
Genus Hylocereus (Dragon fruit)	Fruit	AuNPs	Anticancer (mainly breast cancer)	[48]
Limonia acidissima and conjugated epirubicin	Fruit	AuNPs	Drug delivery that targets breast cancer specifically	[49]
Amomum villosum	Dried fruit	AuNPs	Antioxidant, antimicrobial, anticancer	[50]
Aegle marmelos, Eugenia jambolana, and Soursop	Fruit	AuNPs	Anticancer	[51]

(Continued)

Table 3.1 (Continued)

Extract of compound	Plant parts	Types of MNPs	Pharmacological application	Reference
Chaenomeles sinensis	Fruit	AuNPs, AgNPs	Antioxidant, antimicrobial, anticancer	[52]
Panax ginseng	Root	AgNPs	Anticancer, antiviral	[53]
Glycyrrhiza uralensis	Root	AuNPs AgCl-NPs	Antimicrobial, antioxidant, anticancer	[54]
Rhaphidophora aurea intertwined over Lawsonia inermis and Areca catechu	Aerial root	AuNPs	Anticancer	[55]
Rosmarinus sp. *and Zataria multiflora*	Aerial part	AgNPs	Antioxidant and reducing agent	[56]
Curcuma longa	Rhizomes	AuNP-conjugated graphene oxide	Antioxidant and anti-neoplastic	[57]
Arisaema flavum	Tuber	AgNPs	Antibacterial	[58]
Citrus clementina	Peel	AgNPs	Antimicrobial, anticancer	[59]
Pisum sativum	Peel	AgNPs	Antidiabetic, anticancer, antioxidant, antibacterial	[60]
Ganoderma lucidum – oriental Mushroom	Oriental part	AuNP-Doxorubicin conjugate	Anticancer	[61]
Boswellia serrata	Gum	AgNPs	Antibacterial	[62]
Trigonella-foenum graecum	Seed	AuNPs	Catalytic	[63]
Hydrastis canadensis	Whole	AgNPs	Cytotoxicity	[64]
Gelsemium sempervirens	Whole	AgNPs	Cytotoxicity	[64]
Alternanthera sessilis	Whole	AgNPs	Antioxidant, antimicrobial	[65]
Caria papaya	Fruit	AgNPs	Antimicrobial	[66]

Table 3.1 (Continued)

Extract of compound	Plant parts	Types of MNPs	Pharmacological application	Reference
Cassia fistula	Stem	AuNPs	Antihypoglycemic	[67]
Citrullus colocynthis	Calli	AgNPs	Antioxidant, anticancer	[68]
Citrus sinensis	Peel	AgNPs	Antibacterial	[69]
Dillenia indica	Fruit	AgNPs	Antibacterial	[70]
Dioscorea bulbifera	Tuber	AgNPs	Antimicrobial	[71]
Mirabilis jalapa	Flower	AuNPs	Antimicrobial	[72]

Tinospora cordifolia has been successfully employed to synthesize silver nanoparticles (AgNPs) exhibiting strong antimicrobial and antioxidant properties [19]. AgNPs synthesized from *Tephrosia tinctoria* stem extracts showed the capacity to regulate blood glucose levels [73]. Vanaja et al. [74] created silver nanoparticles with *Cissus quadrangularis* stem extract, which produced good effectiveness against the harmful bacteria *Bacillus subtilis* and *Klebsiella planticola*.

3.3.2 Fruits as a Source for Biogenic Synthesis of Metallic Nanoparticles

Fruits are rich in natural antioxidants, organic acids, vitamins, and phenolic compounds that make them effective stabilizers and reducing agents in the environmentally friendly production of metallic nanoparticles. The use of fruit extracts offers a harmless, sustainable, and eco-friendly substitute for traditional chemical synthesis methods. For instance, *Citrus sinensis* (orange) extract from peel has been utilized to synthesize silver nanoparticles with notable antimicrobial and catalytic properties [75]. Furthermore, the fruit extract from *Rumex hymenosepalus* serves as a stabilizing and reducing agent during the silver nanoparticles synthesis [18].

3.3.3 Seeds as a Source for Green Synthesis of Metallic Nanoparticles

Plant seeds have emerged as a promising biological resource for the environmentally friendly production of metal nanoparticles because of their rich phytochemical composition, including proteins, alkaloids, phenolic chemicals, and flavonoids. Flavonoids and other naturally occurring bioactive components like lignin, vitamins, and saponin are prevalent in the extract from fenugreek seeds. A strong reducing agent, fenugreek seed extract, lowers chloroauric acid and enhances surfactant qualities [76]. Flavonoids have the ability to stabilize the electrostatic stability of gold nanoparticles [76].

3.3.4 Leaves as a Source for the Biogenic Synthesis of Metallic Nanoparticles

Leaves of plants are widely utilized in the green synthesis of NPs due to their abundance of bioactive substances, including flavonoids, terpenoids, phenolic acids, and alkaloids. The process is considered eco-friendly and scalable for industrial applications. Leaf extracts from numerous plant species have been studied by researchers, including *Alternanthera sessilis*, *Murraya koenigii*, and *Centella asiatica*. Recent research indicates that leaves of *Piper nigrum* contain a considerable bioactive component that facilitates the ecologically friendly production of nanoparticles. Extracts from *P. nigrum* have been demonstrated to contain longumine and piper longumine, which serve as a capping agent for the production of AgNPs and may intensify the tumor cells' cytotoxic effects [18, 76]. Researchers generated silver nanoparticles utilizing the leaf extract of *Artemisia nilagirica*, which has been demonstrated to be a good antibacterial agent [18, 77, 78].

3.3.5 Plant Flowers as a Source for the Biogenic Synthesis of Metallic Nanoparticles

Flowers of various plants are rich in phytochemicals like anthocyanins, flavonoids, terpenoids, and phenolics, which make them excellent candidates for the green synthesis of NPs. These compounds serve as natural capping and reducing agents, enabling sustainable and non-toxic methods for synthesizing nanoparticles. For instance, *Hibiscus rosa-sinensis* flower extract has been employed to create gold nanoparticles, using notable catalytic and antimicrobial activities [79]. According to Vankar and Bajpai [72, 80], the aqueous extract of the *Mirabilis jalapa* flower functions as a reducing agent while also producing gold nanoparticles.

3.4 Latex as a Source for Nanoparticle Synthesis

Plant latex has a creamy texture, a viscous fluid released by specialized cells known as laticifers in many plant species. This fluid serves as a defense mechanism and is a complex mixture of various biologically active compounds. Since latexes are known to serve a defense function in plants, they may also have potent antibacterial properties, making plants an excellent source of antimicrobial chemicals [81]. The significance of latex extract from plants in preserving biogenic particles and converting metal ions into nanoparticles was highlighted in a number of studies [81–89].

3.4.1 General Composition of Plant Latex

- **Water (50–90%):** It acts as the primary medium in which all other components are suspended [90].
- **Proteins and Enzymes:** Includes hydrolytic enzymes like proteases (e.g. papain, calotropin), lipases, and oxidases, which are crucial for biological activities and reduction of metal ions [91].
- **Alkaloids:** Nitrogen-containing compounds that contribute to antimicrobial and other bioactive properties [92].
- **Terpenoids and Steroids:** These compounds are known for their reducing ability and stabilizing potential during nanoparticle synthesis [93].
- **Phenolic Compounds and Flavonoids:** Potent antioxidants that donate electrons for the reduction of metal ions to form NPs [94].
- **Polysaccharides, Sugars, and Gums:** Involved in capping and stabilizing nanoparticles by forming a protective layer [95].
- **Resins and Waxes:** Contribute to the sticky nature of latex and provide stabilization during nanoparticle formation [90].

3.4.2 Mechanism of Nanoparticle Synthesis Mediated by Plant Latex

Plant latex is widely utilized in green synthesis of metal nanoparticles like gold (Au), silver (Ag), zinc oxide (ZnO), etc. The mechanism can be summarized as follows:

3.4.2.1 Reduction of Metal Ions

Plant latex contains reducing agents (e.g. flavonoids, enzymes, phenolics) that convert metal ions (e.g. Ag^+, Au^{3+}) into their zero-valent form (e.g. Ag^0, Au^0). This process is eco-friendly and eliminates the need for hazardous chemical reducers [92, 96].

3.4.2.2 Nucleation and Growth

Reduced metal atoms nucleate and aggregate to form tiny clusters, which grow into nanoparticles under controlled conditions (temperature, pH, concentration) [93].

3.4.2.3 Capping and Stabilization

Organic molecules in the latex (such as proteins, polysaccharides, and alkaloids) adhere to the surface of NPs, preventing aggregation and controlling morphology [94].

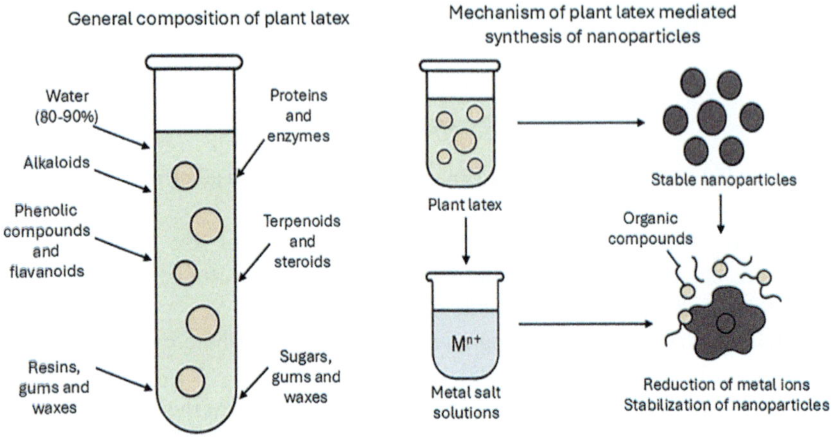

Figure 3.2 General composition of plant latex and mechanism of NPs synthesis mediated by plant latex.

3.4.2.4 Formation of Stable Nanoparticles

The combination of reduction and capping results in stable, uniformly dispersed nanoparticles with enhanced physicochemical properties [91] (Figure 3.2).

3.5 Factors Affecting the Latex-Mediated Synthesis of Metallic Nanoparticles

The green synthesis of metallic NPs utilizing plant latex is influenced by a few biological, physical, and chemical specifications. These factors not only impact the rate of nanoparticle formation but also govern their biological activity, stability, size, structure, and shape.

3.5.1 pH of the Reaction Medium

The pH significantly influences the ionization of functional groups in latex (such as —OH, —COOH, $—NH_2$), which serve as stabilizing and reducing agents. Alkaline pH typically favors the development of smaller and more uniform nanoparticles due to enhanced reduction rates and greater solubility of metal precursors. Example: Silver nanoparticles synthesized using *Jatropha curcas* latex at pH 9 were smaller and more monodispersed than those synthesized at acidic pH [97].

3.5.2 Temperature

Higher temperatures increase the reaction rate and nucleation, potentially leading to smaller nanoparticles with a narrower size distribution. However, excessively high temperatures may denature latex proteins and reduce stability. Example: *Calotropis gigantea* latex-mediated silver nanoparticle synthesis showed optimal size and morphology at 60–80 °C [94].

3.5.3 Reaction Time

The reaction's duration has an impact on the completion of reduction and the growth phase of nanoparticles. Prolonged reaction times can lead to aggregation or overgrowth, increasing particle size. Example: Optimal synthesis of silver NPs using *Euphorbia hirta* latex was observed within three to four hours [98].

3.5.4 Concentration of Metal Salt (Precursor)

The metal salt concentration dictates the quantity of ions that are accessible for reduction, which influences nucleation density. Higher concentrations can lead to particle aggregation, whereas lower concentrations may yield insufficient NPs. Example: a concentration of 1 mM $AgNO_3$ produced well-dispersed silver NPs with *Ficus carica* latex [99].

3.5.5 Concentration and Volume of Latex Extract

The amount of latex controls the availability of reducing and capping agents. A higher concentration may enhance reduction but can also cause excessive capping, leading to larger or aggregated particles. Example: Increased *Hevea brasiliensis* latex volume reduced particle size up to a point, beyond which stabilization was compromised [100].

3.5.6 Stirring Rate and Agitation

Proper mixing enhances the interaction between metal ions and phytochemicals, ensuring uniform reduction and growth. Lack of stirring may cause inhomogeneous particles or incomplete synthesis. Example: Stirring at 500 rpm was optimal for the uniform synthesis of silver NPs using *Calotropis procera* latex [101].

3.5.7 Nature and Composition of Plant Latex

Variability in phytochemical and protein content among different plant species leads to differences in NP morphology, size, and stability. Plants with higher

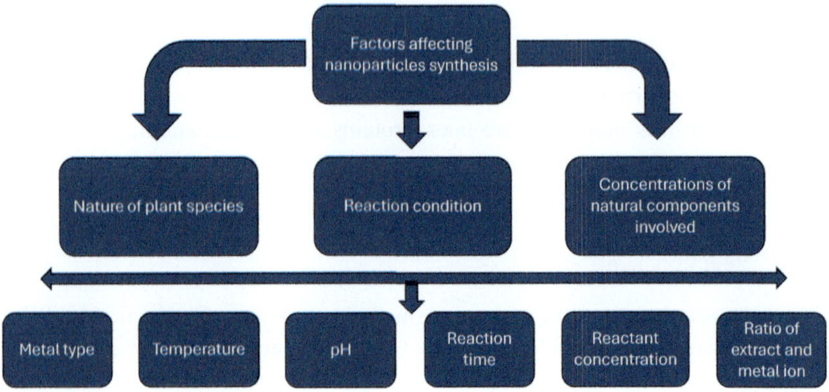

Figure 3.3 Factors influencing the production of metal nanoparticles in plants.

antioxidant content (flavonoids, phenolics) generally yield smaller, more bioactive NPs. Example: latex from *C. procera* and *J. curcas* showed distinct NP morphologies due to differences in their enzymatic and phenolic profiles [90] (Figure 3.3).

3.6 Latex-Mediated Green Nanotherapeutics and Their Applications

3.6.1 Antimicrobial Activity

Silver and copper oxide nanoparticles synthesized from *C. procera* and *E. hirta* exhibit potent antibacterial and antifungal activities by disrupting microbial membranes and generating reactive oxygen species (ROS) [101].

3.6.2 Anticancer Activity

Gold and silver NPs from *C. procera* latex demonstrate cytotoxicity directed against cancerous cells (e.g. MCF-7, HeLa) by inducing apoptosis through mitochondrial pathways [96].

3.6.3 Catalytic Activity

Latex-mediated Ag and ZnO NPs catalyze the reduction of dyes (such as methylene blue) reduction and the degradation of organic pollutants, making them useful in environmental remediation [97].

3.6.4 Antioxidant Activity

Due to phytochemical capping, these NPs exhibit strong free radical scavenging activity, suitable for cosmetic and nutraceutical use [95].

3.6.5 Agricultural Use

ZnO and CuO NPs synthesized using latex have shown plant growth-promoting and pest control effects, indicating potential as nano-fertilizers or biopesticides [97].

3.7 Patents Linked to Latex-Mediated Nanoparticles

Several patents have been filed concerning the synthesis of metallic nanoparticles (NPs) using plant latex or related plant extracts. These patents highlight the growing interest in green synthesis procedures for producing NPs with potential applications in medicine, cosmetics, and environmental remediation. Some notable examples are laid out in Table 3.2.

3.8 Conclusion and Future Perspective

Plant latex is a versatile and efficient bioresource for the green synthesis of NPs, offering an economical, environmentally beneficial, and biologically active alternative to conventional methods. It also holds tremendous potential for the development of green nanotherapeutics due to its biocompatibility, antioxidant potential, and synergistic antimicrobial properties. Latex-derived NPs can be functionalized for targeted drug delivery [90], especially in cancer therapy, by exploiting their natural affinity for cellular membranes. Latex-mediated silver and zinc oxide NPs show promise in topical formulations for treating infectious wounds, burns, and ulcers [94]. Plant latex can be incorporated into biosensors for detecting pathogens or heavy metals, enhancing sensitivity due to its protein content [100]. Latex use aligns with green chemistry principles, making it a sustainable alternative to synthetic stabilizers and reducers in nanotechnology [96]. Integration of latex with polymeric systems or other nanocarriers can enhance the physicochemical properties of drug delivery platforms [100]. Despite its promise, latex-based synthesis presents several challenges, like variability in latex composition [90], limited scalability [96], stability of latex [100], lack of mechanistic understanding [94], etc. There are technical hurdles to overcome, particularly concerning

Table 3.2 List of accessible Patents relating to latex-mediated nanoparticles.

Patent number	Title	Year	Type of NP	Plant material	Remarks	Reference
US9777125B2 (U.S. Patent)	Method for Producing Polymer Latex Particles Coated with Silver NPs	2017	Silver (AgNPs)	Polymer latex (synthetic)	Coated silver NPs on polymer latex particles; antibacterial use	[102]
WO2016106466A1 (WO Patent)	Method for Producing Gold Nanoparticles Using Plant Extracts	2016	Gold (AuNPs)	Various (including latex)	Green synthesis using latex-rich plant extracts	[103]
US20180002189A1 (U.S. Patent)	Method for Producing Gold Nanoparticles from Plant Extracts	2018	Gold (AuNPs)	Leaves, stems, latex	Uses various plant parts, emphasizing latex as a reducing agent	[104]
US9428399B1 (U.S. Patent)	Metal and Metal Oxide Nanoparticle Synthesis Using Plant Extracts	2016	Ag, ZnO, CuO NPs	Seeds (but applicable to latex)	Principles extend to latex; green synthesis platform	[105]
CN104278521A (Chinese Patent)	Method for Preparing Nano-Silver Solution from Natural Plant Latex	2015	Silver (AgNPs)	Natural latex (plant-derived)	Specific use of plant latex; applied in antimicrobial formulations	[106]

standardization, scalability, and mechanistic clarity. Continued research and development could revolutionize its application in nanomedicine, agriculture, and environmental science.

Funding

This study was not supported by outside funds.

Conflicts of Interest

No conflicts of interest are disclosed by the authors. All tables and the figure are original and self-made.

References

1 Antonyraj CA, Jeong J, Kim B, Shin S, Kim S, Lee KY, Cho JK. Selective oxidation of HMF to DFF using Ru/calumina catalyst in moderate boiling solvents toward industrial production. J Ind Eng Chem. 2013;19(3):1056–1059.
2 Sharma D, Kanchi S, Bisetty K. Biogenic synthesis of nanoparticles: a review. Arab J Chem. 2019;12(8):3576–3600.
3 Fu H, Yang X, Jiang X, Yu A. Bimetallic Ag–Au nanowires: synthesis, growth mechanism, and catalytic properties. Langmuir. 2013;29(23):7134–7142.
4 Kwon SJ, Bard AJ. DNA analysis by application of Pt nanoparticle electrochemical amplification with single label response. J Am Chem Soc. 2012;134(26):10777–10779.
5 Xia Y, Yang H, Campbell CT. Nanoparticles for catalysis. Acc Chem Res. 2013;46(8):1671–1672.
6 Banik B, Borkotoky S, Das MK. Biosynthesized colloidal metallic nanoparticles-based nanocosmetic formulations. Nanocosmeceuticals. 2022:369–388.
7 Banik B, Barman J, Das MK. Recent advancement in photo nanotheranostics for cancer treatment. In: Multifunctional Theranostic Nanomedicines in Cancer. 2021 Jan 1. Academic Press, p. 163–176.
8 Rani N, Singh P, Kumar S, Kumar P, Bhankar V, Kumar K. Plant-mediated synthesis of nanoparticles and their applications: a review. Mater Res Bull. 2023;163:112233.
9 Alqarni LS, Alghamdi MD, Alshahrani AA, Nassar AM. Green nanotechnology: recent research on bioresource-based nanoparticle synthesis and applications. J Chem. 2022;2022(1):4030999.

10 Jeevanandam J, Kiew SF, Boakye-Ansah S, Lau SY, Barhoum A, Danquah MK, Rodrigues J. Green approaches for the synthesis of metal and metal oxide nanoparticles using microbial and plant extracts. Nanoscale. 2022;14(7):2534–2571.

11 Manceau A, Nagy KL, Marcus MA, Lanson M, Geoffroy N, Jacquet T, Kirpichtchikova T. Formation of metallic copper nanoparticles at the soil–root interface. Environ Sci Technol. 2008;42(5):1766–1772.

12 Makarov VV, Love AJ, Sinitsyna OV, Makarova SS, Yaminsky IV, Taliansky ME, Kalinina NO. "Green" nanotechnologies: synthesis of metal nanoparticles using plants. Acta Nat. 2014;6(1):35–44.

13 Njagi EC, Huang H, Stafford L, Genuino H, Galindo HM, Collins JB, et al. Biosynthesis of iron and silver nanoparticles at room temperature using aqueous sorghum bran extracts. Langmuir. 2011;27(1):264–271.

14 Harris AT, Bali R. On the formation and extent of uptake of silver nanoparticles by live plants. J Nanopart Res. 2008;10:691–695.

15 Rai M, Yadav A. Plants as potential synthesiser of precious metal nanoparticles: progress and prospects. IET Nanobiotechnol. 2013;7(3):117–124.

16 Shankar SS, Ahmad A, Pasricha R, Sastry M. Bioreduction of chloroaurate ions by geranium leaves and its endophytic fungus yields gold nanoparticles of different shapes. J Mater Chem. 2003;13(7):1822–1826.

17 Shankar SS, Ahmad A, Sastry M. Geranium leaf assisted biosynthesis of silver nanoparticles. Biotechnol Prog. 2003;19(6):1627–1631.

18 Kuppusamy P, Yusoff MM, Maniam GP, Govindan N. Biosynthesis of metallic nanoparticles using plant derivatives and their new avenues in pharmacological applications–an updated report. Saudi Pharm J. 2016;24(4):473–484.

19 Lekkala VDVV, Muktinutalapati AV, Lebaka VR, Lomada D, Korivi M, Li W, Reddy MC. Green synthesis and characterization of silver nanoparticles from *Tinospora cordifolia* leaf extract: evaluation of their antioxidant, anti-inflammatory, antibacterial, and antibiofilm efficacies. Nanomaterials. 2025;15(5):381.

20 Sadalage PS, Patil RV, Havaldar DV, Gavade SS, Santos AC, Pawar KD. Optimally biosynthesized, PEGylated gold nanoparticles functionalized with quercetin and camptothecin enhance potential anti-inflammatory, anti-cancer and anti-angiogenic activities. J Nanobiotechnol. 2021;19(1):84.

21 Dhayalan M, Denison MI, Anitha Jegadeeshwari L, Krishnan K, Nagandra NG. in vitro antioxidant, antimicrobial, cytotoxic potential of gold and silver nanoparticles prepared using Embelia ribes. Nat Prod Res. 2017;31(4):465–468.

22 Saratale RG, Benelli G, Kumar G, Kim DS, Saratale GD. Bio-fabrication of silver nanoparticles using the leaf extract of an ancient herbal medicine, dandelion (Taraxacum officinale), evaluation of their antioxidant, anticancer potential, and antimicrobial activity against phytopathogens. Environ Sci Pollut Res Int. 2018;25(11):10392–10406.

23 Lee SY, Krishnamurthy S, Cho CW, Yun YS. Biosynthesis of gold nanoparticles using Ocimum sanctum extracts by solvents with different polarity. ACS Sustain Chem Eng. 2016;4(5):2651–2659.

24 Ajitha B, Reddy YA, Shameer S, Rajesh KM, Suneetha Y, Reddy PS. Lantana camara leaf extract mediated silver nanoparticles: antibacterial, green catalyst. J Photochem Photobiol B Biol. 2015;149:84–92.

25 Chinnasamy G, Chandrasekharan S, Bhatnagar S. Biosynthesis of silver nanoparticles from Melia azedarach: enhancement of antibacterial, wound healing, antidiabetic and antioxidant activities. Int J Nanomedicine. 2019:9823–9836.

26 Saratale RG, Shin HS, Kumar G, Benelli G, Kim DS, Saratale GD. Exploiting antidiabetic activity of silver nanoparticles synthesized using Punica granatum leaves and anticancer potential against human liver cancer cells (HepG2). Artif Cells Nanomed Biotechnol. 2018;46(1):211–222.

27 Reddy NV, Li H, Hou T, Bethu MS, Ren Z, Zhang Z. Phytosynthesis of silver nanoparticles using Perilla frutescens leaf extract: characterization and evaluation of antibacterial, antioxidant, and anticancer activities. Int J Nanomedicine. 2021:15–29.

28 Devi GK, Sathishkumar K. Synthesis of gold and silver nanoparticles using Mukia maderaspatna plant extract and its anticancer activity. IET Nanobiotechnol. 2017;11(2):143–151.

29 Patra N, Dehury N, Pal A, Behera A, Patra S. Preparation and mechanistic aspect of natural xanthone functionalized gold nanoparticle. Mater Sci Eng C. 2018;90:439–445.

30 Song JY, Kim BS. Biological synthesis of bimetallic Au/Ag nanoparticles using Persimmon (Diopyros kaki) leaf extract. Korean J Chem Eng. 2008;25:808–811.

31 Suriyakalaa U, Antony JJ, Suganya S, Siva D, Sukirtha R, Kamalakkannan S, et al. Hepatocurative activity of biosynthesized silver nanoparticles fabricated using Andrographis paniculata. Colloids Surf B: Biointerfaces. 2013;102:189–194.

32 Krishnaraj C, Jagan EG, Rajasekar S, Selvakumar P, Kalaichelvan PT, Mohan NJ. Synthesis of silver nanoparticles using Acalypha indica leaf extracts and its antibacterial activity against water borne pathogens. Colloids Surf B: Biointerfaces. 2010;76(1):50–56.

33 Singh C, Sharma V, Naik PK, Khandelwal V, Singh H. A green biogenic approach for synthesis of gold and silver nanoparticles using Zingiber officinale. Dig J Nanomater Biostruct. 2011;6(2):535–542.

34 Jayaseelan C, Rahuman AA, Rajakumar G, Vishnu Kirthi A, Santhoshkumar T, Marimuthu S, et al. Synthesis of pediculocidal and larvicidal silver nanoparticles by leaf extract from heartleaf moonseed plant, Tinospora cordifolia Miers. Parasitol Res. 2011;109:185–194.

35 MubarakAli D, Thajuddin N, Jeganathan K, Gunasekaran M. Plant extract mediated synthesis of silver and gold nanoparticles and its antibacterial activity against clinically isolated pathogens. Colloids Surf B: Biointerfaces. 2011;85(2):360–365.

36 Cruz D, Falé PL, Mourato A, Vaz PD, Serralheiro ML, Lino AR. Preparation and physicochemical characterization of Ag nanoparticles biosynthesized by Lippia citriodora (Lemon Verbena). Colloids Surf B: Biointerfaces. 2010;81(1):67–73.

37 Zahir AA, Rahuman AA. Evaluation of different extracts and synthesised silver nanoparticles from leaves of Euphorbia prostrata against Haemaphysalis bispinosa and Hippobosca maculata. Vet Parasitol. 2012;187(3–4):511–520.

38 Sathishkumar M, Sneha K, Won SW, Cho CW, Kim S, Yun YS. Cinnamon zeylanicum bark extract and powder mediated green synthesis of nano-crystalline silver particles and its bactericidal activity. Colloids Surf B: Biointerfaces. 2009;73(2):332–338.

39 Khandanlou R, Murthy V, Saranath D, Damani H. Synthesis and characterization of gold-conjugated Backhousia citriodora nanoparticles and their anticancer activity against MCF-7 breast and HepG2 liver cancer cell lines. J Mater Sci. 2018;53(5):3106–3118.

40 Sulaiman GM, Tawfeeq AT, Jaaffer MD. Biogenic synthesis of copper oxide nanoparticles using olea europaea leaf extract and evaluation of their toxicity activities: an in vivo and in vitro study. Biotechnol Prog. 2018;34(1):218–230.

41 Sankar R, Maheswari R, Karthik S, Shivashangari KS, Ravikumar V. Anticancer activity of Ficus religiosa engineered copper oxide nanoparticles. Mater Sci Eng C. 2014;44:234–239.

42 Kasi SD, Ramasamy JM, Nagaraj D, Santiyagu V, Ponraj JS. Biogenic synthesis of copper oxide nanoparticles using leaf extracts of Cissus quadrangularis and Piper betle and its antibacterial effects. Micro Nano Lett. 2021;16(8):419–424.

43 Hashemi S, Asrar Z, Pourseyedi S, Nadernejad N. Green synthesis of ZnO nanoparticles by Olive (Olea europaea). IET Nanobiotechnol. 2016;10(6):400–404.

44 Zangeneh MM, Ghaneialvar H, Akbaribazm M, Ghanimatdan M, Abbasi N, Goorani S, et al. Novel synthesis of Falcaria vulgaris leaf extract conjugated copper nanoparticles with potent cytotoxicity, antioxidant, antifungal, antibacterial, and cutaneous wound healing activities under in vitro and in vivo condition. J Photochem Photobiol B Biol. 2019;197:111556.

45 Tyavambiza C, Elbagory AM, Madiehe AM, Meyer M, Meyer S. The antimicrobial and anti-inflammatory effects of silver nanoparticles synthesised from Cotyledon orbiculata aqueous extract. Nanomaterials. 2021;11(5):1343.

46 Chahardoli A, Karimi N, Sadeghi F, Fattahi A. Green approach for synthesis of gold nanoparticles from Nigella arvensis leaf extract and evaluation of their antibacterial, antioxidant, cytotoxicity and catalytic activities. Artif Cells Nanomed Biotechnol. 2018;46(3):579–588.

47 Sathishkumar G, Jha PK, Vignesh V, Rajkuberan C, Jeyaraj M, Selvakumar M, et al. Cannonball fruit (Couroupita guianensis, Aubl.) extract mediated synthesis of gold nanoparticles and evaluation of its antioxidant activity. J Mol Liq. 2016;215:229–236.

48 Saravanan M, Vahidi H, Medina Cruz D, Vernet-Crua A, Mostafavi E, Stelmach R, et al. Emerging antineoplastic biogenic gold nanomaterials for breast cancer therapeutics: a systematic review. Int J Nanomedicine. 2020:3577–3595.

49 Kumar CS, Mahesh A, Antoniraj MG, Vaidevi S, Ruckmani K. Ultrafast synthesis of stabilized gold nanoparticles using aqueous fruit extract of Limonia acidissima L. and conjugated epirubicin: targeted drug delivery for treatment of breast cancer. RSC Adv. 2016;6(32):26874–26882.

50 Soshnikova V, Kim YJ, Singh P, Huo Y, Markus J, Ahn S, et al. Cardamom fruits as a green resource for facile synthesis of gold and silver nanoparticles and their biological applications. Artif Cells Nanomed Biotechnol. 2018;46(1):108–117.

51 Vijayakumar S. Eco-friendly synthesis of gold nanoparticles using fruit extracts and in vitro anticancer studies. J Saudi Chem Soc. 2019;23(6): 753–761.

52 Oh KH, Soshnikova V, Markus J, Kim YJ, Lee SC, Singh P, et al. Biosynthesized gold and silver nanoparticles by aqueous fruit extract of Chaenomeles sinensis and screening of their biomedical activities. Artif Cells Nanomed Biotechnol. 2018;46(3):599–606.

53 Sreekanth TV, Nagajyothi PC, Muthuraman P, Enkhtaivan G, Vattikuti SV, Tettey CO, et al. Ultra-sonication-assisted silver nanoparticles using Panax ginseng root extract and their anti-cancer and antiviral activities. J Photochem Photobiol B Biol. 2018;188:6–11.

54 Huo Y, Singh P, Kim YJ, Soshnikova V, Kang J, Markus J, et al. Biological synthesis of gold and silver chloride nanoparticles by Glycyrrhiza uralensis and in vitro applications. Artif Cells Nanomed Biotechnol. 2018;46(2):303–312.

55 Jannathul Firdhouse M, Lalitha P. Cytotoxicity of spherical gold nanoparticles synthesised using aqueous extracts of aerial roots of Rhaphidophora aurea (Linden ex Andre) intertwined over Lawsonia inermis and Areca catechu on MCF-7 cell line. IET Nanobiotechnol. 2017;11(1):2–11.

56 Goodarzi V, Zamani H, Bajuli L, Moradshahi A. Evaluation of antioxidant potential and reduction capacity of some plant extracts in silver nanoparticles' synthesis. Mol Biol Res Commun. 2014;3(3):165.

57 Al-Ani LA, Yehye WA, Kadir FA, Hashim NM, AlSaadi MA, Julkapli NM, Hsiao VK. Hybrid nanocomposite curcumin-capped gold nanoparticle-reduced graphene oxide: anti-oxidant potency and selective cancer cytotoxicity. PLoS One. 2019;14(5):e0216725.

58 Rahman AU, Khan AU, Yuan Q, Wei Y, Ahmad A, Ullah S, et al. Tuber extract of Arisaema flavum eco-benignly and effectively synthesize silver nanoparticles: photocatalytic and antibacterial response against multidrug resistant engineered E. coli QH4. J Photochem Photobiol B Biol. 2019;193:31–38.

59 Saratale RG, Shin HS, Kumar G, Benelli G, Ghodake GS, Jiang YY, et al. Exploiting fruit byproducts for eco-friendly nanosynthesis: citrus × clementina peel extract mediated fabrication of silver nanoparticles with high efficacy against microbial pathogens and rat glial tumor C6 cells. Environ Sci Pollut Res. 2018;25:10250–10263.

60 Patra JK, Das G, Shin HS. Facile green biosynthesis of silver nanoparticles using Pisum sativum L. outer peel aqueous extract and its antidiabetic, cytotoxicity, antioxidant, and antibacterial activity. Int J Nanomedicine. 2019:6679–6690.

61 Kumar DS, Senthilkumar P, Surendran L, Sudhagar B. Ganoderma lucidum oriental mushroom mediated synthesis of gold nanoparticles conjugated with doxorubicin and evaluation of its anticancer potential on human breast cancer MCF-7/DOX cells. Int J Pharm Pharm Sci. 2017;9(9):267–274.

62 Kora AJ, Sashidhar RB, Arunachalam J. Aqueous extract of gum olibanum (Boswellia serrata): a reductant and stabilizer for the biosynthesis of antibacterial silver nanoparticles. Process Biochem. 2012;47(10):1516–1520.

63 Aromal SA, Philip D. Green synthesis of gold nanoparticles using Trigonella foenum-graecum and its size-dependent catalytic activity. Spectrochim Acta A Mol Biomol Spectrosc. 2012;97:1–5.

64 Das RK, Gogoi N, Bora U. Green synthesis of gold nanoparticles using Nyctanthes arbortristis flower extract. Bioprocess Biosyst Eng. 2011;34: 615–619.

65 Niraimathi KL, Sudha V, Lavanya R, Brindha P. Biosynthesis of silver nanoparticles using Alternanthera sessilis (Linn.) extract and their antimicrobial, antioxidant activities. Colloids Surf B: Biointerfaces. 2013;102: 288–291.

66 Jain D, Daima HK, Kachhwaha S, Kothari SL. Synthesis of plant-mediated silver nanoparticles using papaya fruit extract and evaluation of their anti microbial activities. Digest J Nanomater Biostruct. 2009;4(3):557–563.

67 Daisy P, Saipriya K. Biochemical analysis of Cassia fistula aqueous extract and phytochemically synthesized gold nanoparticles as hypoglycemic treatment for diabetes mellitus. Int J Nanomedicine. 2012:1189–1202.

68 Satyavani K, Gurudeeban S, Ramanathan T, Balasubramanian T. Biomedical potential of silver nanoparticles synthesized from calli cells of Citrullus colocynthis (L.) Schrad. J Nanobiotechnol. 2011;9:43.

69 Kaviya S, Santhanalakshmi J, Viswanathan B, Muthumary J, Srinivasan K. Biosynthesis of silver nanoparticles using Citrus sinensis peel extract and its antibacterial activity. Spectrochim Acta A Mol Biomol Spectrosc. 2011;79(3):594–598.

70 Singh S, Saikia JP, Buragohain AK. A novel 'green' synthesis of colloidal silver nanoparticles (SNP) using Dillenia indica fruit extract. Colloids Surf B: Biointerfaces. 2013;102:83–85.

71 Ghosh S, Patil S, Ahire M, Kitture R, Kale S, Pardesi K, et al. Synthesis of silver nanoparticles using Dioscorea bulbifera tuber extract and evaluation of its synergistic potential in combination with antimicrobial agents. Int J Nanomedicine. 2012:483–496.

72 Vankar PS, Bajpai D. Preparation of gold nanoparticles from *Mirabilis jalapa* flowers. Indian J Biochem Biophys. 2010;47(3):157–160.

73 Rajaram K, Aiswarya DC, Sureshkumar P. Green synthesis of silver nanoparticle using Tephrosia tinctoria and its antidiabetic activity. Mater Lett. 2015;138:251–254.

74 Vanaja M, Rajeshkumar S, Paulkumar K, Gnanajobitha G, Malarkodi C, Annadurai G. Phytosynthesis and characterization of silver nanoparticles using stem extract of Coleus aromaticus. Int J Mater Biomater Appl. 2013;3(1):1–4.

75 Ibrahim HM. Green synthesis and characterization of silver nanoparticles using banana peel extract and their antimicrobial activity against representative microorganisms. J Radiat Res Appl Sci. 2015;8(3):265–275.

76 Mittal AK, Chisti Y, Banerjee UC. Synthesis of metallic nanoparticles using plant extracts. Biotechnol Adv. 2013;31(2):346–356.

77 Jacob SJ, Finub JS, Narayanan A. Synthesis of silver nanoparticles using *Piper longum* leaf extracts and its cytotoxic activity against Hep-2 cell line. Colloids Surf B: Biointerfaces. 2012;91:212–214.

78 Vijayakumar M, Priya K, Nancy FT, Noorlidah A, Ahmed AB. Biosynthesis, characterisation and anti-bacterial effect of plant-mediated silver nanoparticles using Artemisia nilagirica. Ind Crop Prod. 2013;41:235–240.

79 Rao BL, Gouda GP, Shivananda CS. Green synthesis of silver nanoparticles using Hibiscus Rosa Sinensis flower extract. In AIP Conference Proceedings 2020 May 4 (Vol. 2220, No. 1). AIP Publishing.

80 Banik B, Das MK. Nano Medicine and Nano Safety: Recent Trends and Clinical Evidences. Nanomaterials for alternative antibiotic therapy. 2020, p. 85–96.

81 Bhagyashri AC, Jogendra CH, Avinash VP. Plant latex: an inherent spring of phsarmaceuticals. World J Pharm Sci. 2015;4:1781–1796.

82 Mesquita ML, Desrivot J, Fournet A, Paula JE, Grellier P, Espindola LS. Antileishmanial and trypanocidal activity of Brazilian Cerrado plants. Mem Inst Oswaldo Cruz. 2005;100:783–787.

83 Domsalla A, Melzig MF. Occurrence and properties of proteases in plant latices. Planta Med. 2008;74(7):699–711.

84 Kitajima S, Kamei K, Taketani S, Yamaguchi M, Kawai F, Komatsu A, Inukai Y. Two chitinase-like proteins abundantly accumulated in latex of mulberry show insecticidal activity. BMC Biochem. 2010;11:1–7.

85 Fernandez-Arche A, Saenz MT, Arroyo M, De la Puerta R, Garcia MD. Topical anti-inflammatory effect of tirucallol, a triterpene isolated from Euphorbia lactea latex. Phytomedicine. 2010;17(2):146–148.

86 De Marino S, Gala F, Zollo F, Vitalini S, Fico G, Visioli F, Iorizzi M. Identification of minor secondary metabolites from the latex of Croton lechleri (Muell-Arg) and evaluation of their antioxidant activity. Molecules. 2008;13(6):1219–1229.

87 Mendonça RJ, Maurício VB, de Bortolli TL, Lachat JJ, Coutinho-Netto J. Increased vascular permeability, angiogenesis and wound healing induced by the serum of natural latex of the rubber tree Hevea brasiliensis. Phytother Res. 2010;24(5):764–768.

88 Chandhru M, Logesh R, Kutti Rani S, Ahmed N, Vasimalai N. Green synthesis of silver nanoparticles from plant latex and their antibacterial and photocatalytic studies. Environ Technol. 2022;43(20):3064–3074.

89 Asare GA, Bugyei K, Sittie A, Yahaya ES, Gyan B, Adjei S, et al. Genotoxicity, cytotoxicity and toxicological evaluation of whole plant extracts of the medicinal plant Phyllanthus niruri (Phyllanthaceae). Genet Mol Res. 2012;11(1):100–111.

90 Rai M, Pandey S. Plant latex as a novel source for the synthesis of nanoparticles: a review. Appl Nanosci. 2020;10(8):2867–2881.

91 Siddiqui S, Al-Naqeb G, Kausar MA. Biogenic synthesis of nanoparticles from latex-bearing plants and their antimicrobial properties. J King Saud Univ Sci. 2022;34(3):101902.

92 Gade A, Bonde P, Ingle A, Marcato PD, Duran N, Rai M. Exploitation of *Azadirachta indica* A. Juss seed kernels for synthesis of silver nanoparticles. Colloids Surf B: Biointerfaces. 2008;66(1):140–146.

93 Bhambure R, Bule M, Shaligram N, Kamat M, Kanekar P. Extracellular biosynthesis of silver nanoparticles using Klebsiella pneumoniae. Biotechnol Bioeng. 2009;104(5):984–991.

94 Kumar V, Yadav SK, Yadav SC. Green synthesis and characterization of silver nanoparticles using latex of Calotropis gigantea and its antimicrobial efficacy. Mater Sci Eng. 2021;118:111478.

95 Rajendran K, Gnanavel V, Suganthi R. Latex-mediated biosynthesis of silver nanoparticles and their antimicrobial properties. J Mol Liq. 2021;329:115560.

96 Ahmed S, Ahmad M, Swami BL, Ikram S. A review on plants extract-mediated synthesis of silver nanoparticles for antimicrobial applications: a green expertise. J Adv Res. 2016;7(1):17–28.

97 Heera P, Sharma Y, Anand V, Kumar A, Kumar R. Preparation of Zno nanoparticles by green synthesis using jatropha latex as reducing agent. Available at SSRN 5052267.

98 Elumalai EK, Prasad TN, Venkata Kambala VK, Nagajyothi PC, David E. Green synthesis of silver nanoparticle using *Euphorbia hirta* L and their antifungal activities. Arch Appl Sci Res. 2010;2(6):76–81.

99 Alagar M, Rajagopal K. Green synthesis of silver nanoparticles using Ficus carica latex and their antibacterial activity. Asian J Pharm Clin Res. 2017;10(7):353–356.

100 Ashraf S, ul Ain Q, Raza A, Riaz S. Latex from Hevea brasiliensis: a novel reducing agent for the synthesis of gold and silver nanoparticles. Mater Lett. 2016;181:166–169.

101 Singh R, Shedbalkar UU, Wadhwani SA, Chopade BA. Bacteriagenic silver nanoparticles: synthesis, mechanism, and applications. Appl Microbiol Biotechnol. 2020;104(2):457–471.

102 LG Household & Health Care Ltd. Method for producing polymer latex particles coated with silver nanoparticles (U.S. Patent No. US9777125B2). U.S. Patent and Trademark Office. 2017. https://patents.google.com/patent/US9777125B2.

103 University of Malaya. Method for producing gold nanoparticles using plant extracts (WO Patent No. WO2016106466A1). World Intellectual Property Organization. 2016. https://patents.google.com/patent/WO2016106466A1.

104 Chakrabarty AM, et al. Method for producing gold nanoparticles from plant extracts (U.S. Patent Application No. US20180002189A1). United States Patent Application Publication. 2028. https://patents.justia.com/patent/20180002189.

105 Katti KV. Synthesis of nanoparticles of metals and metal oxides using plant seed extracts (U.S. Patent No. US9428399B1). U.S. Patent and Trademark Office. 2016. https://patents.google.com/patent/US9428399B1.

106 Guangdong University of Technology. Method for preparing nano-silver solution from natural plant latex (Chinese Patent No. CN104278521A). China National Intellectual Property Administration. 2015. https://patents.google.com/patent/CN104278521A.

4

Biological Synthesis of Nanoparticles by Bacteria and Biogenic Sustainable Nanotherapeutics

Junmoni Nath[1], *Bhanu P. Sahu*[1], *and Malay K. Das*[2]

[1]*Department of Pharmaceutics, School of Pharmaceutical Sciences, Girijananda Chowdhury University, Guwahati, Assam, India*
[2]*Drug Delivery Research Laboratory, Department of Pharmaceutical Sciences, Dibrugarh University, Dibrugarh, Assam, India*

4.1 Introduction

Designing and modifying materials on an extraordinarily small scale, between 1 nm and 100 nm, is the focus of the cutting-edge scientific discipline known as nanotechnology. This technique enables the creation of materials with special qualities and functions that are suited for particular uses by operating at the atomic, molecular, and supramolecular levels. Nanoscience has become well-known worldwide in recent years for its creative methods of making nanoparticles in a sustainable manner. By providing solutions for sophisticated electronics, efficient energy systems, medical diagnostics, and illness treatment, these microscopic particles have transformed a variety of industries, including healthcare, energy, electronics, agriculture, and even space exploration. A new use of nanotechnology that has a broad influence in the medical field is nanotherapeutics. With the development of nanotherapeutics and diagnostics, drugs can become more effective, with fewer or no adverse issues [1–5].

Two main approaches can be used to produce nanomaterials: The first is a bottom-up approach, which involves combining single atoms or molecules through chemical methods like electrochemical, photochemical, hydrothermal, microwave, Sono chemical, redox, co-precipitation, pyrolysis, and sol-gel methods [6], and the second is a top-down strategy (physical techniques like lithography, evaporation-condensation ball milling, vapor and gas phase, arc discharge, electro-deposition, pulsed laser ablation, sonication, and spray pyrolysis), where nanomaterials are created by shrinking large molecules to nanoscale size [7].

Biogenic Synthesis of Nanoparticles and Sustainable Nanotherapeutics,
First Edition. Edited by Malay K. Das.

Furthermore, nanomaterials are created using a bottom-up biological method, which utilizes viruses, bacteria, biomolecules, plants, fungus, and yeasts [8].

Although industries generally favor the chemical and physical approaches, their drawbacks outweigh their benefits. Typically, they are costly, time- and energy-intensive, involve intricate processes, and produce hazardous byproducts [9, 10]. Therefore, compared to traditional chemical and physical methods, biological synthesis of nanoparticles offers a promising and sustainable alternative. This method is environmentally beneficial because it uses biological sources such as microbes and plants. By utilizing easily accessible, renewable resources, it also lowers expenses. The potential of biogenic nanoparticles to transform nanoparticle synthesis for a range of applications is drawing increasing attention [11–14]. The ability of bacteria to undergo genetic mutations and their distinct growth patterns make them particularly useful for the generation of biogenic and industrial nanoparticles. These qualities enable them to be customized for certain applications. Furthermore, bacteria can absorb and digest metals, including those present in environmental contaminants, which makes them useful for the production of sustainable nanoparticles and environmental cleanup projects [15]. The production of nanoparticles utilizing a variety of biological organisms, such as bacteria, actinobacteria, yeast, microalgae, fungi, and viruses, has been the subject of numerous studies. These living organisms provide sustainable, organic processes for creating nanoparticles with particular qualities. This biogenic technique is preferred because it is environmentally friendly and adaptable [10, 16–21]. Microbial cells grow easily, flourish under a variety of conditions, and proliferate quickly. Exopolysaccharides, nanocellulose, and nanowires are special nanomaterials that bacteria and microalgae can produce. Owing to their unique architectures and characteristics, these materials are useful for a wide range of applications [22, 23].

The capacity of some bacterial species, including *Bacillus licheniformis, Pseudomonas deceptionensis,* and *Pyrococcus furiosus,* to produce metal nanoparticles of silver, gold, iron, nickel, and zinc has recently been investigated. These microorganisms convert metal ions into nanoparticles via their own metabolic mechanisms. Nanoparticles with certain physical, chemical, and biological characteristics can be created by genetically altering their metabolic pathways. This method enables the creation of customized nanoparticles for a range of applications in medicine, electronics, and environmental remediation [10, 17–19, 21, 24].

4.2 Mechanisms of Nanoparticle Synthesis by Bacteria

The biological synthesis of nanoparticles by bacteria has emerged as a significant area of research in green nanotechnology because of its eco-friendly, cost-effective, and scalable nature. Unlike chemical and physical methods, which often involve hazardous chemicals, high energy consumption, and environmental risks, bacterial synthesis of nanoparticles operates under mild ambient conditions [25]. This approach utilizes the natural metabolic and enzymatic machinery of bacteria to reduce metal ions to stable nanoparticles, offering a sustainable alternative for nanoparticle production. Nanoparticles are synthesized by bacteria through intracellular and extracellular mechanisms, primarily as a defense mechanism against the toxic effects of metal ions. The process begins with the interaction of metal ions with bacterial cells, followed by their uptake, reduction, and eventual nucleation into nanoparticles. These steps are mediated by bacterial enzymes, reducing agents, and biomolecules, which act as capping agents to stabilize the nanoparticles [26]. A schematic representation of these mechanisms is shown in Figure 4.1.

4.2.1 Mechanisms of Metal Ion Uptake and Transport

Bacteria interact with the metal ions in their environment through adsorption, active transport, and passive diffusion. Metal ions, such as Fe^{2+}, Cu^{2+}, and Zn^{2+}, are essential for various metabolic processes and are transported into bacterial

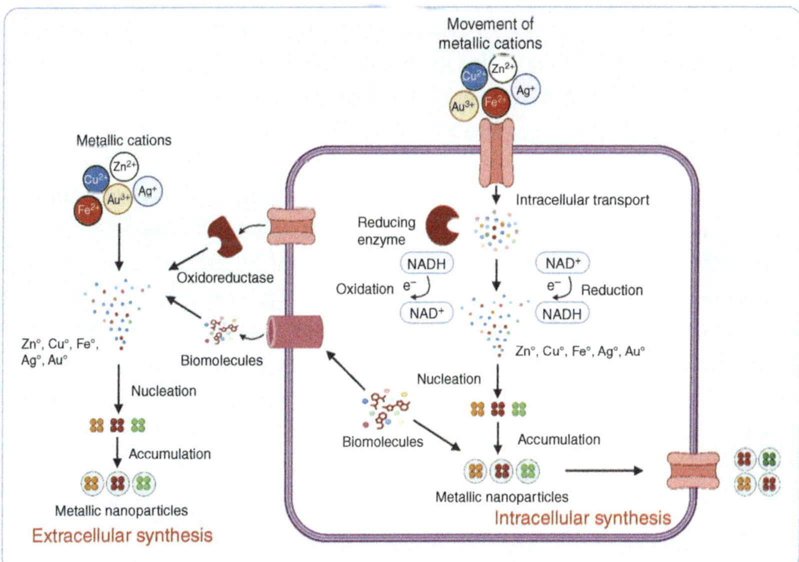

Figure 4.1 Schematic representation of extracellular and intracellular biosynthesis of metallic nanoparticles (NPs) by bacterial cells. This process involves the reduction of metallic cations to metallic nanoparticles facilitated by biomolecules either outside (extracellular) or inside (intracellular) the cell.

cells through highly specific systems, including ATP-binding cassette (ABC) transporters and metal ion channels. Toxic and non-essential metal ions like Ag^+ and Au^{3+}, on the other hand, enter bacterial cells through non-specific transport mechanisms or by mimicking essential ions [27]. For example, in *Cupriavidus metallidurans*, genes in the gold-induced gig operon are strongly upregulated upon exposure to gold ions, facilitating their uptake and detoxification. This system reduces Au^{3+} to elemental Au nanoparticles, which accumulate within or on the bacterial surface as a protective measure against metal toxicity [28].

4.2.2 Enzymatic and Nonenzymatic Reduction of Metal Ions

Once metal ions are internalized or adsorbed onto the bacterial surface, their reduction to elemental states is facilitated by enzymatic and nonenzymatic processes. The mediating agents and electron transfer mechanisms used in these two processes are different. Specific bacterial enzymes, such as hydrogenase, nitrate reductase, or cytochromes, function as catalysts in enzymatic reduction, transforming metal ions into their neutral metallic forms using electron donors such as NADH or NADPH. Conversely, non-enzymatic reduction involves physicochemical interactions between metal ions and bacterial cell surface functional groups (such as carboxyl, hydroxyl, and amine groups), which act as organic stabilizing and reducing agents. Enzymes such as nitrate reductase, hydrogenase, and cytochromes play crucial roles in this process [29]. These enzymes utilize electron donors such as NADH and NADPH to transfer electrons to metal ions, reducing them to their elemental states [30]. For example, nitrate reductase in *B. licheniformis* reduces $Ag+$ ions to Ag^0, resulting in the formation of stable silver nanoparticles (AgNPs) [31, 32]. Similarly, *Shewanella oneidensis* utilizes membrane-bound hydrogenases and cytochromes to reduce Pd^{2+} and Pt^{2+} ions to their metallic forms, enabling the synthesis of Pd and Pt nanoparticles [33, 34]. Nonenzymatic reduction is mediated by the functional groups present on bacterial cell surfaces, such as carboxyl, hydroxyl, and amine groups. These groups interact with the metal ions, facilitating the reduction and stabilization of the resulting nanoparticles. For example, the cell wall of *Pseudomonas aeruginosa* contains negatively charged groups that bind to Au ions, enhancing their reduction and subsequent nanoparticle formation [35].

4.2.3 Nucleation and Growth of Nanoparticles

The reduced metal atoms nucleate to form clusters that grow into nanoparticles. This process is influenced by environmental factors, such as pH, temperature, and metal ion concentration, as well as by the presence of bacterial biomolecules that act as capping agents [36]. These agents prevent aggregation and control the size and shape of the nanoparticles. Intracellular nucleation occurs in the cytoplasm, where reducing agents and enzymes regulate the nanoparticle size and homogeneity [37]. For example, *Magnetospirillum magneticum* synthesizes magnetite (Fe_3O_4) nanoparticles within specialized organelles called magnetosomes [38]. These organelles provide a controlled environment for the biomineralization of nanoparticles, resulting in a uniform size and shape. On the other hand, extracellular nucleation involves enzymes or metabolites secreted by bacteria into the surrounding medium. *Pseudomonas stutzeri* produces extracellular AgNPs stabilized by proteins and biomolecules in the periplasmic space [39]. This extracellular approach offers advantages, such as easier recovery and purification of nanoparticles.

4.3 Influence of Environmental Factors

The synthesis of nanoparticles by bacteria is profoundly influenced by environmental parameters, including pH, temperature, and metal ion concentration. Each of these factors plays a critical role in determining the size, morphology, and stability of the produced nanoparticles. The pH of the medium affects the reduction potential of metal ions and the activity of bacterial enzymes involved in nanoparticle synthesis. For example, acidic conditions typically favor the formation of smaller spherical nanoparticles owing to increased enzymatic activity and higher reduction rates [34]. In contrast, neutral or alkaline conditions promote diverse morphologies such as rods, hexagons, and plates. *Shewanella algae* produce AuNPs with different shapes depending on pH, ranging from spherical at acidic pH to triangular and hexagonal at neutral or alkaline pH [40]. Temperature significantly affects the kinetics of enzymatic reactions and the rate of metalion reduction. Elevated temperatures enhance the reaction rate, leading to faster synthesis and smaller nanoparticles owing to reduced aggregation. However, excessively high temperatures may denature bacterial enzymes or destabilize biomolecules, thereby reducing their capping efficiency. Studies on *P. stutzeri* indicate that AgNPs synthesized at higher temperatures exhibit improved crystallinity and homogeneity [26]. The concentration of metal ions in the medium determines the availability of substrates for bacterial enzymes and influences nanoparticle size. Low concentrations often result in smaller nanoparticles with a uniform size distribution, whereas higher concentrations can lead to larger, polydisperse particles owing to increased nucleation events. *P. aeruginosa* demonstrates this phenomenon, producing AgNPs of varying sizes depending on the initial Ag^+ ion concentration [41, 42].

Additional environmental parameters, such as light, ionic strength, and nutrient availability, also influence nanoparticle synthesis. Light has been shown to enhance the rate of nanoparticle production by promoting photoreduction processes in some bacterial systems. For example, *Bacillus amyloliquefaciens* synthesizes silver nanoparticles more efficiently under solar irradiation, with light intensity directly correlated with production yield [43]. Similarly, ionic strength affects the electrostatic interactions between metal ions and bacterial cells, altering adsorption and reduction dynamics [44].

4.4 Types of Nanoparticles Synthesized by Bacteria

The ability of bacteria to produce a wide variety of nanoparticles (NPs) with special qualities has drawn much interest because of their economical and environmentally friendly production processes. Silver nanoparticles (AgNPs), which are among the most often synthesized nanoparticles, are produced by bacterial species, such *as Bacillus* spp., *Pseudomonas fluorescens*, and Streptomyces sp. They are typically 5–100 nm in size [45–47]. Owing to their strong antibacterial properties, AgNPs can be applied in both medicinal and environmental settings. Bacteria like *Streptomyces griseoruber* and *Rhodopseudomonas capsulata* are prominent makers of gold nanoparticles (AuNPs), another form that is synthesized extensively [48, 49]. When AuNPs are synthesized, they typically produce particles between 5 and 50 nm in size. These particles have important catalytic and biological characteristics such as anticancer activity. Apart from silver and gold, bacteria such as *Streptomyces zaomyceticus* also produce copper oxide nanoparticles with diameters ranging from 78 to 80 nm and have antibacterial and biocontrol properties [50]. Bacteria such as *Bacillus subtilis* produce titanium dioxide nanoparticles (TiO_2NPs), which are widely recognized for their environmental remediation and photocatalytic capabilities [51]. Table 4.1 provides a comprehensive overview of the bacterial species utilized in the biosynthesis of various nanoparticles, highlighting their sizes, morphologies, and types of nanoparticles synthesized.

4.5 Biogenic Nanoparticles in Sustainable Nanotherapeutics and Its Applications

Biogenic sustainable nanotherapeutics, which use biologically produced nanoparticles for therapeutic purposes, are a novel and environmentally responsible approach to nanomedicine. These nanoparticles are produced by biological processes that use natural extracts, microbes, or plants as stabilizing and reducing agents. By minimizing the usage of dangerous chemicals, this approach also supports sustainable development objectives by lessening its negative effects

Table 4.1 Bacterial species used for the biosynthesis of various nanoparticles, along with their types, sizes, shapes and references.

Bacteria Used	Type of Nanoparticle	Size	Shape	References
Aeromonas hydrophila	Zinc Oxide	42–64 nm	Spherical, Oval	[52]
Bacillus licheniformis	Zinc Oxide Nanoflowers	3.8 nm	Flower-like	[53]
Rhodococcus pyridinivorans	Zinc Oxide	100–130 nm	Spherical	[54]
Lactobacillus fermentum	Silver	11.2 nm	Spherical	[55]
Paracoccus haeundaensis BC74171T	Gold	21 nm	Spherical	[56]
Micrococcus yunnanensis J2	Gold	54 nm	Spherical	[57]
Corynebacterium glutamicum	Gold	5–50 nm	Irregular	[58]
Lactobacillus plantarum	Zinc Oxide	7–19 nm	Roughly Spherical	[59]
Pseudomonas fluorescens	Silver	5–50 nm	Myriad Shapes	[45]
Lactobacillus plantarum	Silver	4.7–24.3 nm	Spherical	[60]
Streptomyces sp	Silver	5–3.9 nm	Spherical	[46]
Bacillus subtilis	Titanium Dioxide	66–77 nm	Spherical, Oval	[51]
Arthrobacter nitroguajacolicus	Gold	40 nm	Spherical	[61]
Lactobacillus casei	Iron Oxide	10–15 nm	Spherical	[62]
Bacillus sp.	Silver	5–15 nm	Spherical	[47]
Bacillus brevis	Silver	41–68 nm	Spherical	[19]
Pseudomonas stutzeri	Silver	200 nm	Equilateral Triangles, Hexagons	[39]
Pseudomonas aeruginosa	Gold	15–30 nm	Spherical	[63]
Rhodopseudomonas capsulate	Gold	10–20 nm	Spherical	[48]

(Continued)

Table 4.1 (Continued)

Bacteria Used	Type of Nanoparticle	Size	Shape	References
Halomonas elongata	Copper Oxide	57–79 nm	Rectangular	[64]
Klebsiella pneumonia, Escherichia coli, Enterobacter cloacae	Silver	52.5 nm	Spherical	[65]
Bacillus licheniformis	Silver	50 nm	Spherical	[31]
Klebsiella pneumonia	Silver	1–6 nm	Uniform	[66]
Bacillus cereus	Silver	10–30 nm	Spherical	[67]
Bacillus cereus	Silver	20–40	Spherical	[68]
Arthrobacter sp. B4	Silver	9–72 nm	Spherical	[69]
Escherichia coli	Cadmium Sulfide	2–5 nm	Spherical	[70]
Streptomyces zaomyceticus Oc-5	Copper Oxide	78–80 nm	Spherical, Crystalline	[50]
Streptomyces sp. MHM38	Copper Oxide	1.72–13.49 nm	Spherical	[71]
Paenibacillus polymyxa Sx3	Zinc Oxide (ZnO), Manganese Dioxide (MnO_2), Magnesium Oxide (MgO)	ZnO: 62.8 nm, MnO_2: 18.8 nm, MgO: 10.9 nm	Spherical	[72]
Pseudomonas aeruginosa JP-11	CdS (Cadmium Sulfide)	20–40 nm	Spherical	[73]
Streptomyces rochei MHM13	Silver	22–85 nm	Spherical	[74]
Streptomyces sp. OSIP1 and OSNP14	Silver	8–15 nm	Spherical	[75]
Rhodopseudomonas capsulate	Gold	10–20 nm	Spherical	[48]
Streptomyces griseoruber	Gold	5–50 nm	—	[49]
Streptomyces capillispiralis (Ca-1)	Copper	3.6–5.9 nm	Spherical	[76]
Clostridium thermoaceticum	Cadmium Sulfide	2–5 nm	Amorphous	[77]
Corynebacterium sp. SH09	Silver	10–15 nm	Spherical	[78]
Desulfobacteraceae	Zinc Sulfide	2–5 nm	Spherical	[79]

Table 4.1 (Continued)

Bacteria Used	Type of Nanoparticle	Size	Shape	References
Desulfovibrio desulfuricans	Palladium and Selenium	5–10 nm	Spherical	[80]
Desulfovibrio vulgaris	Gold, Uranium, and Chromium	10–20 nm	Spherical	[81]
Enterobacter cloacae	Silver and Selenium	10–20 nm	Wurtzite Crystal	[82]
Escherichia coli DH5α	Silver	8–9 nm	Spherical	[83]
Escherichia coli MC4100	Gold	Less than 10–50 nm	Spherical, Triangular, Hexagonal, and Rod-shaped	[84]
Aeromonas sp. SH10	Silver	6.4 nm	Spherical	[85]
Bacillus cereus	Silver	20–40 nm	Spherical	[70]
Bacillus megatherium D01	Gold	1.9 ± 0.8 nm	Spherical	[86]
Bacillus subtilis 168	Gold	5–25 nm	Octahedral	[87]
Bacillus subtilis	Silver	5–50 nm	Spherical and Triangular	[88]
Nocardiopsis sp. MBRC-1	Silver	45 nm	Spherical	[89]
Pseudomonas putida NCIM 2650	Silver	70 nm	Spherical	[90]
Pseudomonas stutzeri AG259	Silver	35–46 and up to 200 nm	Hexagonal, Equilateral Triangle, Crystalline Silver	[91]
Rhodobacter sphaeroides	Zinc Sulfide	5–8 nm	Spherical	[92]
Rhodopseudomonas capsulata	Gold	10–20 nm	Nanoplate and Spherical	[48]
Geobacillus sp.	Gold	5–50 nm	Quasi-hexagonal	[93]
Geovibrio ferrireducens	Gold	10–20 nm	Spherical	[94]

Table 4.1 (Continued)

Bacteria Used	Type of Nanoparticle	Size	Shape	References
Klebsiella aerogenes	Cadmium Sulfide	20–200 nm	Crystalline	[95]
Rhodopseudomonas palustris	Cadmium Sulfide	8.01 ± 0.25 nm	Crystalline, Face-centered Cubic	[96]
Serratia nematodiphila	Silver	10–31 nm	Spherical and Crystalline	[97]
Shewanella algae	Gold	5 nm	Elemental	[98]
Shewanella algae Strain BRY	Magnetite	Various Sizes Changed with pH	Fine-grained Crystal	[99]

on the environment. Because of their distinct physicochemical characteristics, which include improved biocompatibility, stability, and bioavailability, biogenic nanoparticles are an excellent choice for antibacterial applications, drug administration, and cancer treatment. Biogenic nanotherapeutics are paving the way for efficient and sustainable healthcare solutions by fusing the benefits of green chemistry with cutting-edge medical applications [100, 101].

4.5.1 Anticancer Therapy

Due to a confluence of genetic and environmental variables, cancer continues to rank among the leading causes of death, globally. Significant progress has been made in the treatment of cancer in recent years with nanoparticles produced using biogenic techniques. Among these, *Lactobacillus rhamnosus* GG-derived AgNPs have shown exceptional effectiveness against the human colorectal cancer cell line HT-29. Similarly, Lactobacillus species-derived zinc oxide nanoparticles (ZnONPs) have demonstrated effectiveness in the treatment of colon cancer (HT-29). Moreover, *Lactobacillus casei-derived* selenium nanoparticles (SeNPs) have demonstrated remarkable efficacy against colon cancer cells. Copper oxide nanoparticles (CuONPs) biosynthesized from *L. casei* were particularly efficient against HT-29 colon carcinoma cell line and human gastric carcinoma cells (AGS). Streptomyces species-derived platinum nanoparticles have demonstrated potent anti-breast cancer action against MCF-7 breast cancer cells [102, 103]. Additionally, important antibacterial, antioxidant, and anticancer properties were demonstrated using green nanotechnology employing silver nanoparticles

(Ag@CTNPs) from *Clitoriaternatea* extract integrated into polymer beads. Using MTT and apoptosis assays, Ag@CT-SA/GEL beads have exhibited potent anticancer effects against lung cancer cell lines [104]. Furthermore, Enterococcus species-mediated gold nanoparticles caused human colorectal cancer cells to become cytotoxic, and they may be employed as a pro-apoptotic drug to treat colon cancer [105]. These results highlight the therapeutic potential of biogenically produced nanoparticles as efficient and targeted medicines for the treatment of cancer.

4.5.2 Bioanalysis

Humanity has greatly benefited from the application of nanoparticles for biological material sensing, which has shown itself to be a useful strategy for enhancing medical care and diagnostics [106]. Because of their special qualities, namely their large surface area and capacity to interact with biological molecules, many kinds of nanoparticles are used in biosensing applications [107]. As an illustration of the potential of nanoparticles in cancer detection, the green synthesis of gold-silver (Au-Ag) alloy nanoparticles mediated by chloroplasts can be considered, which has been applied to cancer analysis [108]. Furthermore, *hypopnea valencia*-mediated gold nanoparticles (AuNPs) have been employed as a noninvasive diagnostic technique for detecting pregnancy in women [109]. Additionally, silver nanoparticles (AgNPs) mediated by *Noctiluca scintillans* were assessed to identify oral health problems like gum infections and oral discharge, proving their usefulness in the diagnosis of oral health diseases [110]. In addition, platinum nanoparticles (PtNPs) mediated by *Streptomyces myriocystum* were used to identify allergies and asthma, suggesting that they could be used to diagnose respiratory diseases [111]. These illustrations demonstrate the various uses of biogenic nanoparticles in diagnostics, highlighting their contribution to the development of medical technology and the enhancement of human health.

4.5.3 Antimicrobial

The potent antibacterial qualities of silver nanoparticles are well known in the medical community. They also work well in fighting viral illnesses, lessening inflammation, and treating fungal infections. Silver nanoparticles demonstrate strong antibacterial action, particularly against *Escherichia coli* and other fecal coliform bacteria, which are very sensitive to them [112]. The antibacterial and biocompatible qualities of gold (Au) nanoparticles are well established. Nevertheless, Au nanoparticles require the presence of other biomolecules to increase their effectiveness; they do not show antibacterial action on their own.

Au nanoparticles can be cross-linked with materials like collagen, chitosan, or gelatin to enhance their interaction with other macromolecules [113]. Because of their capacity to interact with cell wall macromolecules and disrupt membrane proteins, gold (Au) nanoparticles can be used as potent antifungal agents. Instability results from this contact, which compromises the fungal cell wall's structural integrity. In particular, the enzyme cell wall glucan synthase, which is responsible for producing glucan, a crucial element of the fungal cell wall, is inhibited by Au nanoparticles. The stability of the cell wall is compromised by the inhibition of this enzyme by Au NPs, which ultimately damages fungal cells [114, 115]. By disrupting the synthesis of virulence factors, such as exopolysaccharides, and altering metabolic processes, such as surface hydrophobicity, which are essential for bacterial–host cell interactions and the integrity of biofilms in microorganisms, Au nanoparticles prevent Proteus spp. from forming biofilms. Furthermore, Au nanoparticles can stop the growth of microbial cells by a variety of methods by reducing ATP synthase activity, which modifies the cell's membrane potential and results in cell death by collapsing the energy metabolism. For example, 2 nm-sized Au nanoparticles that were synthesized showed improved antibacterial capabilities [116]. Moreover, studies have shown that iron (Fe) nanoparticles, which may be produced from a variety of source extracts, have strong antibacterial activity and may be used to treat a range of illnesses. The relationship between particular bacterial strains and nanoparticles has been the subject of research. Currently, iron-based nano systems are used in medical devices to successfully prevent the growth of microbial colonies [117–120]. Additionally, copper nanoparticles (CuNPs) synthesized using *Shewanella lihinica* PV-4 were tested for antibacterial properties. At a concentration of 100 μg/mL, the CuNPs demonstrated an antibacterial efficiency of $86.3 \pm 0.2\%$ against a suspension of *E. coli* at 10^5 CFU/ml within 12 hours [121]. Furthermore, using *Bacillus cereus*, extracellular synthesis of spherical silver nanoparticles (AgNPs) with a diameter of 10–30 nm was carried out. These nanoparticles demonstrated antibacterial efficacy against Gram-positive Streptococcus and Gram-negative *E. coli*, successfully suppressing bacterial growth at concentrations as low as 50 ppm [69].

4.5.4 Wound Healing

An important process in wound healing is the anti-inflammatory response, which promotes the synthesis of immune-regulatory chemicals including cytokines and interleukins. T-cells, B-cells, and macrophages are among the immune cells and keratinocytes that are activated by these substances. In wound treatment, copper is commonly used to speed up the healing process and prevent microbiological infections at the wound site. It helps increase vascular endothelial growth factor (VEGF) expression, which promotes angiogenesis. Wound healing is greatly improved when angiogenesis, anti-inflammatory reactions, and immune activation are all stimulated simultaneously [122, 123]. Again, when applied on a 10% ointment basis, silver nanoparticles (AgNPs) biosynthesized from *Euphorbia milii* leaf extract with a size range of 20–30 nm exhibit good wound-healing properties in rats with albinism [124].

In a study by Tiwari et al. [125], *P. aeruginosa* was used to biosynthesize copper nanoparticles (CuNPs), which showed a notable increase in wound-healing speed. In comparison to copper in its native or bulk form, the study found that copper nanoparticles were more successful in accelerating wound healing. This demonstrates the increased therapeutic potential of biologically synthesized copper nanoparticles [125]. In a similar vein, *Penicillium chrysogenum* was used to biosynthesize silver nanoparticles (AgNPs) and assess their antibacterial capabilities. Strong antibacterial action was observed in these silver nanoparticles, particularly against *Staphylococcus aureus*, a frequent bacterium linked to wound infections. Ag NPs were found to be effective against a variety of different microbes, such as *B. cereus, P. aeruginosa*, and *E. coli*, indicating their broad-spectrum antibiotic capability. This illustrates even more how beneficial biosynthesized nanoparticles are for preventing infections and accelerating the healing process [126].

4.5.5 Anti-Diabetic

The potential therapeutic uses of biologically produced nanoparticles (NPs) have drawn a lot of interest, especially in the treatment of diseases like diabetes mellitus. Li et al. [127] synthesized biogenic platinum nanoparticles (PtNPs) from the leaves of *Withania somnifera*, also referred to as ashwagandha. These nanoparticles were spherical in shape and 12 nm in size. To assess the synthetic PtNPs' anti-diabetic potential, they were given to streptozotocin-induced diabetic rats. The findings demonstrated a noteworthy decrease in the rats' plasma glucose levels, suggesting that the biogenic PtNPs had an anti-diabetic impact. This work demonstrated the potential of biosynthesized nanoparticles as useful tools for diabetes management, offering a safe, natural substitute for traditional therapies [127]. In another study, by altering several biochemical pathways linked to diabetes mellitus, silver nanoparticles (AgNPs) produced through biological synthesis demonstrated exceptional efficacy in the treatment of the condition [128]. Similarly, it has been discovered that gold nanoparticles (AuNPs) biosynthesized from *Sargassum swartzii*, a brown alga, exhibit anti-diabetic qualities, providing a viable natural method for managing diabetes [129]. Furthermore, in diabetic rats, the green-synthesized silver nanoparticles (AgNPs) from *Phagnalonniveum* enhanced body weight, lipid profiles, and liver and kidney functioning while also dramatically lowering blood glucose levels, suggesting that AgNPs may be a useful anti-diabetic medication [130]. These naturally occurring

NPs have the potential to affect insulin secretion, glucose metabolism, and lessen diabetes-related consequences.

4.5.6 Other Applications

Nanoparticles (NPs) have many industrial, environmental, and biomedical uses due to their small size at the nanoscale, which raises the surface-to-volume ratio. Their potential applications are augmented by the more effective interactions with other chemicals made possible by their larger surface area. Gold nanoparticles (AuNPs) and zinc oxide nanoparticles (ZnONPs) from *S. griseoruber* and Cordyceps have been studied for their environmental applications, particularly in the degradation of methylene blue dyes [49, 131]. These biogenic nanoparticles have high surface area, ease of functionalization, and reactivity, making them effective in breaking down synthetic dyes. They also have potential for environmental remediation, detoxifying toxic compounds like lead, making them suitable for industrial waste cleaning and water treatment [132, 133]. TiO_2 nanoparticles from *Bacillus mycoides* are used in green solar cell development due to their high surface area, photocatalytic activity, and stability, enhancing efficiency and sustainability, and reducing environmental impact compared to traditional methods [134]. Additionally, Cadmium selenide (CdSe) quantum dots (QDs) produced from *Chlorella pyrenoidosa* have been used in biosensing applications to detect the cancer medication imatinib. The adaptability of iron oxide (Fe_2O_3) nanoparticles in a variety of biosensing systems has also been demonstrated by their use in the detection of glucose and hydrogen peroxide [135, 136]. Furthermore, biogenic nanoparticles penetrate biological membranes, making them promising for anticancer therapy and drug delivery. Bacterial magnetosomes loaded with doxorubicin showed increased tumor toxicity and superior MRI contrast, offering higher relaxivity than traditional contrast agents [137, 138]. As a result, it is clear that microbial nanoparticles have a wide range of potential uses in several scientific domains.

4.6 Advantages of Bacterial Nanoparticle Synthesis [42, 139–144]

Inexpensive to Cultivate: Cultivating bacteria is not as expensive as cultivating other biological systems, such as plant or mammalian cell cultures. Their ability to flourish in low-cost, basic media makes them a cost-effective option for producing nanoparticles on a wide scale.

High Growth Rate: Bacteria reproduce quickly; their numbers frequently double in a matter of hours. This quick expansion makes the technique effective for industrial or research applications by guaranteeing a steady and scalable output of nanoparticles.

Genetic Manipulability: Through very simple genetic engineering, scientists can alter the metabolic pathways of bacteria or add particular genes to improve the creation of nanoparticles. The size, shape, and content of the produced nanoparticles can be highly controlled thanks to this genetic plasticity.

4.6.1 Disadvantages

Polydispersity: Using microbes to synthesize nanoparticles frequently presents the problem of polydispersity. It is challenging to attain consistency because of the size and shape variances in the particles created using this procedure. The performance of nanoparticles may be impacted by this irregularity, especially in applications that call for exact properties.

Agglomeration: Microorganism-based nanoparticles have a tendency to group or adhere to one another, which compromises their stability and usefulness. The success of agglomerated particles in a variety of applications depends on their ability to maintain their nanoscale characteristics.

Risks of Contamination: Unwanted microorganism contamination during bacterial production is a risk. The characteristics of nanoparticles can be changed by contaminants, and the introduction of pathogens in biomedical applications may be harmful to human health. The need for meticulous monitoring and strict aseptic conditions might raise production costs and complexity.

Uncontrolled Particle Size and Shape: Achieving consistency in nanoparticle size and morphology is one of the main obstacles. The functionality and application of nanoparticles can be impacted by the variety of sizes and forms that bacterial systems frequently create. Without improving the culture conditions or utilizing further processing methods, it is challenging to obtain exact control over these characteristics.

4.7 Future Perspectives

Bacterial nanoparticle synthesis holds great potential in green nanotechnology, producing environmentally sustainable solutions. However, the full efficiency of natural resources like waste materials and microorganisms has not been fully explored. Future research should focus on understanding fundamental biosynthesis mechanisms, chemical agents, biological elements, and environmental factors to optimize bacterial nanoparticle production. Advancements in nanotechnology should also develop precise monitoring techniques, such as electron microscopy and spectroscopy, to track nanoparticle synthesis in real-time and provide insights into consistency and reproducibility.

Enhancing knowledge of the biosynthesis process, refining synthesis methods, and creating scalable, environmentally friendly production systems are key to the future of bacterial biogenic nanoparticle synthesis. This will open the door for bacterial nanoparticles to be widely used in a variety of fields, such as environmental remediation and medicine. Natural sources of biogenic sustainable nanotherapeutics have the potential to transform medicine by providing low-toxic, effective, and environmentally friendly treatments for conditions including bacterial infections and cancer.

4.8 Conclusion

The process by which bacteria produce nanoparticles has been thoroughly explained here. The purpose of this chapter is to introduce the idea and stress the necessity of creating biogenic nanoparticles from bacterial sources. We have also concentrated on the difficulties that arise when synthesizing nanoparticles. We have also included the many types of nanoparticles that bacteria can produce. Additionally, in light of recent discoveries, we have discussed the uses of biogenic sustainable nanotherapeutics, including their antibacterial, anticancer, wound-healing, antidiabetic, and other various applications. Lastly, we have briefly discussed the prospects for biogenic nanoparticles produced by bacteria in the future, along with some strong advice and ideas for enhancements to help them become safe and biocompatible substances. All things considered, taking into account the scientific benefits and drawbacks of bacterial biogenic nanoparticles, we urge scientists to focus their efforts on streamlining the synthesis procedure and creating these nanoparticles as potent anti-disease treatments.

References

1 Elfeky AS, Salem SS, Elzaref AS, Owda ME, Eladawy HA, Saeed AM, et al. Multifunctional cellulose nanocrystal/metal oxide hybrid, photo-degradation, antibacterial and larvicidal activities. Carbohydr Polym. 2020;230:115711.

2 Salem SS, Fouda A. Green synthesis of metallic nanoparticles and their prospective biotechnological applications: an overview. Biol Trace Elem Res. 2021;199(2):344–370.

3 Gahlawat G, Choudhury AR. A review on the biosynthesis of metal and metal salt nanoparticles by microbes. RSC Adv. 2019;9(15):12944–12967.

4 Grasso G, Zane D, Dragone R. Microbial nanotechnology: challenges and prospects for green biocatalytic synthesis of nanoscale materials for sensoristic and biomedical applications. Nanomaterials. 2020;10(11):11.

5 Prasad M, Lambe UP, Brar B, Shah I, Manimegalai J, Ranjan K, Prasad G. Nanotherapeutics: an insight into healthcare and multi-dimensional applications in the medical sector of the modern world. Biomed Pharmacother. 2018;97:1521–1537.

6 Andreo J, Ettlinger R, Zaremba O, Peña Q, Lächelt U, de Luis RF, et al. Reticular nanoscience: bottom-up assembly nanotechnology. J Am Chem Soc. 2022;144(21):7531–7550.

7 Sikiru S, Ayodele OA, Sanusi Y, Adebukola SY, Soleimani H, Yekeen N, Haslija AA. A comprehensive review on nanotechnology application in wastewater treatment: a case study of metal-based using green synthesis. J Environ Chem Eng. 2022;10:108065.

8 Saratale RG, Karuppusamy I, Saratale GD, Pugazhendhi A, Kumar G, Park Y, et al. A comprehensive review on green nanomaterials using biological systems: recent perception and their future applications. Colloids Surf B: Biointerfaces. 2018;170:20–35.

9 Ngoepe NM, Hato MJ, Modibane KD, Hintsho-Mbita NC. Biogenic synthesis of metal oxide nanoparticle semiconductors for wastewater treatment. In: Tade M (ed.). Photocatalysis and Advanced Oxidation Processes for Wastewater Treatment. 2020, p. 1–31.

10 Castro L, Blazquez ML, Gonzalez FG, Ballester A. Mechanism and applications of metal nanoparticles prepared by bio-mediated process. Rev Adv Sci Eng. 2014;3(3):199–216.

11 Jacob JM, Ravindran R, Narayanan M, Samuel SM, Pugazhendhi A, Kumar G. Microalgae: a prospective low-cost green alternative for nanoparticle synthesis. Curr Opin Environ Sci Health. 2020;10:1–8.

12 Kim TY, Kim MG, Lee JH, Hur HG. Biosynthesis of nanomaterials by Shewanella species for application in lithium ion batteries. Front Microbiol. 2018;9:2817.

13 Khatoon N, Ahmad R, Sardar M. Robust and fluorescent silver nanoparticles using Artemisia annua: biosynthesis, characterization and antibacterial activity. Biochem Eng J. 2015;102:91–97.

14 Meenakshi S. Bacterial cell–a bioreactor for the synthesis of nanoparticles. Chettinad Health City Med J. 2020;9:130–136.

15 Nancharaiah YV, Mohan SV, Lens PNL. Biological and bioelectrochemical recovery of critical and scarce metals. Trends Biotechnol. 2016;34:137–155.

16 Fariq A, Khan T, Yasmin A. Microbial synthesis of nanoparticles and their potential applications in biomedicine. J Appl Biomed. 2017;15:241–248.

17 Jang HD, Kim SK, Chang H, Choi JH, Cho BG, Jo EH, et al. Three-dimensional crumpled graphene based platinum–gold alloy nanoparticle composites as superior electrocatalysts for direct methanol fuel cells. Carbon. 2015;93:869–877.

18 Singh P, Kim YJ, Zhang D, Yang DC. Biological synthesis of nanoparticles from plants and microorganisms. Trends Biotechnol. 2016;34:588–599.

19 Saravanan M, Barik SK, MubarakAli D, Prakash P, Pugazhendhi A. Synthesis of silver nanoparticles from Bacillus brevis (NCIM 2533) and their antibacterial activity against pathogenic bacteria. Microb Pathog. 2018;116: 221–226.

20 Buszewski B, Railean-Plugaru V, Pomastowski P, Rafinska K, Szultka-Mlynska M, Golinska P, et al. Antimicrobial activity of biosilver nanoparticles produced by a novel *Streptacidiphilus durhamensis* strain. J Microbiol Immunol Infect. 2018;51(1):45–54.

21 Sanaeimehr Z, Javadi I, Namvar F. Antiangiogenic and antiapoptotic effects of green-synthesized zinc oxide nanoparticles using *Sargassum muticum* algae extraction. Cancer Nanotechnol. 2018;9:1–16.

22 Li C, Zhou L, Yang H, Lv R, Tian P, Li X, et al. Self-assembled exopolysaccharide nanoparticles for bioremediation and green synthesis of noble metal nanoparticles. ACS Appl Mater Interfaces. 2017;9(26):22808–22818.

23 Golmohammadi H, Morales-Narvaez E, Naghdi T, Merkoci A. Nanocellulose in sensing and biosensing. Chem Mater. 2017;29(15):5426–5446.

24 Shivaji S, Madhu S, Singh S. Extracellular synthesis of antibacterial silver nanoparticles using psychrophilic bacteria. Process Biochem. 2011;46(9): 1800–1807.

25 Tsekhmistrenko S et al. Bacterial synthesis of nanoparticles: a green approach. Biosyst Divers. 2020;28(1):9–17.

26 Iravani S. Bacteria in nanoparticle synthesis: current status and future prospects. Int Sch Res Notices. 2014;2014:359316.

27 Zhang S, Lu J, Wang Y, Verstraete W, Yuan Z, Guo J. Insights of metallic nanoparticles and ions in accelerating the bacterial uptake of antibiotic resistance genes. J Hazard Mater. 2022;421:126728.

28 Reith F, Etschmann B, Grosse C, Moors H, Benotmane MA, Monsieurs P, et al. Mechanisms of gold biomineralization in the bacterium Cupriavidus metallidurans. Proc Natl Acad Sci. 2009;106(42):17757–17762.

29 Durán M, Silveira CP, Durán N. Catalytic role of traditional enzymes for biosynthesis of biogenic metallic nanoparticles: a mini-review. IET Nanobiotechnol. 2015;9(5):314–323.

30 Wypij M, Golińska P. Mechanism of microbial synthesis of nanoparticles. In: Microbial Nanotechnology. 2020. CRC Press, p. 100–116.

31 Kalimuthu K, Babu RS, Venkataraman D, Bilal M, Gurunathan S. Biosynthesis of silver nanocrystals by Bacillus licheniformis. Colloids Surf B: Biointerfaces. 2008;65(1):150–153.

32 Karley D, Shukla SK, Rao TS. Biosynthesis of silver nanoparticle using *Bacillus licheniformis* culture-supernatant for combating pathogenic biofilms. Microb Pathog. 2024;194:106833.

33 Ng CK, Tan TK, Song H, Cao B. Reductive formation of palladium nanoparticles by Shewanella oneidensis: role of outer membrane cytochromes and hydrogenases. RSC Adv. 2013;3(44):22498–22503.

34 Rajput VD, Minkina T, Kimber RL, Singh VK, Shende S, Behal A, et al. Insights into the biosynthesis of nanoparticles by the genus Shewanella. Appl Environ Microbiol. 2021;87(22):e01390–e01321.

35 Lahiri D, Nag M, Sheikh HI, Sarkar T, Edinur HA, Pati S, Ray RR. Microbiologically-synthesized nanoparticles and their role in silencing the biofilm signaling cascade. Front Microbiol. 2021;12:636588.

36 Park TJ, Lee KG, Lee SY. Advances in microbial biosynthesis of metal nanoparticles. Appl Microbiol Biotechnol. 2016;100:521–534.

37 Hulkoti NI, Taranath T. Biosynthesis of nanoparticles using microbes—a review. Colloids Surf B: Biointerfaces. 2014;121:474–483.

38 Singappuli-Arachchige D et al. The magnetosome protein, mms6 from *Magnetospirillum magneticum* strain AMB-1, is a lipid-activated ferric reductase. Int J Mol Sci. 2022,23(18):10305.

39 Klaus T, Joerger R, Olsson E, Granqvist CG. Silver-based crystalline nanoparticles, microbially fabricated. Proc Natl Acad Sci. 1999;96(24): 13611–13614.

40 Ogi T, Saitoh N, Nomura T, Konishi Y. Room-temperature synthesis of gold nanoparticles and nanoplates using Shewanella algae cell extract. J Nanopart Res. 2010;12:2531–2539.

41 Fabrega J, Fawcett SR, Renshaw JC, Lead JR. Silver nanoparticle impact on bacterial growth: effect of pH, concentration, and organic matter. Environ Sci Technol. 2009;43(19):7285–7290.

42 Narayanan KB, Sakthivel N. Biological synthesis of metal nanoparticles by microbes. Adv Colloid Interf Sci. 2010;156(1–2):1–13.

43 Jara N, Milán NS, Rahman A, Mouheb L, Boffito DC, Jeffryes C, Dahoumane SA. Photochemical synthesis of gold and silver nanoparticles—a review. Molecules. 2021;26(15):4585.

44 Malinowska I, Ryżyńska Z, Mrotek E, Klimczuk T, Zielińska-Jurek A. Synthesis of $CoFe_2O_4$ nanoparticles: the effect of ionic strength, concentration, and precursor type on morphology and magnetic properties. J Nanomater. 2020;2020(1):9046219.

45 Syed B, Prasad N, Dhananjaya BL, Yallappa S, Satish S. Synthesis of silver nanoparticles by endosymbiont *Pseudomonas fluorescens* CA 417 and their bactericidal activity. Enzym Microb Technol. 2016;95:128–136.

46 Karthik L, Kumar G, Kirthi AV, Rahuman AA, Bhaskara Rao KV. *Streptomyces* sp. LK3 mediated synthesis of silver nanoparticles and its biomedical application. Bioprocess Biosyst Eng. 2014;37:261–267.

47 Pugazhenthiran N, Anandan S, Kathiravan G, Udaya Prakash NK, Crawford S, Ashokkumar M. Microbial synthesis of silver nanoparticles by *Bacillus* sp. J Nanopart Res. 2009;11:1811–1815.

48 He S, Guo Z, Zhang Y, Zhang S, Wang J, Gu N. Biosynthesis of gold nanoparticles using the bacteria *Rhodopseudomonas capsulata*. Mater Lett. 2007;61:3984–3987.

49 Ranjitha VR, Rai VR. Actinomycetes mediated synthesis of gold nanoparticles from the culture supernatant of *Streptomyces griseoruber* with special reference to catalytic activity. 3 Biotech. 2017;7:299.

50 Hassan SE-D, Fouda A, Radwan AA, Salem SS, Barghoth MG, Awad MA, et al. Endophytic actinomycetes *Streptomyces* spp mediated biosynthesis of copper oxide nanoparticles as a promising tool for biotechnological applications. JBIC. J Biol Inorg Chem. 2019;24:377–393.

51 Vishnu Kirthi A, Abdul Rahuman A, Rajakumar G, Marimuthu S, Santhoshkumar T, Jayaseelan C, et al. Biosynthesis of titanium dioxide nanoparticles using bacterium *Bacillus subtilis*. Mater Lett. 2011;65:2745–2747.

52 Jayaseelan C, Rahuman AA, Kirthi AV, Marimuthu S, Santhoshkumar T, Bagavan A, et al. Novel microbial route to synthesize ZnO nanoparticles using *Aeromonas hydrophila* and their activity against pathogenic bacteria and fungi. Spectrochim Acta A Mol Biomol Spectrosc. 2012;90:78–84.

53 Tripathi R, Bhadwal AS, Gupta RK, Singh P, Shrivastav A, Shrivastav B. ZnO nanoflowers: novel biogenic synthesis and enhanced photocatalytic activity. J Photochem Photobiol B Biol. 2014;141:288–295.

54 Kundu D, Hazra C, Chatterjee A, Chaudhari A, Mishra S. Extracellular biosynthesis of zinc oxide nanoparticles using *Rhodococcus pyridinivorans* NT2: multifunctional textile finishing, biosafety evaluation, and in vitro drug delivery in colon carcinoma. J Photochem Photobiol B Biol. 2014;140: 194–204.

55 Sintubin L, De Windt W, Dick J, Mast J, Van Der Ha D, Verstraete W, Boon N. Lactic acid bacteria as reducing and capping agents for the fast and efficient production of silver nanoparticles. Appl Microbiol Biotechnol. 2009;84:741–749.

56 Patil MP, Kang M-J, Niyonizigiye I, Singh A, Kim J-O, Seo Y-B, Kim G-D. Extracellular synthesis of gold nanoparticles using the marine bacterium *Paracoccus haeundaensis* BC74171T and evaluation of their antioxidant activity and antiproliferative effect on normal and cancer cell lines. Colloids Surf B: Biointerfaces. 2019;183:110455.

57 Jafari M, Rokhbakhsh-Zamin F, Shakibaie M, Moshafi MH, Ameri A, Rahimi HR, Forootanfar H. Cytotoxic and antibacterial activities of biologically synthesized gold nanoparticles assisted by *Micrococcus yunnanensis* strain J2. Biocatal Agric Biotechnol. 2018;15:245–253.

58 Sneha K, Sathishkumar M, Mao J, Kwak I, Yun Y-S. *Corynebacterium glutamicum*-mediated crystallization of silver ions through sorption and reduction processes. Chem Eng J. 2010;162:989–996.

59 Selvarajan E, Mohanasrinivasan V. Biosynthesis and characterization of ZnO nanoparticles using *Lactobacillus plantarum* VITES07. Mater Lett. 2013;112:180–182.

60 Yusof HM, Rahman NA, Mohamad R, Zaidan UH. Microbial mediated synthesis of silver nanoparticles by *Lactobacillus plantarum* TA4 and its antibacterial and antioxidant activity. Appl Sci. 2020;10:6973.

61 Dehnad A, Hamedi J, Derakhshan-Khadivi F, Abusov R. Green synthesis of gold nanoparticles by a metal resistant *Arthrobacter nitroguajacolicus* isolated from gold mine. IEEE Trans Nanobiosci. 2015;14:393–396.

62 Fatemi M. Biosynthesis of iron oxide nanoparticles by cytoplasmic extracts of bacteria *Lactobacillus casei*. Asian J Green Chem. 2018;2:181–188.

63 Husseiny M, El-Aziz MA, Badr Y, Mahmoud M. Biosynthesis of gold nanoparticles using *Pseudomonas aeruginosa*. Spectrochim Acta A Mol Biomol Spectrosc. 2007;67:1003–1006.

64 Rad M, Taran M, Alavi M. Effect of incubation time, $CuSO_4$ and glucose concentrations on biosynthesis of copper oxide (CuO) nanoparticles with rectangular shape and antibacterial activity: Taguchi method approach. Nano Biomed Eng. 2018;10:25–33.

65 Shahverdi AR, Minaeian S, Shahverdi HR, Jamalifar H, Nohi A-A. Rapid synthesis of silver nanoparticles using culture supernatants of enterobacteria: a novel biological approach. Process Biochem. 2007;42:919–923.

66 Mokhtari N, Daneshpajouh S, Seyedbagheri S, Atashdehghan R, Abdi K, Sarkar S, et al. Biological synthesis of very small silver nanoparticles by culture supernatant of *Klebsiella pneumonia*: the effects of visible-light irradiation and the liquid mixing process. Mater Res Bull. 2009;44:1415–1421.

67 Prakash A, Sharma S, Ahmad N, Ghosh A, Sinha P. Synthesis of AgNPs by *Bacillus cereus* bacteria and their antimicrobial potential. J Biomater Nanobiotechnol. 2011;2:155.

68 Sunkar S, Nachiyar CV. Biogenesis of antibacterial silver nanoparticles using the endophytic bacterium *Bacillus cereus* isolated from *Garcinia xanthochymus*. Asian Pac J Trop Biomed. 2012;2:953–959.

69 Yumei L, Yamei L, Qiang L, Jie B. Rapid biosynthesis of silver nanoparticles based on flocculation and reduction of an exopolysaccharide from *Arthrobacter* sp. B4: its antimicrobial activity and phytotoxicity. J Nanomater. 2017;2017:9703614.

70 Sweeney RY, Mao C, Gao X, Burt JL, Belcher AM, Georgiou G, Iverson BL. Bacterial biosynthesis of cadmium sulfide nanocrystals. Chem Biol. 2004;11:1553–1559.

71 Bukhari SI, Hamed MM, Al-Agamy MH, Gazwi HS, Radwan HH, Youssif AM. Biosynthesis of copper oxide nanoparticles using *Streptomyces* MHM38 and its biological applications. J Nanomater. 2021;2021:6693302.

72 Ogunyemi SO, Zhang M, Abdallah Y, Ahmed T, Qiu W, Ali M, et al. The bio-synthesis of three metal oxide nanoparticles (ZnO, MnO_2, and MgO) and their antibacterial activity against the bacterial leaf blight pathogen. Front Microbiol. 2020;11:3099.

73 Raj R, Dalei K, Chakraborty J, Das S. Extracellular polymeric substances of a marine bacterium mediated synthesis of CdS nanoparticles for removal of cadmium from aqueous solution. J Colloid Interface Sci. 2016;462:166–175.

74 Abd-Elnaby HM, Abo-Elala GM, Abdel-Raouf UM, Hamed MM. Antibacterial and anticancer activity of extracellular synthesized silver nanoparticles from marine *Streptomyces rochei* MHM13. Egypt J Aquat Res. 2016;42:301–312.

75 Bakhtiari-Sardari A, Mashreghi M, Eshghi H, Behnam-Rasouli F, Lashani E, Shahnavaz B. Comparative evaluation of silver nanoparticles biosynthesis by two cold-tolerant *Streptomyces* strains and their biological activities. Biotechnol Lett. 2020;42:1985–1999.

76 Hassan SELD, Salem SS, Fouda A, Awad MA, El-Gamal MS, Abdo AM. New approach for antimicrobial activity and bio-control of various pathogens by biosynthesized copper nanoparticles using endophytic actinomycetes. J Radiat Res Appl Sci. 2018;11:262–270.

77 Cunningham DP, Lundie LL Jr. Precipitation of cadmium by *clostridium thermoaceticum*. Appl Environ Microbiol. 1993;59(1):7–14.

78 Mohanta YK, Hashem A, EF Abd_Allah, Jena SK, Mohanta TK. Bacterial synthesized metal and metal salt nanoparticles in biomedical applications: an up and coming approach. Appl Organomet Chem. 2020;34(9):e5810.

79 Agarwal S, Darbar S. Microorganism assisted synthesized metal and metal oxide nanoparticles for removal of heavy metal ions from the wastewater effluents. In: Environmental Applications of Microbial Nanotechnology. 2023. Elsevier, p. 127–148.

80 Omajali JB et al. Characterization of intracellular palladium nanoparticles synthesized by *Desulfovibrio desulfuricans* and *Bacillus benzeovorans*. J Nanopart Res. 2015;17:1–17.

81 Kashefi K, Tor JM, Nevin KP, Lovley DR. Reductive precipitation of gold by dissimilatory Fe (III)-reducing bacteria and archaea. Appl Environ Microbiol. 2001;67(7):3275–3279.

82 Kessi J, Ramuz M, Wehrli E, Spycher M, Bachofen R. Reduction of selenite and detoxification of elemental selenium by the phototrophic bacterium Rhodospirillum rubrum. Appl Environ Microbiol. 1999;65(11):4734–4740.

83 Gabrielyan L, Badalyan H, Gevorgyan V, Trchounian A. Comparable antibacterial effects and action mechanisms of silver and iron oxide nanoparticles on Escherichia coli and Salmonella typhimurium. Sci Rep. 2020;10(1):13145.

84 Du L, Jiang H, Liu X, Wang E. Biosynthesis of gold nanoparticles assisted by Escherichia coli DH5α and its application on direct electrochemistry of hemoglobin. Electrochem Commun. 2007;9(5):1165–1170.

85 Rai A, Singh A, Ahmad A, Sastry M. Role of halide ions and temperature on the morphology of biologically synthesized gold nanotriangles. Langmuir. 2006;22(2):736–741.

86 Wen L, Lin Z, Gu P, Zhou J, Yao B, Chen G, Fu J. Extracellular biosynthesis of monodispersed gold nanoparticles by a SAM capping route. J Nanopart Res. 2009;11:279–288.

87 Southam G, Beveridge TJ. The in vitro formation of placer gold by bacteria. Geochim Cosmochim Acta. 1994;58(20):4527–4530.

88 Saifuddin N, Wong C, Yasumira AN. Rapid biosynthesis of silver nanoparticles using culture supernatant of bacteria with microwave irradiation. J Chem. 2009;6(1):61–70.

89 Manivasagan P, Venkatesan J, Senthilkumar K, Sivakumar K, Kim SK. Biosynthesis, antimicrobial and cytotoxic effect of silver nanoparticles using a novel Nocardiopsis sp. MBRC-1. Biomed Res Int. 2013;2013(1):287638.

90 Thamilselvi V, Radha K. Synthesis of silver nanoparticles from *Pseudomonas putida* NCIM 2650 in silver nitrate supplemented growth medium and optimization using response surface methodology. Dig J Nanomater Biostruct. 2013;8(3).

91 Singh PK, Kundu S. Biosynthesis of gold nanoparticles using bacteria. Proc Natl Acad Sci India Sect B Biol Sci. 2014;84:331–336.

92 Xiao X, Ma XB, Yuan H, Liu PC, Lei YB, Xu H, et al. Photocatalytic properties of zinc sulfide nanocrystals biofabricated by metal-reducing bacterium Shewanella oneidensis MR-1. J Hazard Mater. 2015;288:134–139.

93 Correa-Llantén DN et al. Gold nanoparticles synthesized by *Geobacillus* sp. strain ID17, a thermophilic bacterium isolated from Deception Island, Antarctica. Microb Cell Factories. 2013;12:1–6.

94 Rizki IN, Okibe N. Size-controlled production of gold bionanoparticles using the extremely acidophilic Fe (III)-reducing bacterium, *Acidocella aromatica*. Minerals. 2018;8(3):81.

95 Alsaggaf MS, Elbaz AF, El Badawy- S, Moussa SH. Anticancer and antibacterial activity of cadmium sulfide nanoparticles by Aspergillus niger. Adv Polym Technol. 2020;2020(1):4909054.

96 Bai HJ, Zhang ZM, Guo Y, Yang GE. Biosynthesis of cadmium sulfide nanoparticles by photosynthetic bacteria Rhodopseudomonas palustris. Colloids Surf B: Biointerfaces. 2009;70(1):142–146.

97 Malarkodi C, Rajeshkumar S, Paulkumar K, Vanaja M, Jobitha GD, Annadurai G. Bactericidal activity of bio mediated silver nanoparticles synthesized by Serratia nematodiphila. Drug Invent Today. 2013;5(2):119–125.

98 Konishi Y, Tsukiyama T, Tachimi T, Saitoh N, Nomura T, Nagamine S. Microbial deposition of gold nanoparticles by the metal-reducing bacterium Shewanella algae. Electrochim Acta. 2007;53(1):186–192.

99 Yang J, Ju P, Dong X, Duan J, Xiao H, Tang X, et al. Green synthesis of functional metallic nanoparticles by dissimilatory metal-reducing bacteria "Shewanella": a comprehensive review. J Mater Sci Technol. 2023;158:63–76.

100 Kulkarni D, Sherkar R, Shirsathe C, Sonwane R, Varpe N, Shelke S, et al. Biofabrication of nanoparticles: sources, synthesis, and biomedical applications. Front Bioeng Biotechnol. 2023;11:1159193.

101 Patil S, Chandrasekaran R. Biogenic nanoparticles: a comprehensive perspective in synthesis, characterization, application, and its challenges. J Genet Eng Biotechnol. 2020;18(1):67.

102 Abdoli M, Mohammadi G, Mansouri K, Khaledian S, Taran T, Martinez F. A review on anticancer, antibacterial, and photocatalytic activity of various nanoparticles synthesized by probiotics. J Biotechnol. 2022;354:63–71.

103 Kah G, Chandran R, Abrahamse H. Biogenic silver nanoparticles for targeted cancer therapy and enhancing photodynamic therapy. Cells. 2023;12(15):2012.

104 Veetil VT, Jayakrishnan V, Aravindan V, Rajeeve AD, Koolath S, Yamuna R. Biogenic silver nanoparticles incorporated hydrogel beads for anticancer and antibacterial activities. Sci Rep. 2024;14(1):27269.

105 Vairavel M, Devaraj E, Shanmugam R. An eco-friendly synthesis of *Enterococcus sp.*–mediated gold nanoparticle induces cytotoxicity in human colorectal cancer cells. Environ Sci Pollut Res. 2020;27:8166–8175.

106 Michael H, Alan LG, Serge C. Nanomaterials for biosensing applications: a review. Front Chem. 2014;2:63.

107 Ramesh M, Janani R, Deepa C, Rajeshkumar L. Nanotechnology-enabled biosensors: a review of fundamentals, design principles, materials, and applications. Biosensors. 2023;13:40.

108 Zhang Y, Gao G, Qian Q, Cui D. Chloroplasts-mediated biosynthesis of nanoscale Au-Ag alloy for 2-butanone assay based on electrochemical sensor. Nanoscale Res Lett. 2012;7:475.

109 Kuppusamy P, Mashitah MY, Maniam GP, Govindan N. Biosynthesized gold nanoparticle developed as a tool for detection of HCG hormone in pregnant women urine sample. Asian Pac J Trop Dis. 2014;4:237.

110 Elgamouz A, Idriss H, Nassab C, Bihi A, Bajou K, Hasan K, et al. Green synthesis, characterization, antimicrobial, anti-cancer, and optimization of colorimetric sensing of hydrogen peroxide of algae extract capped silver nanoparticles. Nanomaterials. 2020;10:1861.

111 Sharma D, Kanchi S, Bisetty K. Biogenic synthesis of nanoparticles: a review. Arab J Chem. 2019;12:3576–3600.

112 Ali HR, Emam AN, Koraney NF, Hefny EG, Ali SF. Combating the prevalence of water-borne bacterial pathogens using anisotropic structures of silver nanoparticles. J Nanopart Res. 2020;22:1–15.

113 Singh A, Gautam PK, Verma A, Singh V, Shivapriya PM, Shivalkar S, et al. Green synthesis of metallic nanoparticles as effective alternatives to treat antibiotics resistant bacterial infections: a review. Biotechnol Rep. 2020;25:e00427.

114 Gutiérrez JA, Caballero S, Díaz LA, Guerrero MA, Ruiz J, Ortiz CC. High antifungal activity against *Candida* species of monometallic and bimetallic nanoparticles synthesized in nanoreactors. ACS Biomater Sci Eng. 2018;4:647–653.

115 Yu Q, Li J, Zhang Y, Wang Y, Liu L, Li M. Inhibition of gold nanoparticles (AuNPs) on pathogenic biofilm formation and invasion to host cells. Sci Rep. 2016;6:1–14.

116 Li X, Robinson SM, Gupta A, Saha K, Jiang Z, Moyano DF, et al. Functional gold nanoparticles as potent antimicrobial agents against multi-drug-resistant bacteria. ACS Nano. 2014;8:10682–10686.

117 Perveen S, Nadeem R, ur Rehman S, Afzal N, Anjum S, Noreen S, et al. Green synthesis of iron (Fe) nanoparticles using *Plumeria obtusa* extract as a reducing and stabilizing agent: antimicrobial, antioxidant and biocompatibility studies. Arab J Chem. 2022;15:103764.

118 Naseem T, Farrukh MA. Antibacterial activity of green synthesis of iron nanoparticles using *Lawsonia inermis* and *Gardenia jasminoides* leaves extract. J Chem. 2015;2015:912342.

119 Dana E, Taha A, Afkar E. Green synthesis of iron nanoparticles by *Acacia nilotica* pods extract and its catalytic, adsorption, and antibacterial activities. Appl Sci. 2018;8:1922.

120 Shejawal KP, Randive DS, Bhinge SD, Bhutkar MA, Wadkar GH, Jadhav NR. Green synthesis of silver and iron nanoparticles of isolated proanthocyanidin: its characterization, antioxidant, antimicrobial, and cytotoxic activities against COLO320DM and HT29. J Genet Eng Biotechnol. 2020;18:1–11.

121 Lv Q, Zhang B, Xing X, Zhao Y, Cai R, Wang W, Gu Q. Biosynthesis of copper nanoparticles using *Shewanella loihica* PV-4 with antibacterial activity: novel approach and mechanisms investigation. J Hazard Mater. 2018;347:141–149.

122 Sen CK, Khanna S, Venojarvi M, Trikha P, Ellison EC, Hunt TK, Roy S. Copper-induced vascular endothelial growth factor expression and wound healing. Am J Physiol Heart Circ Physiol. 2002;282(5):H1821–H1827.

123 Borkow G, Gabbay J, Dardik R, Eidelman AI, Lavie Y, Grunfeld Y, et al. Molecular mechanisms of enhanced wound healing by copper oxide-impregnated dressings. Wound Repair Regen. 2010;18(2):266–275.

124 Gong CP, Li SC, Wang RY. Development of biosynthesized silver nanoparticles-based formulation for treating wounds during nursing care in hospitals. J Photochem Photobiol B. 2018;183:137–141.

125 Tiwari DK, Behari J, Sen P. Application of nanoparticles in wound healing. World J Nano Sci Eng. 2014;4(2):33–39.

126 Akila S, Nanda A, Salai RG. In-vivo wound healing activity of silver nanoparticles: an investigation. Int J Sci Res. 2014;3(7):1208–1212.

127 Li Y, Zhang J, Gu J, Chen S. Biosynthesis of polyphenol-stabilised nanoparticles and assessment of anti-diabetic activity. J Photochem Photobiol B. 2017;169:96–100.

128 Prabhu S, Vinodhini S, Elanchezhiyan C, Rajeswari D. Retracted: evaluation of antidiabetic activity of biologically synthesized silver nanoparticles using *Pouteria sapota* in streptozotocin-induced diabetic rats. J Diabetes. 2018;10(1):28–42.

129 Dhas TS, Kumar VG, Karthick V, Vasanth K, Singaravelu G, Govindaraju K. Effect of biosynthesized gold nanoparticles by *Sargassum swartzii* in alloxan induced diabetic rats. Enzym Microb Technol. 2016;95:100–106.

130 Ul Haq MN, Shah GM, Gul A, Foudah AI, Alqarni MH, Yusufoglu HS, et al. Biogenic synthesis of silver nanoparticles using *Phagnalon niveum* and its in vivo anti-diabetic effect against alloxan-induced diabetic Wistar rats. Nanomaterials. 2022;12(5):830.

131 Li JF, Rupa EJ, Hurh J, Huo Y, Chen L, Han Y, et al. *Cordyceps militaris* fungus mediated zinc oxide nanoparticles for the photocatalytic degradation of methylene blue dye. Optik. 2019;183:691–697.

132 Lok CN, Ho CM, Chen R, He QY, Yu WY, Sun H, et al. Proteomic analysis of the mode of antibacterial action of silver nanoparticles. J Proteome Res. 2006;5:916–924.

133 Shen W, Qu Y, Pei X, Li S, You S, Wang J, et al. Catalytic reduction of 4-nitrophenol using gold nanoparticles biosynthesized by cell-free extracts of *aspergillus* sp. WL-Au. J Hazard Mater. 2017;321:299–306.

134 Ordenes-Aenishanslins NA, Saona LA, Duran-Toro VM, Monrãs JP, Bravo DM, Perez-Donoso JM. Use of titanium dioxide nanoparticles biosynthesized by *Bacillus mycoides* in quantum dot sensitized solar cells. Microb Cell Factories. 2014;13:90.

135 Mishra A, Ahmad R, Perwez M, Sardar M. Reusable green synthesized biomimetic magnetic nanoparticles for glucose and H_2O_2 detection. BioNanoScience. 2016;6:93–102.

136 Zhang Z, Chen J, Yang Q, Lan K, Yan Z, Chen J. Eco-friendly intracellular microalgae synthesis of fluorescent CdSe QDs as a sensitive nanoprobe for determination of imatinib. Sensors Actuators B Chem. 2018;263:625–633.

137 Gonzalez-Ballesteros N, Prado-Lopez S, Rodraguez-Gonzalez JB, Lastra M, Rodraguez-Arguelles MC. Green synthesis of gold nanoparticles using brown algae *Cystoseira baccata*: its activity in colon cancer cells. Colloids Surf B: Biointerfaces. 2017;153:190–198.

138 Sun JB, Wang ZL, Duan JH, Ren J, Yang XD, Dai SL, et al. Targeted distribution of bacterial magnetosomes isolated from *Magnetospirillum gryphiswaldense* MSR-1 in healthy Sprague-Dawley rats. J Nanosci Nanotechnol. 2009;9:1881–1885.

139 Prasad R, Pandey R, Barman I. Engineering tailored nanoparticles with microbes: quo vadis? Wiley Interdiscip Rev Nanomed Nanobiotechnol. 2016;8(2):316–330.

140 Singh NA, Narang J, Garg D, Jain V, Payasi D, Suleman S, Swami RK. Nanoparticles synthesis via microorganisms and their prospective applications in agriculture. Plant Nano Biol. 2023;Sep 3:100047.

141 Das RK, Pachapur VL, Lonappan L, Naghdi M, Pulicharla R, Maiti S, et al. Biological synthesis of metallic nanoparticles: plants, animals and microbial aspects. Nanotechnol Environ Eng. 2017;2:1–21.

142 Bloch K, Pardesi K, Satriano C, Ghosh S. Bacteriogenic platinum nanoparticles for application in nanomedicine. Front Chem. 2021;9:624344.

143 Nyabadza A, McCarthy É, Makhesana M, Heidarinassab S, Plouze A, Vazquez M, Brabazon D. A review of physical, chemical and biological synthesis methods of bimetallic nanoparticles and applications in sensing, water treatment, biomedicine, catalysis and hydrogen storage. Adv Colloid Interf Sci. 2023;103010.

144 Seabra AB, Durán N. Nanotoxicology of metal oxide nanoparticles. Metals. 2015;5(2):934–975.

5

Biological Synthesis of Nanoparticles by Fungi and Biogenic Sustainable Nanotherapeutics

Jyoti Kumari Mishra, Jack Patgiri, Priyanka Das, Moumita Gope, and Malay K. Das

Drug Delivery Research Laboratory, Department of Pharmaceutical Sciences, Dibrugarh University, Dibrugarh, Assam, India

5.1 Introduction

Fungi are heterotrophic eukaryotic organisms whose cells contain a nucleus (Figure 5.1), similar to that of animal cells [1]. Unlike plants, they lack chlorophyll, do not carry out photosynthesis, and have a cell wall made of chitin (a polysaccharide made of *N*-acetylglucosamine and glucans), which serves as a substitute for cellulose, providing mechanical strength and cellular integrity. They can be unicellular or multicellular, forming networks of thread-like structures called hyphae that together make up the mycelium (Figure 5.2). Inside, their cellular machinery includes a nucleus, mitochondria, endoplasmic reticulum, and Golgi apparatus. Unlike animals, their plasma membrane contains ergosterol (a steroid molecule that replaces cholesterol), which is a key target for many antifungal medications [2]. Many fungi also produce extracellular enzymes and secondary metabolites, which help in nutrient breakdown and are used in pharmaceutical and industrial applications. As fungi are heterotrophs, they cannot make their own food; instead, they release acids and enzymes to digest organic matter externally and absorb the resulting nutrients through their hyphae. They help recycle nutrients and energy in ecosystems by breaking down matter and forming key symbiotic relationships. Some fungi can switch between unicellular and multicellular forms during their life cycles. The scientific study of fungi, called mycology, has shown that "fungi are actually more closely related to animals than to plants or bacteria" [3].

Fungi are grouped into five main phyla depending on how they reproduce and their evolutionary traits (Figure 5.3) [3]. The first one is Chytridiomycota

Biogenic Synthesis of Nanoparticles and Sustainable Nanotherapeutics,
First Edition. Edited by Malay K. Das.

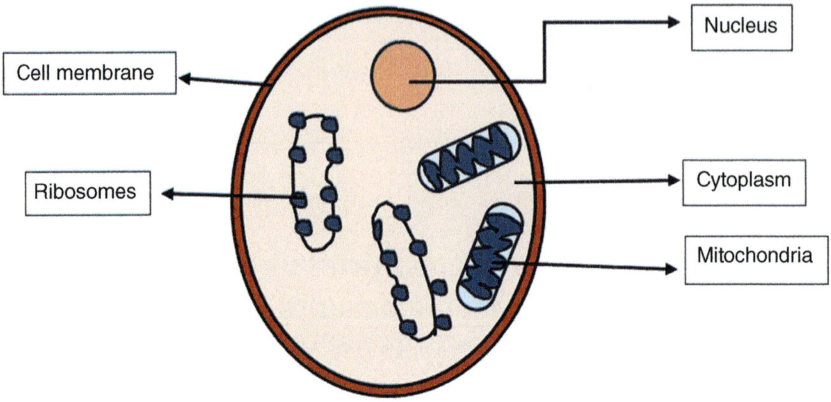

Figure 5.1 A basic fungal cell.

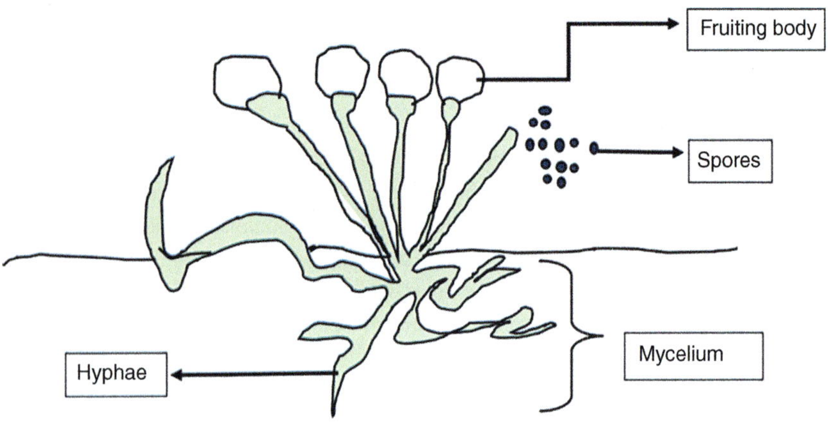

Figure 5.2 Multicellular fungus.

(Chytrids). They are considered as the most ancient fungal group and are found mainly in aquatic environments due to the presence of flagella on their spores and gametes [4]. The second phyla, Zygomycota (Conjugated Fungi), reproduce sexually by forming thick-walled, multinucleate structures called zygosporangia. The third type produces seed-like asci contained in fruiting bodies called ascocarps, most often reproduces asexually, and is known by the term Ascomycota, which is also known as Sac Fungi. Basidiomycota (Club Fungi) include familiar mushrooms, puffballs, and shelf fungi with club-shaped basidia forming spores typically on gills under the mushroom cap (basidiocarps) and are known for their ecological diversity and productivity [4]. Finally, Glomeromycota comprises fungi that form

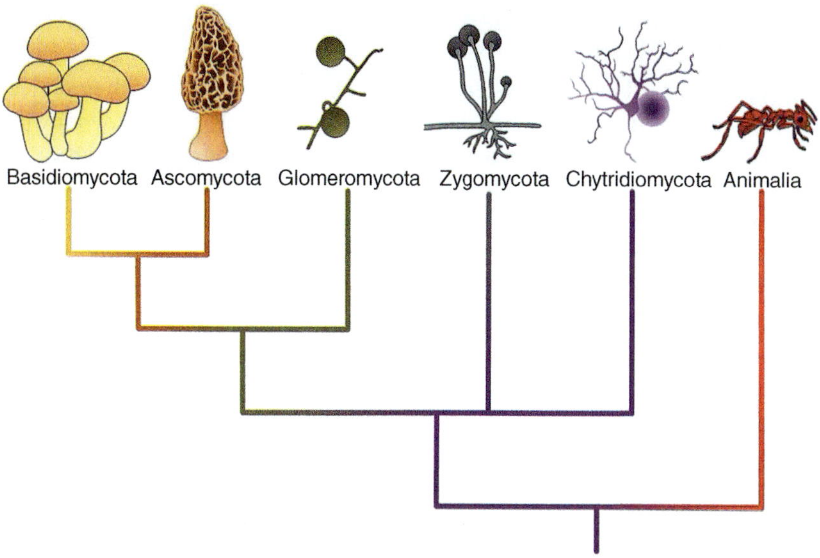

Figure 5.3 Classification of fungi. *Source:* Holmberg and Prud'homme-Genereux [3] / LibreTexts / CC BY 4.0.

important mycorrhizal associations with plant roots, aiding in nutrient and water uptake, featuring coenocytic hyphae and lacking sexual reproduction, including possible accumulation in ecosystems and human exposure pathways [5].

The life cycle of fungi is highly versatile, and it may include both sexual and asexual reproductive phases, enabling them to thrive in diverse environments [6]. In the asexual phase, fungal filaments or hyphae develop specialized structures, such as sporangia or conidiophores, which produces and releases asexual spores. These spores are dispersed with the help of wind, water, or animals and germinate under favorable conditions to produce new hyphal networks that are genetically identical to the parent, thus allowing for rapid colonization [7]. The sexual phase introduces genetic variation through the fusion of compatible hyphae or gametes. This process begins with plasmogamy and then karyogamy occurs. In plasmogamy, the cytoplasm of two mating types merges, resulting in a dikaryotic stage with two distinct nuclei. Karyogamy occurs due to the fusion of two nuclei, forming a diploid zygote. Meiosis allows the zygote to produce genetically diverse haploid spores. These spores then spread and develop into new fungal organisms. While this general cycle applies broadly, specific fungal groups exhibit unique reproductive features; for example, Zygomycota form zygospores, Ascomycota develop asci containing ascospores, and Basidiomycota produce basidia bearing basidiospores [8].

5.2 Fungi and the Biogenic Synthesis of Nanoparticles

The application of biological systems such as plants, microbes, and their derivatives for nanoparticle production has become increasingly popular worldwide due to its fast, eco-friendly approach in contrast to conventional physico-chemical processes [5]. Among these biological routes, microorganisms have emerged as efficient bio-nanofactories capable of producing nanoparticles with desirable sizes and uniform distribution. One specific branch of this field focuses on fungi-mediated nanoparticle synthesis, also known as mycosynthesis, while the wider field of study is referred to as myconanotechnology. This area holds vast promise owing to the remarkable diversity of fungi [9]. Fungal systems have recently become important platforms in producing metallic nanoparticles, e.g. gold, silver, cadmium sulfide (CdS), and platinum. Fungi are particularly well-suited for biological nanoparticle production because of their inherent tolerance to metals and their capacity to accumulate metal ions. This accumulation occurs through various mechanisms, including extracellular interactions with metabolites and polymers or through specific binding with polypeptides [10]. These mechanisms often involve metabolism-dependent uptake and transformation of metal ions. Additionally, fungi are widely cultivated, making them attractive for large-scale applications. Their scalability, including the use of solid-state fermentation techniques, offers a clear advantage for commercial and industrial production. Another benefit is the simplicity of using fungal species as they grow quickly, are readily cultivable, and can be maintained under standard laboratory conditions [8].

5.3 Mechanism of Fungi-Mediated Nanoparticle Synthesis

Fungi are found almost everywhere in nature and come in many forms from microscopic organisms to large mushrooms. They are eukaryotic and their cells consist of nucleus and other organelles. Since fungi cannot make their own food, they rely on organic matter, either by breaking it down as saprophytes or by living off other organisms as parasites [11]. This is mainly because they are easy to grow in the laboratory, and they can tolerate and interact with metals quite well. They also release a large number of proteins outside their cells, which helps stabilize the nanoparticles they form. Compared to plant-based systems, fungal mycelium is more efficient due to its better metal-binding ability and stronger resistance to physical stress [4].

When fungi are used to make nanoparticles, the process usually involves exposing fungal biomass to metal salt solutions. Two common approaches are extracellular and intracellular synthesis (Figure 5.4). In the extracellular method,

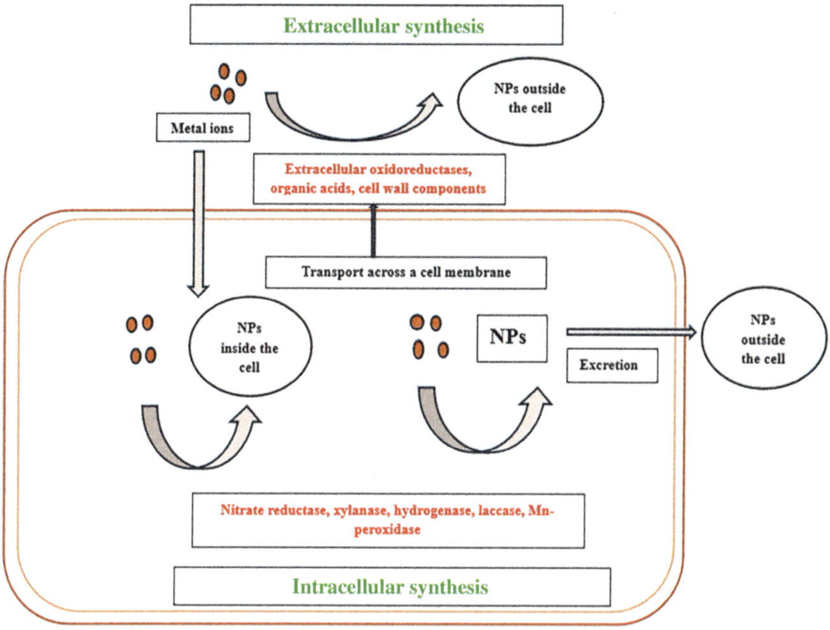

Figure 5.4 Mechanism of fungal nanoparticle synthesis.

fungi secrete helpful compounds like enzymes, proteins, polysaccharides, and flavonoids into their surroundings [4]. These compounds act as natural reducing agents and stabilizers, helping convert metal ions like silver and gold into nanoparticles. Enzymes such as *Nitrate reductase* that use NADH as a cofactor, play a key role by donating electrons to these ions, turning them into their neutral metallic forms. Other fungal compounds like hydroxyquinolines and anthraquinones also support this process. This method is preferred by many researchers because it is simple, scalable, and makes it easy to recover the nanoparticles afterward [8].

On the other hand, the intracellular method (Figure 5.4) consists of introducing metal ions directly into fungal cultures [8]. These ions first attach to the fungal cell wall through electrostatic interactions. After that, they are transported inside the cell, where fungal enzymes like *Nitrate reductase, Xylanase, Hydrogenase, Laccase,* etc. convert them into neutral atoms. These atoms join to create nanoparticles inside or under the cell membrane. For instance, *Fusarium oxysporum* has been widely studied for producing silver nanoparticles inside its cells using this method. Similar results have been seen with gold ions as well, where the reduction happens inside the cytoplasm with the help of redox-active enzymes [10].

The details of various fungi-derived nanoparticles have been elaborated in the Table 5.1.

Table 5.1 Various fungi-derived nanoparticles and their applications.

Fungus Species	Type of Nanoparticle	Synthesis Site	Particle Size (nm)	Key Applications	References
Aspergillus fumigatus	Zinc Oxide (ZnO)	Extracellular	1.2–6.8	Industrial Processes, Medical Treatments, Agricultural	[2, 12, 13]
Aspergillus oryzae	Iron-based ($FeCl_3$)	Extracellular	10–24.6	Farming, Biomedical, and Engineering Uses	[14–16]
Aspergillus tubingensis	Calcium Phosphide ($Ca_3P_2O_8$)	Extracellular	28.2	Used in Agricultural, Biomedical and Engineering Sectors	[14]
Rhizopus oryzae	Gold (Au)	Cell Surface	10	Pesticide Development	[1]
Helminthosporum solani	Gold (Au)	Extracellular	2–70	Anti-cancer Formulations	[9]
Aspergillus niger	Silver (Ag)	Extracellular	3–30	Antibacterial and Antifungal Uses	[17, 18]
Aspergillus fumigatus	Silver (Ag)	Not Specified	15–46	Antiviral Potential Against HIV-1	[5]
Pleurotus sajor caju	Silver (Ag)	Extracellular	30.5	Antibacterial Properties	[9]
Aspergillus flavus	Silver (Ag)	On Cell Wall	~8.92	Not Specified	[19]
Verticillium luteoalbum	Gold (Au)	Intracellular	<10	Not Specified	[20, 21]
Trichoderma viride	Silver (Ag)	Not Specified	5–40	Antibacterial Activity	[22]
Penicillium fellutanum	Silver (Ag)	Cell Surface Adsorption	5–25	Not Specified	[23]
Fusarium oxysporum	Iron Oxide (Fe_3O_4)	Extracellular	20–50	Medical Application	[19]
Soil-derived Fungal Isolates	Zinc, Magnesium, Titanium	Extracellular	Various Sizes	Not Specified	[4]

Aspergillus flavus TFR7	Titanium Dioxide (TiO_2)	Extracellular	12–15	Plant Nutrient Enhancement	[4, 14]
Aspergillus terreus CZR-1	Silver (Ag)	Extracellular	2.5	Agriculture, Biomedical and Engineering Uses	[4, 11]
Verticillium species	Iron Oxide (Fe_3O_4)	Extracellular	20–50, 100–400	Not Specified	[15]
Trichoderma viride	Silver (Ag)	Extracellular	2–4	Biosensor Technology, Bio Imaging	[11, 24, 25]
Penicillium strain J3	Silver (Ag)	Extracellular	10–100	Not Specified	[8, 26]
Fusarium semitectum	Gold (Au)	Not Specified	25	Optoelectronic Applications	[5, 9]
Aureobasidium pullulans	Gold (Au)	Intracellular	23–35	Not Specified	[10, 19]
Volvariella volvacea	Gold (Au)	Not Specified	20–150	Therapeutic Applications	[9]
Fusarium oxyporum	Silver (Ag)	Extracellular	20–50	Antibacterial Properties	[19]
Candida albicans	Gold (Au)	Not Specified	20–40, 60–80	Liver Cancer Detection	[17]

5.4 Factors Influencing the Synthesis of Fungi-Mediated Nanoparticles

5.4.1 Species and Strain

The type of fungal species is a key factor influencing nanoparticle synthesis, as each fungus produces unique enzymes and metabolites that affect the characteristics of the resulting particles [12]. For example, fungi like *F. oxysporum*, *Aspergillus niger*, and *Trichoderma harzianum* have been widely used because they are capable of tolerating high concentrations of metal ions and secrete large quantities of extracellular proteins. These proteins help reduce and stabilize nanoparticles. Research has shown that *F. oxysporum* was capable of producing silver nanoparticles as small as 5–13 nm under certain conditions, such as 50 °C temperature, pH 6, and 10 mM silver nitrate ($AgNO_3$) concentration [10].

5.4.2 pH

The acidity or alkalinity of the reaction environment plays a significant role in myconanotecnology. The pH affects the activity of the enzymes influencing the interaction of fungal functional groups with metal ions. The alkaline pH conditions enhance the reduction rate and lead to the formation of smaller and more uniformly dispersed nanoparticles, while acidic conditions slow down the process and often produce aggregated or irregularly shaped particles [27]. In a study using *Colletotrichum* sp., formed nanoparticles in just 20 minutes at an alkaline pH and 50 °C [10]. Similarly, *Epicoccum nigrum* showed smaller particle sizes at pH 11 compared to pH 6 [9].

5.4.3 Temperature

Temperature directly influences the rate of fungal metabolism and enzyme activity [28]. A moderate increase in temperature generally speeds up nanoparticle synthesis, but temperatures that are too high can deactivate enzymes and denature proteins, leading to poor quality or no nanoparticle formation. For instance, *T. harzianum* showed optimal silver nanoparticle synthesis at 40 °C, while *R. stolonifer* failed to synthesize nanoparticles at 80 °C due to enzyme denaturation [10]. Another example is *Aspergillus oryzae*, which completed nanoparticle synthesis in just 20 minutes at 90 °C, compared to 6 hours at 30 °C [10].

5.4.4 Metal Precursor Concentration

The concentration of the metal salt used in synthesis, such as silver nitrate ($AgNO_3$), affects nanoparticle size and size distribution [29]. Low concentrations tend to produce smaller, more uniform nanoparticles, whereas high concentrations can cause particle aggregation. For example, *A. oryzae* produced nanoparticles ranging from 7 nm to 27 nm at 1 mM $AgNO_3$ but increasing the concentration to 10 mM resulted in much larger particles, up to 62 nm [10]. In another study, *R. stolonifer* synthesized nanoparticles as small as 2.86 nm using 10 mM $AgNO_3$, highlighting the need to optimize concentration for each fungal system [10].

5.4.5 Reaction Time

Reaction time influences nanoparticle size and morphology. Shorter reaction times typically result in smaller, spherical particles, while longer reaction periods may cause additional growth or aggregation. For instance, *A. niger* synthesized 2–5 nm nanoparticles in 24 hours at pH 7 and 30 °C but extending the reaction time to 72 hours led to particles as large as 80 nm and changed their shapes to include rods and triangles [10, 30]. Thus, by carefully controlling the reaction time, the desired nanoparticle properties can be obtained.

5.4.6 Aeration and Agitation

Oxygen supply and agitation during synthesis can affect fungal respiration and metabolism, which further influences enzyme secretion and nanoparticle formation. Proper aeration improves homogeneity and promotes more consistent size and size distribution [14]. For example, *Trichoderma longibrachiatum* synthesized stable nanoparticles (24.4 nm) without agitation with appropriate biomass and temperature control, suggesting that though agitation may not always be required, it can enhance results depending on the fungal system [10].

5.4.7 Biomass Concentration

The amount of fungal biomass used in the synthesis reaction affects the availability of reducing and stabilizing agents. A higher biomass generally increases the rate of reduction and particle yield, but it may also lead to unwanted aggregation. For example, *Guignardia Mangifera* produced nanoparticles of 5–30 nm using 10 g of biomass at 30 °C and pH 7, while *F. oxysporum* achieved optimal synthesis with 11 g of biomass under different conditions [10].

5.4.8 Nature of Fungal Metabolites

Fungi secrete a variety of biomolecules like enzymes, proteins, and polysaccharides that help reduce metal ions and keep the particles stable [31]. These capping agents help in preventing particle agglomeration and also contribute to the biological activity of the nanoparticles. In the case of *F. oxysporum*, enzymes like *Nitrate reductase* and secondary metabolites such as anthraquinones reduce silver ions [10]. Similarly, studies confirmed the use of proteins as capping nanoparticles through FTIR and UV-VIS spectroscopy in *A. niger*-mediated synthesis [10].

5.4.9 Metal Ion Type and Redox Potential for Multimetallic Systems

In multimetallic nanoparticle synthesis, the type of metal ion and its redox potential influence which metals are reduced first, affecting the structure, such as whether the particles form alloys or core-shell configurations. Metals with high reduction potential usually reduce before others. In a study involving *Candida* species, Cd^{2+} and Hg^{2+} ions were reduced sequentially to form Cd_4HgS_5 nanoparticles of 10 nm size inside fungal cells [9].

5.4.10 Synergistic Effects in Multimetallic Systems

Synergistic interactions can improve catalytic activity, stability, and antimicrobial potential compared to single-metal nanoparticles [14]. For example, *A. niger* has been used to make silver-gold nanoparticles capped with proteins, which showed promising anticoagulant and thrombolytic activities in blood assays [9]. Another case involved the synthesis of copper-selenium-zinc oxide nanoparticles (Tri-CSZNPs) by *A. niger*, which demonstrated strong antifungal effects against multiple pathogenic fungi [9].

5.5 Life Cycle Assessment and Toxicity of Fungal-Derived Nanoparticles

A life cycle assessment (LCA) provides a structured approach to evaluate the environmental and health impacts of fungal-derived nanoparticles across all stages of their production and use. This process begins with sourcing raw materials, including growth media like sugars, salts, vitamins, and minerals. Choosing substrates from agricultural waste instead of synthetic sources can improve environmental sustainability [20]. The assessment also accounts for chemicals and solvents used during the synthesis and purification steps, while considering the impacts from growing and transporting plant materials for fungal cultures [32]. Cultivation and extraction, including maintaining proper incubation time and temperatures, require energy and resources. These stages also evaluate how efficiently and safely nanoparticles are produced, as well as the energy and chemicals used, and waste materials generated. Environmental impacts from operations such as centrifugation, filtration, and advanced analytical methods which include microscopy and spectroscopy are also analyzed during the purification and characterization processes of nanoparticle synthesis [8]. Transportation and storage assessments consider the environmental costs of shipping, fuel consumption, packaging materials, and energy required for providing specific storage conditions such as controlled temperature and humidity. The application phase evaluates both the benefits and potential drawbacks of nanoparticle use in agriculture (effects on soil health and crop yield), medicine (safety and effectiveness), electronics (energy-efficiency), and environmental uses [9]. Finally, waste management and recycling are key parts of the LCA. It involves examining ways to recover nanoparticles from used products, the environmental advantages and disadvantages of disposal methods like landfilling and incineration, and the potential for environmental accumulation and human exposure over time [20]. Long-term impacts are studied to understand the degradation, transformation, and potential risks of nanoparticles for ecosystems and human health [4].

Fungal-mediated nanoparticles are considered eco-friendly in production. However, they can pose significant toxicity risks that can be adjusted with careful assessment. Types of toxicity include environmental and ecotoxic effects on soil and aquatic organisms, bioaccumulation in plants and animals, cytotoxic and genotoxic impacts in human and animal cells, soil contamination with heavy metals, unintended genetic changes in plants and microbes, and the development of resistance in pathogens [33].

Detection methods involve the use of plant growth inhibition assays to measure reduced germination and root/shoot length, microbiological tests for assessing soil microbial biomass and diversity, cytotoxicity assays (for example, MTT assay) to evaluate cell viability, genotoxicity assays (for example, comet assay) to detect DNA damage, and spectroscopic and microscopic analyses (for example TEM, SEM, UV–VIS analyses) to characterize nanoparticle properties related to toxicity [20]. Environmental monitoring assesses nanoparticle accumulation in soil, water, and biological tissues, while ecotoxicological tests on aquatic organisms such as algae, invertebrates, and fish measure mortality, developmental abnormalities, and bioaccumulation potential [8].

Reported results include reduced germination rates and stunted root and shoot growth in crops like wheat, lettuce, soybean, and barley; alterations in soil microbial community composition and enzyme activities; degradation of overall soil health; and mortality, reduced growth, and bioaccumulation in soil invertebrates like earthworms. In aquatic systems, evidence shows toxic effects such as oxidative stress, membrane disruption, mitochondrial damage, and silver ion release causing cellular damage in algae, invertebrates like *Daphnia magna*, and fish such as zebrafish and rainbow trout [8]. Observed outcomes include bioaccumulation in fish liver and gills and earthworm tissues. However, biomagnification through food chains is generally low due to transformations like sulfidation that reduce silver ion availability. Human and animal toxicity risks highlighted include potential cytotoxicity, genotoxicity, inflammation, organ damage, and allergic responses [15].

5.6 Sustainability and Green Chemistry Aspects of Fungi-Mediated Nanoparticles

Growing awareness of environmental issues has pushed researchers to look for cleaner, more sustainable ways to produce nanoparticles. One promising approach is using fungi to make these particles in fields like medicine, catalysis, and environmental cleanup [5]. This "green" method is increasingly viewed as a strong alternative to traditional chemical processes, because it offers better safety, efficiency, and environmental compatibility [20].

Fungi are especially attractive for this application, because they naturally produce a wide variety of bioactive molecules, such as proteins and enzymes. These substances aid in converting metal ions to their elemental form and help stabilize the formed nanoparticles [1, 34]. These microorganisms, which include both filamentous fungi and yeasts, are not only easy to culture and handle but also exhibit significant biomass production. Common species used for biogenic nanoparticle synthesis include *Aspergillus* sp., *Beauveria bassiana, Candida albicans, F. oxysporum, Penicillium* sp., *R. stolonifer, and Trichoderma* sp., among others. Their diverse biology makes them excellent candidates for eco-friendly nanoparticle production [5].

Unlike conventional chemical methods, which often depend on harsh reducing agents like sodium borohydride or Tollen's reagent, fungal systems use natural, biologically produced compounds. Fungal metabolites like proteins, enzymes, and other organic molecules work as both reducing agents (turning metal ions into nanoparticles) and stabilizers (preventing clumping). This approach avoids toxic chemicals entirely, making the process safer for people and better for the environment [35].

The sustainability of fungi-mediated nanoparticle synthesis is particularly noteworthy [8]. Unlike chemical processes, which often result in noxious by-products and require substantial energy inputs, fungal synthesis utilizes natural metabolic processes that occur at ambient temperatures and pressures [20]. The minimal use of harmful chemicals and the reduced environmental impact of this approach are key reasons for its growing popularity in the field of green chemistry. This approach avoids toxic chemicals entirely, making the process safer for people and better for the environment [33].

Beyond being environmentally friendly, this biological route offers practical advantages for real-world use. Nanoparticles made with fungi tend to have good biocompatibility, making them suitable for use in medicine, drug delivery, and cosmetics. Researchers can also fine-tune their size, shape, and surface properties, which is vital for targeted applications ranging from treating diseases to cleaning up environmental contaminants [1].

Overall, producing nanoparticles with fungi represents an important move toward greener, more sustainable nanotechnology. By using natural processes and avoiding hazardous chemicals, this approach not only reduces the environmental footprint but also brings new possibilities for creating safe, effective, and eco-friendly nanomaterials for a wide range of industries.

5.7 Challenges and Limitations

The biogenic synthesis of nanoparticles using fungi is often viewed as simple, cost-effective, and scalable due to the rapid growth of fungal cultures, their natural reducing agents (such as enzymes and proteins), and the generally non-toxic nature of their extracts [5]. However, there are still several important challenges and limitations to consider. Key challenges include choosing the right type of nanoparticle, controlling its structure, size, stability, dosage, and method of application. Other issues involve batch-to-batch variation, potential toxicity, and environmental and health risks [36]. One major hurdle is selecting suitable metal precursors and ensuring they are compatible with each other, especially when working with mixtures of metals and metal oxides. Achieving uniform mixing and reaction under controlled conditions like precise pH and temperature is difficult and essential for consistent results. Another significant limitation is the need for advanced characterization techniques. Appropriate analyses of the nanoparticles' size, shape, composition, and properties is necessary, which can be technically challenging and expensive [9]. Additionally, different metals have unique physical and chemical properties that can complicate the synthesis of nanoparticles. These properties might interfere with processes like crystallization, cause unwanted aggregation, or affect how well the nanoparticles reach their intended target. Cultivating fungal strains in appropriate growth conditions is also critical, as the production of bioactive compounds acting as reducing and stabilizing agents depends on factors like pH, temperature, and nutrient availability. These compounds help control the uniformity, shape, and size of the nanoparticles produced [9]. Further, batch-to-batch variability remains a persistent challenge. Environmental factors such as fluctuations in pH, temperature, and nutrient levels can impact fungal metabolism and the characteristics of the nanoparticles [8]. To address this, researchers are exploring genetic engineering approaches to develop metabolically stable fungal strains, as well as process optimization strategies to improve reproducibility [14]. Hence for this, more research is needed to refine the selection of metal precursors and develop a better understanding of their physical and chemical properties. Greater attention should also be paid to thoroughly studying nanoparticle toxicities. For successful development and application of fungus-mediated nanoparticles, there is a need for clear regulatory approvals, safety and handling guidelines, and careful assessment of economic viability [5].

5.8 Future Perspective

Fungi-mediated nanoparticles show a lot of promise for the future, especially for more eco-friendly and safer technologies. Going forward, research could focus on refining fungal strains to achieve increased nanoparticle production with more precise control over their dimensions and stability [35]. It would also be useful to study edible and medicinal fungi for producing nanoparticles that can safely be used in medicine and healthcare. More research is required to understand how using fungi, metal ions turn into nanoparticles, and which enzymes or compounds are responsible for this process. Another important step is to create clear and repeatable methods for large-scale production, so these nanoparticles can be produced in large quantities without losing quality [14]. Future studies should also be conducted to investigate the stability, safety, and environmental perspective of the nanoparticles [37]. Moreover, combining fungal strains with other green technologies like using plant extracts or friendly bacteria might also yield better results. The use of artificial intelligence with fungal biotechnology could make myconanotechnology faster, smarter, more efficient, and more eco-friendly. All these ideas can help in taking this green nanotechnology from small lab experiments to large-scale production for various biomedical applications.

5.9 Conclusion

Among various natural systems, fungi stand out as excellent agents for synthesizing nanoparticles. They are easy to cultivate, robust, environmentally friendly, and their extracellular production methods are rapid and cost-effective. Thanks to their natural ability to produce bioactive compounds, fungi can reduce metal ions to stable nanoparticles that hold significant promise for use in medicine, agriculture, cosmetics, and environmental applications. Metal nanoparticles have found wide-ranging applications in the pharmaceutical, agriculture, and textile industries. While this approach is eco-friendly and sustainable, it also includes challenges like improper molecule estimates, steadiness, and the need for standardized strategies. Future research should prioritize optimizing fungal strains, clarifying the biochemical mechanisms underlying nanoparticle formation, and ensuring product safety through rigorous toxicological studies. Overcoming these challenges is essential to unlocking the full potential of fungal-mediated nanoparticle production as a truly green, sustainable nanotechnology suitable for broad industrial and medical applications.

Conflict of Interest: The authors declare that there is no conflict of interest. Figures 5.1, 5.2, and 5.4 and Table 5.1 are self-made and original.

References

1 Loshchinina EA, Vetchinkina EP, Kupryashina MA. Diversity of biogenic nanoparticles obtained by the fungi-mediated synthesis: a review. Biomimetics. 2022;8(1):1.

2 Raliya R, Tarafdar JC. ZnO nanoparticle biosynthesis and its effect on phosphorous mobilizing enzyme secretion and gum contents in clusterbean (Cyamopsis tetragonoloba L.). Agric Res. 2013;2(1):48–57.

3 Holmberg TJ, Prud'homme-Genereux A. Principles of Biology II, OL ed. 2023. Thompson Rivers University, LibreTexts.

4 Anjum S, Vyas A, Sofi T. Fungi-mediated synthesis of nanoparticles: characterization process and agricultural applications. J Sci Food Agric. 2023;103(10):4727–4741.

5 Chauhan A, Anand J, Parkash V, Rai N. Biogenic synthesis: a sustainable approach for nanoparticles synthesis mediated by fungi. Inorg Nano-Met Chem. 2023;53(5):460–473.

6 Bafana A, Kumar SV, Temizel-Sekeryan S, Dahoumane SA, Haselbach L, Jeffryes CS. Evaluating microwave-synthesized silver nanoparticles from silver nitrate with life cycle assessment techniques. Sci Total Environ. 2018;636:936–943.

7 Finnveden G, Hauschild MZ, Ekvall T, Guinée J, Heijungs R, Hellweg S, et al. Recent developments in life cycle assessment. J Environ Manag. 2009;91(1):1–21.

8 Adeleke BS, Olowe OM, Ayilara MS, Fasusi OA, Omotayo OP, Fadiji AE, et al. Biosynthesis of nanoparticles using microorganisms: a focus on endophytic fungi. Heliyon. 2024;10(21):e39636.

9 Sidhu AK, Agrawal SB, Verma N, Kaushal P, Sharma M. Fungal-mediated synthesis of multimetallic nanoparticles: mechanisms, unique properties, and potential applications. Front Nanotechnol. 2025;7:1549713.

10 Guilger-Casagrande M, Lima RD. Synthesis of silver nanoparticles mediated by fungi: a review. Front Bioeng Biotechnol. 2019;7:287.

11 Carris LM, Little CR, Stiles CM. Introduction to Fungi. Plant Health Instr. 2012. doi:10.1094/PHI-I-2012-0426-01.

12 Kadam VV, Ettiyappan JP, Balakrishnan MR. Mechanistic insight into the endophytic fungus mediated synthesis of protein capped ZnO nanoparticles. Mater Sci Eng B. 2019;243:214–221.

13 Sharma JL, Dhayal V, Sharma RK. White-rot fungus mediated green synthesis of zinc oxide nanoparticles and their impregnation on cellulose to develop environmental friendly antimicrobial fibers. 3 Biotech. 2021;11(6):269.

14 Vijayaraghavan K, Ashokkumar T. Plant-mediated biosynthesis of metallic nanoparticles: a review of literature, factors affecting synthesis, characterization techniques and applications. J Environ Chem Eng. 2017;5(5):4866–4883.

15 Tarafdar JC, Raliya R. Rapid, low-cost, and ecofriendly approach for iron nanoparticle synthesis using Aspergillus oryzae TFR9. J Nanopart. 2013;2013:1–4.

16 Miller KM, Reddy MB, Quashie D Jr, Velez FJ, Ali J, Singh P, et al. Iron-enriched Aspergillus oryzae as an alternative to iron sulphate to limit iron accumulation, growth and motility of the enteric pathogen *S. Typhimurium*. Br J Nutr. 2023;130(3):411–416.

17 Owais M, Chauhan A, Tufail M, Sherwani A, Sajid M, Suri CR, et al. Fungus-mediated biological synthesis of gold nanoparticles: potential in detection of liver cancer. Int J Nanomedicine. 2011;6:2305–2317.

18 Boroumand Moghaddam A, Namvar F, Moniri M, Md Tahir P, Azizi S, Mohamad R. Nanoparticles biosynthesized by fungi and yeast: a review of their preparation, properties, and medical applications. Molecules. 2015;20(9):16540–16565.

19 Riddin TL, Gericke M, Whiteley CG. Analysis of the inter- and extracellular formation of platinum nanoparticles by Fusarium oxysporum f. sp. lycopersici using response surface methodology. Nanotechnology. 2006;17(14):3482–3489.

20 Jangid H, Kumar G. Ecotoxicity of fungal-synthesized silver nanoparticles: mechanisms, impacts, and sustainable mitigation strategies. 3 Biotech. 2025;15(4):101.

21 Sardul SS, Harshita S, Shyamji S. Biosynthesis of silver nanoparticles by endophytic fungi: its mechanism, characterization techniques and antimicrobial potential. Afr J Biotechnol. 2017;16(14):683–698.

22 Yadav A, Kon K, Kratosova G, Duran N, Ingle AP, Rai M. Fungi as an efficient mycosystem for the synthesis of metal nanoparticles: progress and key aspects of research. Biotechnol Lett. 2015;37(11):2099–2120.

23 Kathiresan K, Manivannan S, Nabeel MA, Dhivya B. Studies on silver nanoparticles synthesized by a marine fungus, Penicillium fellutanum isolated from coastal mangrove sediment. Colloids Surf B: Biointerfaces. 2009;71(1):133–137.

24 Eissing T. A computational systems biology software platform for multiscale modeling and simulation: integrating whole-body physiology, disease biology, and molecular reaction networks. Front Physiol. 2011;2:1–11.

25 Popli D, Anil V, Subramanyam AB, Namratha MN, Ranjitha VR, Rao SN, et al. Endophyte fungi, Cladosporium species-mediated synthesis of silver nanoparticles possessing in vitro antioxidant, anti-diabetic and anti-Alzheimer activity. Artif Cells Nanomed Biotechnol. 2018;46(Suppl 1):676–683.

26 Sabri MA, Umer A, Awan GH, Hassan MF, Hasnain A. Selection of suitable biological method for the synthesis of silver nanoparticles. Nanomater Nanotechnol. 2016;6:29.

27 Zhang H, Zhou H, Bai J, Li Y, Yang J, Ma Q, et al. Biosynthesis of selenium nanoparticles mediated by fungus Mariannaea sp. HJ and their characterization. Colloids Surf A Physicochem Eng Asp. 2019;571:9–16.

28 Edman M, Hagos S, Carlsson F. Warming effects on wood decomposition depend on fungal assembly history. J Ecol. 2021;109(4):1745–1757.

29 Sati A, Ranade TN, Mali SN, Yasin HKA, Pratap A. Silver nanoparticles (AgNPs): comprehensive insights into biosynthesis, key influencing factors, multifaceted applications, and toxicity—a 2024 update. ACS Omega. 2025;10(10):7549–7582.

30 Šebesta M, Vojtková H, Cyprichová V, Ingle AP, Urík M, Kolenčík M. Mycosynthesis of metal-containing nanoparticles—synthesis by Ascomycetes and Basidiomycetes and their application. Int J Mol Sci. 2023;24:304. doi:10.3390/ijms24010304.

31 Priyadarshini E, Priyadarshini SS, Cousins B, Pradhan N. Metal–fungus interaction: review on cellular processes underlying heavy metal detoxification and synthesis of metal nanoparticles. Chemosphere. 2021:129976. doi:10.1016/j.chemosphere.2021.129976.

32 Nwe N, Furuike T, Tamura H. Production, properties and applications of fungal cell wall polysaccharides: chitosan and glucan. In: Jayakumar R, Prabaharan M, Muzzarelli RAA (eds). Chitosan for Biomaterials II. 2011. Springer, Berlin, Heidelberg, p. 187–207. (Advances in Polymer Science; vol. 244).

33 Iravani S. Green synthesis of metal nanoparticles using plants. Green Chem. 2011;13(10):2638–2650.

34 Marslin G, Siram K, Maqbool Q, Selvakesavan RK, Kruszka D, Kachlicki P, Franklin G. Secondary metabolites in the green synthesis of metallic nanoparticles. Materials. 2018;11(6):940. doi:10.3390/ma11060940.

35 Narayanan KB, Sakthivel N. Biological synthesis of metal nanoparticles by microbes. Adv Colloid Interf Sci. 2010;156(1–2):1–13.

36 Dhanjal DS, Mehra P, Bhardwaj S, Singh R, Sharma P, Nepovimova E, et al. Mycology nanotechnology interface: applications in medicine and cosmetology. Int J Nanomedicine. 2022;17:2505–2533.

37 Oberdörster G, Oberdörster E, Oberdörster J. Nanotoxicology: an emerging discipline evolving from studies of ultrafine particles. Environ Health Perspect. 2005;113(7):823–839.

6

Biological Synthesis of Nanoparticles by Microalgae and Biogenic Sustainable Nanotherapeutics

Ranu Biswas, Md Ahesan Ansari, Arjun Kumar Sen, and Emdad Hossain

Department of Pharmaceutical Technology, Jadavpur University, Kolkata, West Bengal, India

6.1 About Algae and their Use

Many people are unaware of the vital role of algae in our ecosystem. This lack of appreciation is particularly common among larger populations, who often refer to algae using unflattering terms like "seaweed" and even "frog spittle." Most people's encounters with algae tend to be negative—whether it's dealing with algal blooms in local lakes, finding insect-covered seaweed on the beach, or contending with red tides that can poison seafood and harm human health. Given these associations, it's no surprise that few have taken the time to study phycology, the science of algae, despite their essential contributions to our lives [1]. Oceans, rivers, lakes, and numerous other aquatic environments are inhabited by an extensive variety of photosynthetic organisms identified as algae. These organisms are not just limited to the green varieties often associated with pond scum; they encompass a wide range of forms, colors, and functions. As primary producers and the foundation of the food web, they are essential to aquatic habitats [1–3].

Based on their structure, reproductive strategies, and pigmentation, algae are segregated into a number of types. The major categories include Chlorophyta (green algae, e.g., *Chlamydomonas*), Rhodophyta (red algae, e.g., *Porphyra*), Phaeophyta (brown algae, e.g., *Macrocystis*), Bacillariophyta (diatoms), Dinoflagellates (e.g., *Karenia brevis*), and Cyanobacteria. Algae are fundamental to ecosystem functioning for several reasons. As primary producers, they use photosynthesis to transform sunlight into chemical energy. Algae contribute significantly to global primary production, with marine algae—particularly phytoplankton—accounting for about half of the earth's total photosynthetic output [2]. For a variety of aquatic creatures, such as fish, zooplankton, and larger marine animals, they provide an essential source of food. Higher trophic

Biogenic Synthesis of Nanoparticles and Sustainable Nanotherapeutics,
First Edition. Edited by Malay K. Das.

levels are supported by the energy that algae absorb and distribute throughout the food chain. Underwater forests formed by larger algae, like kelp, offer a variety of marine creatures' refuge and habitat. These ecosystems support high biodiversity and are essential for the survival of many marine organisms. Through photosynthesis, algae produce oxygen, which is vital for the survival of aerobic organisms. It is estimated that marine phytoplankton contribute about 50% of the oxygen in the earth's atmosphere. Additionally, algae play a significant role in carbon cycling. Through photosynthesis, they absorb carbon dioxide, and when they die, some of this carbon is sequestered in ocean sediments, contributing to the long-term removal of carbon from the atmosphere [4].

Algae have substantial economic significance and are utilized in various industries, including food, biofuels, and pharmaceuticals. They are also used in agriculture and environmental monitoring. Agar, carrageenan, and alginate are processed from algae and utilized as thickening and gelling elements in cuisine. Algae can also be eaten raw (e.g., seaweed in sushi). Algal biofuel research has accelerated as an environmentally friendly alternative replacing fossil fuels. Certain species, like *Chlorella* and *Spirulina*, are high in lipid content. They can be converted into biodiesel. Furthermore, bioactive elements with possible health advantages can be found in algae. Because of its high protein and antioxidant content, *Spirulina*, for instance, is sold as a dietary supplement. In agriculture, algal fertilizers and soil conditioners can enhance soil health and crop yields. They are rich in nutrients and can improve the microbial community in the soil. For environmental monitoring, algae serve as bioindicators of water quality. Changes in algal communities can indicate shifts in environmental conditions, such as nutrient loading and pollution. Because of their nourishing, antiaging, and skin-protecting qualities, algae extracts are frequently used in cosmetology and personal care products. Products containing seaweed or microalgae are popular due to their perceived natural benefits. In recent years, nanoparticles (NPs) have been synthesized with the support of cyanobacteria [4–7].

6.2 Spectrum of Algae-Mediated Nanoparticle's Green Synthesis

In the last few decades, nanotechnology has quickly developed because NPs and nanomaterials have outstanding physical and chemical properties. This technology affects almost every area of science by creating new solutions to various research problems [8]. Chemical and physical methods are used for making NPs; however, these processes can be harmful to human health and the environment. Currently, there is a growing trend toward the synthesis of NPs using alternative, environmentally friendly, nontoxic, and cost-effective green methods [9, 10]. Algae, fungi, bacteria, and plant-based systems enable the green synthesis of NPs without impurities. A wide range of biological entities within these systems act as both reducing and capping agents involved in NP synthesis, either inside the biological entity or in the external environment. The reduction of precursor salts leads to the formation of NPs, with these agents acting as reducing or stabilizing agents. These components reduce precursor salts into NPs, serving as reducing or stabilizing agents [8].

There is a growing trend toward utilizing algae for green synthesis, as these organisms are enriched with natural constituents that function as nanobio factories [11]. They grow rapidly, can be easily harvested, and can be scaled up at low cost, making them key players in nanobio synthesis. Found in a variety of habitats, algae are the most common photosynthesis-producing organisms on earth. Algae are an ideal option for green synthesis because they are capable of converting metals into NPs. Metallic and metal oxide NPs are generated from a variety of algae, such as blue-green (*Cyanophycean*), brown (*Phaeophyceae*), green (*Chlorophyceae*), and red (*Rhodophyceae*) algae [12]. For a very long time, algae have been utilized in industry and commerce as fertilizers, additives, cosmetics, medications, and bioremediation agents [8].

6.3 Pathway of NPs Biosynthesis Mediated by Algae

Heavy-metal ions have been observed to hyper-assemble in algae, and they are remarkably adept at reforming them into more malleable forms. Because of these encouraging characteristics, algae have been considered model organisms enabling the creation of various nanomaterials, particularly metallic NPs [12]. Algae include a variety of biological compounds, including antioxidant-rich substances, pigments, chlorophylls, oils, minerals, proteins, carbohydrates, lipids, polyunsaturated fatty acids, and other phytochemicals, which aid in making metal ions less charged or zero valency forms. The three-phase process of bio-reduction includes an activation phase mediated by algal enzymes, as indicated by the color change of the solution. Nucleated metal elements combine during the development phase to form thermodynamically stable NPs of various sizes and shapes. NPs acquire their final shape during the last termination phase [13]. Algal NP biosynthesis (Figure 6.1) can occur either intracellularly or extracellularly, based on the location of NP generation. In the dose-dependent intracellular process, NP biosynthesis occurs within the algal cell. Reducing agents are nicotinamide adenine dinucleotide phosphate (NADPH) or NADPH-dependent reductase, which are produced during metabolic processes like as respiration, photosynthesis, and nitrogen fixation [14]. *Ulva intestinalis* and *Rhizoclonium fontinale* algae were incubated with chloroauric acid for 72 hours at 20 °C to

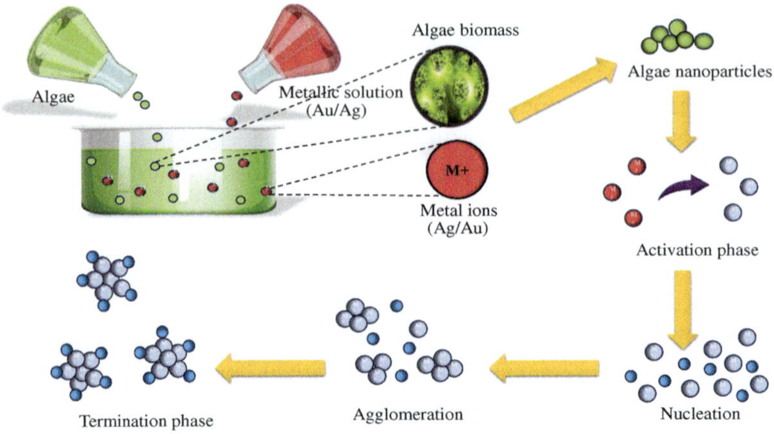

Figure 6.1 General mechanism of synthesis of microalgae-mediated NPs.

perform intracellular generation of AuNPs. The color of the thallus changing from green to purple indicated the biosynthesis of gold nanoparticles (AuNPs). The incubation of the gold solution with the biomass did not change its color, indicating the absence of enzymes or metabolites in the process. However, *Klebsormidium flaccidum* in the silica gel suspension changes color, with the chloroplasts turned to purple from green. This change suggests that the cells can reduce the gold precursor, chloroauric acid. Dark patches within the thylakoid membrane were also visible in transmission electron microscope investigation, suggesting the existence of gold precursor salt in reduced form, which is made feasible by the NADPH-dependent reductase enzyme, or NADPH [8]. The inside of cells generation of AuNPs by the cell wall of the green microalga, *Tetraselmisko chinensis*, was demonstrated by Senapati and colleagues in their 2012 study, highlighting its potential for nanotechnology applications. Ultraviolet–visible spectroscopy confirmed that there was no extracellular synthesis. AuNPs were more concentrated near the cell wall as compared to the cytoplasm, due to involvement of bioactive moieties in bio-reduction [15].

Metal ions adhere to algae's cell surfaces during the extracellular process. Metabolites interact with these ions, aiding in their reduction at the cell surface. This process demonstrates the significant impact of algal cells on their chemical environment [16]. At higher pH, the aggregation of the NPs is minimized due to the enhanced reducing capability of the functional groups. Additionally, it stabilizes the NPs by facilitating interactions with the amine groups. This dual action promotes a more stable NP formulation [17, 18]. A new direction for nanotechnology research is made possible by *Spirulina platensis's* extracellular generation of AuNPs using chloroauric acid [19].

6.4 Critical Review on Synthesis of Various NPs Using Various Algae

6.4.1 Biosynthesis of NPs Through Brown Algae

Gold, silver, and copper NPs are the most common and widely produced metallic NPs. Because of their special qualities and prospective uses in a variety of industries, including environmental research, electronics, and medicine, these particles are made from brown algae. The synthesis process often leverages the natural compounds found within algae, which act as reducing along with stabilizing agents that make NPs environmentally friendly and effective for multiple uses. Various species of brown algae have been reported in the literature for the biosynthesis of AgNPs, including *Gelidiella acerosa*, *Turbinaria conoides*, *Padina pavonica*, and *Cystophora moniliformis*. Extracellular production of 96-nm-sized spherical silver nanoparticles (AgNPs) from *T. conoides*, demonstrating significant antibacterial activity against various species [20]. *Turbinaria conoides* is notably one of the key brown algae species traditionally used to produce AuNPs. Brown algae can biosynthesize metal oxide NPs, such as zinc oxide (ZnO NPs) and titanium dioxide (TiO_2 NPs), alongside metallic NPs [21].

6.4.2 Nps' Biosynthesis Employing Red Algae

Recently, the NPs synthesis using red algae has gained success. Self aggregation can affect particle uniformity and slow crystallization growth impacts efficiency. However, ongoing research is addressing these stability issues. There are several advantages of using red algae to mediate the synthesis of AgNPs. This approach is not only efficient and eco-friendly, but it is also more time-efficient and cost-effective compared to traditional physicochemical methods. Red algae strains like *Kappaphycus alvarezii* have shown potential for biosynthesizing AgNPs, indicating a promising approach for sustainable production [22]. Studies indicate that AgNPs produced by using red algae strains are primarily spherical and range in size from 20 to 60 nm. These dimensions play a vital role in various fields, impacting their properties and effectiveness. Understanding these characteristics can help optimize synthesis for specific uses. Few studies have examined the biosynthesis of AuNPs using red algae, especially in comparison to the many studies on AgNPs. Along with *Lemanea fluviatilis*, other species like *Kappaphycus alvarezi* can also easily produce AuNPs [23].

6.4.3 Biosynthesis of NPs Through Blue-Green Algae

In contrast to brown and red algae, blue-green algae are widely exploited for the generation of a diverse spectrum of NPs. Of the blue-green algae, *S. platensis* is the primary generator of AgNPs, a species known for its rich nutrient profile and potential therapeutic properties. Because of its capacity to biosynthesize AgNPs, this microalga has drawn interest in the domains of nanotechnology, agriculture, and medicine. Production of NP involves the reduction of silver ions, resulting in the formation of well-dispersed NPs that exhibit unique antibacterial and antiviral properties. *Spirulina platensis*, with its high growth rate and ease of cultivation, offers a sustainable and environmentally friendly method for synthesizing these valuable nanomaterials [24]. *Spirulina platensis* is a free-floating, filamentous cyanobacterium with multicellular trichomes forming a left-handed helix. It contains 60%–70% vegetable protein, and fatty acids that aid in reducing and capping NPs. With peptides as well as proteins serving as reducing agents, scientists have discovered that *S. platensis* can mediate the extracellular fabrication of sphere-like, octahedral, and cubic AuNPs. *Phormidium valderianum* can produce AuNPs in different shapes. This process happens outside the cell and uses substances from the cytoplasm to help reduce the gold. *Chlamydomonas reinhardtii*, a significant species of freshwater green algae, plays a crucial role in the production of cadmium sulfide bimetallic nanoparticles (CdSNPs). This microalga has the unique ability to interact with and influence the formation of these NPs, which are composed of cadmium and sulfur. Its involvement in this process highlights the potential of *C. reinhardtii* in bioremediation and nanotechnology applications, as it helps to facilitate the synthesis and stabilization of CdSNPs in aquatic environments [25].

6.4.4 NPs' Biosynthesis Employing Green Algae

Microgreen algae and macrogreen algae are the two varieties of green algae. Among the diverse types of monometallic NPs, AgNPs hold significant promise as they can be effectively synthesized *in vitro* using various species of green microalgae. More than 20 different species are used to synthesize AgNPs [26].

Most species of micro green algae used for extracellular applications produce AgNPs that vary in size and shape. For example, *Pithophora oedogonia* produces cubical and hexagonal NPs measuring 24–55 nm, while *Chlorococcum humicola* generates spherical NPs of 16 nm. *Chlorella vulgaris* creates triangular NPs sized at 28 nm, and *C. reinhardtii* produces either rectangular or rounded NPs that range from 1 to 15 nm. Lastly, *Enteromorpha flexuosa* produces circular NPs measuring 15 nm [27].

Ulva fasciata, widely used green macroalgae, is used to develop nanosized silver colloids and applied to cotton fabric both with and without citric acid. This study aimed to assess the antimicrobial efficacy of the treated fabrics. The amide, carboxylic, and nitro chemicals found in *Gracilaria edulis* have been employed to create spherical AgNPs and octahedral ZnONPs [28].

6.5 Application of Algae-Based Nanoparticles

Green synthesized NPs are biocompatible and free from toxicity as they are devoid of reducing agents. This results in lower toxicity compared to chemically synthesized NPs. Because algae species have natural biomolecules that give little to no toxicity, they can be used primarily in a variety of biomedical applications (Figure 6.2 and Table 6.1) without the need for hazardous compounds to reduce and stabilize NPs [20, 29].

6.5.1 Antibacterial Activity

Self-medication and excessive antibiotic use give rise to the antibiotic resistance. One of the biggest worldwide health challenges is treating drug-resistant strains of bacteria in an effective and secure way. Since NPs have effective and better antibacterial activity, there seems to be a movement toward their use as a substitute antibacterial agent. NPs have broad-spectrum activity because they kill bacteria by rupturing their cell membranes and producing reactive oxygen species [21, 28]. The biosynthesized AgNPs from algae are studied for antimicrobial action against a few strains of bacteria. The growth rate of *Pseudomonas aeruginosa* and *Bacillus* species is effectively retarded by the AgNPs derived from the brown seaweed *P. tetrastromatica* [22, 30]. The aqueous extract of *P. oedogonia* that was

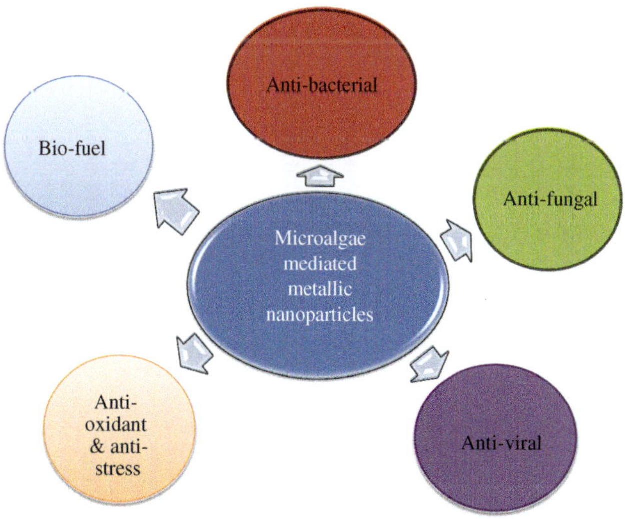

Figure 6.2 Different applications of metallic nanoparticles synthesized using micro-algae.

Table 6.1 Application of nanoparticles synthesized by various microalgae.

Microalgae species	Nanoparticle type	Size range	Synthesis method	Biomedical applications	References
Pithophora oedogonia	Ag	10–50 nm	Extracellular process	Significant antibacterial effectiveness against various bacteria, including *E. coli, V. cholerae, P. aeruginosa*, and *S. flexneri*.	[24, 29]
Sargassum longifolium	Ag	10–50 nm	Extracellular process (biomass extract)	Demonstrated exceptional fungicidal activity, tested at various concentrations of pathogenic species like *Candida albicans*.	[29, 30]
Gelidiella acerosa	Ag	10–50 nm	Extracellular process (proteins, polysaccharides, and other metabolites)	Antifungal effect against *Trichoderma reesei, Mucor indicus, Humicolainsolens*, and *Fusarium dimerum*	[30, 31]
Spirulina platensis	CuO, Ag	1–100 nm	Extracellular process	Selective antifouling agent and biofilm preventing agent	[36]
Sargassum vulgare	Ag	1–20 nm	Extracellular process (proteins, polysaccharides, and other metabolites)	Strong anticancer effects on HeLa cells and human myeloblastic leukemic cells (HL60)	[40]
Sargassum muticum	Ag	10–50 nm	Extracellular process (proteins, polysaccharides, and other metabolites)	Anticancer activity via ROS generation., showed *in vitro* cytotoxicity to the MCF7 breast cancer cell line	[41]
Acanthophora spicifera	Au	10–50 nm	Extracellular process (proteins, polysaccharides, and other metabolites)	Effective against the colorectal adenocarcinoma HT-29 cell line in one investigation	[42]

Ulva lactuca	Ag	20–40 nm	Extracellular process (proteins, polysaccharides, and other metabolites)	Used as bioremediation agents; for instance, photocatalytic degradation of methyl orange dye, antimalarial activity	[43, 44]
Chlorella ellipsoidea	Ag	7–12 nm	Both extracellular and intracellular methods	Potential photocatalytic activity on the hazardous pollutant dyes	[47]
Noctiluca scintillans	Ag	261–1121 μm	Extracellular process	Used as a biosensing agent for colorimetric detection (for the identification) of hydrogen peroxide, shows antimicrobial activity	[49]
Sargassum myriocystum	Ag	10–50 nm	Extracellular process	Employed for the biosensing of the body's adrenaline levels	[50, 51]

used to formulate AgNPs showed significant antibacterial effectiveness against various bacteria, including *Escherichia coli, Vibrio cholerae, P. aeruginosa*, and *Shigella flexneri. Pseudomonas aeruginosa* showed maximum zone of inhibition (17.2 mm), demonstrating exceptional antibacterial efficacy of AgNPs against greater resistance to gram-positive rods [24, 29]. For *P. aeruginosa*, the most extensive zone of inhibition documented was 17.2 mm, suggesting that AgNPs have exceptional antibacterial capabilities against resistant gram-negative rods. Yet another report focused on the evaluation of *Neodesmuspupukensis*-mediated AgNPs and AuNPs with reference to the different bacterial strains against which they exerted antibacterial activity.

6.5.2 Antifungal Activity

Fungal infections are considered to be increasingly threatening public health concerns with the increasing scarcity and resistance to antifungal drugs. NPs have demonstrated exceptional fungicidal activity, suggesting that they may be an innovative remedy for fungal infections [29]. It has been demonstrated that green synthesized AgNPs are currently the most effective antifungal agent. According to a recent study, *Sargassum longifolium* is used to synthesize AgNPs and tested for antifungal properties at various concentrations of pathogenic species like *Candida albicans*. The study's findings showed that, in proportion to the dosages administered, each fungus species' growth was significantly inhibited [30]. Another study examined the antifungal effects of AgNPs made from an aqueous extract of the red seaweed *G. acerosa* against *Trichoderma reesei, Mucor indicus, Humicolainsolens*, and *Fusarium dimerum*. The AgNPs are effective against fungal infection in comparison to standard antifungal medication. Moreover, AgNPs developed by using *Hypnea musciformis* and *U. latica* demonstrated efficacy in preventing the growth of fungal strains such *C. parapsilosis* [31]. AuNPs have also been developed from algae to combat fungal infection, but very little literature is available concerning this aspect. Aqueous extract of brown seaweed *Dictyota bartayresiana* is used to prepare AuNPs to treat fungal infection against *Humicola insolens* [32].

6.5.3 Antifouling Agent and Biofilm Prevention

This undesired development on submerged areas is known as biofouling, and it presents serious health hazards in addition to financial losses in the industrial, medical, and maritime sectors [33, 34]. Antifouling methods include a variety of approaches, from biocides to toxic chemicals. Unfortunately, these substances often accumulate and contribute to environmental pollution [35]. The AgNP-capped biofilm formation is effective against *E. coli* and *Micrococcus* species. It can be utilized as a selective antifouling agent against specific species [36]. CuNPs have also been explored as antibiofilm agents like AgNPs. Studies showed that CuNPs decreased the hydrophobicity of *P. aeruginosa* on the outside of cells and extracellular polymeric substances in addition to suppressing the formation of biofilms [8]. Recently, the effectiveness of biofilm-forming bacterial strains *Sargassum myriocystum* in producing AgNPs at different concentrations was assessed against *Staphylococcus epidermidis* and *P. aeruginosa*, and maximum inhibition was achieved at 50 μg/mL. In contrast, *S. epidermidis* treated with 50 μg/ml of AgNPs showed an inhibition rate of 48.34%. When *P. aeruginosa* was exposed to the same higher dose of 50 μg/ml, the suppression of biofilm development reached to 55.49% [37, 38].

6.5.4 Anticancerous Activity

The application of NPs to the treatment of cancer is a new field of study in nanobiotechnology [39]. Recently, many studies have been reported on NPs developed from algae. AgNPs derived from *S. vulgare* and measuring 10 nm in size showed strong anticancer effects on human myeloblastic leukemic cells (HL60) along with HeLa cells [40]. Furthermore, AgNPs generated from *Sargassum muticum* showed *in vitro* cytotoxicity to the breast cancer cell line named as MCF7. These AgNPs have caused intracellular intracellular reactive oxygen species (ROS) that leads to apoptosis and further destroys cancer cells. Using the 3-(4,5-dimethylthiazole-2-yl)-2,5-diphenyltetrazolium bromide (MTT) screening method, AgNPs produced by *S. myriocystum* are cytotoxic to the HeLa cell line at multiple concentrations (0, 2, 4, 8, 16, 32, 64, 128, 256, and 512 μg/ml). Results showed 50% inhibition and apoptotic activity, and that the total cytotoxicity rose as the medium's AgNP concentration rose [37]. Similarly, algal-mediated AgNPs' capacity to cause cytotoxicity against the breast cancer MCF-7 cell line was assessed *in vitro* at doses varying from 0 to 100 μg/ml at 24-, 36-, and 48-hour periods. Anticancer potential of AgNPs was confirmed at the MIC of 20 μg/ml against cancer cells in the breast that had nuclear disintegration, apoptosis, also cell death [41].

Additionally, it has been found that algae-mediated AuNPs exhibit strong anticancer properties against a variety of cell types. AuNPs mediated by *Acanthophora spicifera* are effective against the colorectal adenocarcinoma HT-29 cell line in one investigation [42].

6.5.5 Nano-Bioremediation

Recently, there has been a lot of interest in using NPs as a novel method for decontaminating contaminated locations. NPs synthesized via algae have been used as bioremediation agents; for instance, photocatalytic degradation of methyl orange dye was achieved by visible light illumination using AgNPs synthesized by *U. lactuca*. Furthermore, population of *Plasmodium falciparum*, which is known to be chloroquine resistant, was greatly reduced with a smaller amount of *U. lactuca*-mediated AgNPs [43, 44]. When compared to cyanobacterial extract, microchaete AgNPs showed superior decolorization capabilities against methyl red azo dyes in comparative research [45]. AuNPs made from the water-based extracts of brown algae *S. tenerrimum*and *T. conoides* were tested for their catalytic activity against the organic dyes rhodamine *B*, sulforhodamine, and aromatic nitro compounds [46]. Research showed that AgNPs loaded with *C. ellipsoidea*-mediated biomatrix are potential photocatalytic on the hazardous pollutant dyes throughout three reduction cycles [47]. Lipid-cadmium sulfide NPs were synthesized using the green alga *Scenedesmus obliquus*. As synthesized, a significant increase in the adsorption kinetics of Cd^{2+} ions into the biomass of algae was observed, which irreversibly bound to the chemisorbed monolayer of Cd^{2+}. For the bioremediation of Cd^{2+} ions, *Scenedesmus* acts as an efficient model alga owing to its high retention properties [8, 48]. In this way, several forms of algae-mediated NPs are contributing to the remediation of organic and aromatic chemicals, heavy metals, and dyes of all kinds.

6.5.6 Biosensing

It has demonstrated exceptional optical qualities that make it suitable for a wide range of biosensing applications that detect hormones qualitatively and quantitatively in the human body [48]. A colorimetric detector for the identification of hydrogen peroxide, an antimicrobial, was recently developed using AgNPs mediated by *Noctiluca scintillans*. Patients are administered it for minor skin scratches, gum disease, tooth pain, teeth whitening, and oral discharge. Temperature, time, and pH all affect how quickly hydrogen peroxide breaks down on the AgNPs' catalytic surface [49]. *Sargassum myriocystum* is used to produce platinum nanoparticles (PtNPs), a hormone-based drug used to treat allergies, asthma, and heart attacks. PtNPs are also used for biosensing the body's adrenaline levels [50, 51].

6.6 Algae-Mediated Biogenic Nanotherapeutics: Pre-Clinical and Clinical Evidences

Micro algae or microphytes are organisms that are mainly found in freshwater and marine systems and are microscopic in nature having the capacity of photosynthesis. They are unicellular in nature, which can stay single or in chain forms, and their size ranges from few microns to few hundred microns. They have strong adaptive capacity and thus can stay in almost all ecosystems [52]. These microbes are widely used in pharmaceuticals, food, cosmetics, and nutraceutical industries due to their high concentration of bioactive resources, comprising fatty acids, proteins, as well as vitamins [53]. It is seen that the micro-algae can effectively adsorb NPs, distinct metal ions via physical and electrical interactions depositions and thus can be used as an ideal carrier of drug molecules (anticancer agents, antibiotics). Thus, their use in the drug delivery process has already shown excellent results [54]. In 2011, Aw and co-researchers worked on the possibility of loading diatoms using a drug model that is poorly soluble in water (nonsteroidal anti-inflammatory drug [NSAID]) Indomethacin, with a loading effectiveness till 22% (percentage by weight). Initial drug release was seen within six hours (drugs adsorbed on the surface) and the remaining drug was released throughout the length of two weeks (drugs entrapped within the pores of diatoms) [55]. The sustained release and accelerated permeation of mesalamine and prednisone by diatom microparticles in the gastrointestinal system were examined by Zhang et al. in 2013. At concentrations of up to 1000 μg/ml, they demonstrated very little cytotoxicity of bare diatoms on HT-29, HT-116, and Caco-2-colon cancer cell lines. Studies conducted *in vitro* revealed that the medications are released over an extended period of time in various bodily fluids [56].

In 2017, Khalid et al. synthesized AgNPs employing ethanolic extracts from the following three different freshwater microalgae species: *Dixonosphaerium* sp. strain HM1 (DHM1), *Dixonosphaerium* sp. strain HM2 (DHM2), and *Pectinodesmus* sp. strain HM3 (PHM3). They then tested the antibacterial, antiviral, and antitumor activities of these NPs. As an antibacterial agent, these AgNPs showed effective antibacterial activity against 14 different gram-positive and gram-negative bacterial strains. Studies revealed that PHM3 AgNPs were ovoid and larger in shape and thus showed better activity than the other two strains' AgNPs. None of the three types of algae inhibited *S. aureus* or *P. aeruginosa*; by contrast, the basic ethanolic extract of the algal strains exhibited negligible antibacterial activity. As an antifungal agent, again PHM3 AgNPs showed better results in comparison with the other two against *C. albicans,* and maximum activity was seen at a concentration of 1 mg/mL. A test of cytotoxicity was also made against HepG2 cells with the AgNPs of the three algal strains, taking Taxol as the standard drug. All three AgNPs showed positive cytotoxic activity against HepG2 cells [57]. Utilizing the DPPH model, Chokshi et al. in 2016 investigated the production of AgNPs utilizing the de-oiled biomass of *Acutodesmus dimorphus (Turpin)*, a thermally resistant microalga with the potential for producing biofuel, and found that it exhibited notable antioxidant properties [58].

The antibacterial properties of gold and AgNPs were investigated against methicillin-resistant *Staphylococcus aureus* (MRSA), natural *S. aureus*, and *Streptococcus* sp. by Aldayel et al. in 2022 utilizing the microalgae *C. vulgaris*. *Streptococcus* sp. had the least amount of antibacterial activity, while methicillin-resistant *S. aureus* exhibited the most. Between AgNPs and AuNPs,

AgNPs showed greater antibacterial property over AuNPs [59]. Sudha et al. in 2012 synthesized AgNPs using microalgae, and their antibacterial properties were also tested. Microalgal water was collected from the mangrove. *Corynebacterium* sp. *V. cholera*, *P. vulgaris*, and *Salmonella typhi* were used as bacterial strains. Microalgae *Microcoleus* sp. was utilized for green synthesis of AgNPs. *Microcoleus* sp. of microalgae gave the maximum biological activities when compared to the remaining microalgae and were found to synthesize AgNPs and produce bactericidal effects against test bacterial strains [60]. Sathavisam et al. in 2024 synthesized copper ferrite ($CuFe_2O_4$) NPs using *Chlorella* green microalgae biomass extract and tested for antibacterial properties as a function of pH, temperature, and time of reaction. It was seen that these NPs were effective against both *E. coli* and *S. aureus* at pH 10 and pH 12 when the NPs were synthesized for eight hours at 180 °C [61].

6.7 Challenges

The production of NPs from biological sources has recently become a more environmentally friendly option than traditional chemical processes. Given their ease of development and ability to withstand harsh pH and temperature conditions, microalgae stand out among the various green sources investigated as viable, low-cost, environmentally friendly technologies for large-scale NP production. Although NP manufacturing is quick, chemical processes are not ecologically friendly since they employ hazardous chemical reducing agents and consume a lot of energy. In contrast, green synthesis uses reducing agents from biological sources such as algae, viruses, bacteria, and plant extracts, and is carried out under normal conditions, thus reducing energy costs. Therefore, this green synthesis has emerged as an alternative approach for the production of metallic NPs, though it is still in its infancy. In 2020, Wong et al. evaluated the toxicity of various metallic NPs and ranked them according to how well they could stop marine diatom growth of *Skeletonema costatum* as ZnO nanoparticle (*n*ZnO)s > rutile TiO_2 > MgO > annealed MgO > anatase TiO_2 > γ-$Al2O_3$ > In_2O_3 > α-Al_2O_3 > SnO_2 [62]. Despite their many uses in various industries, green synthesized NPs have several drawbacks, such as morphology (size, shape, and crystal behavior) that is affected by the conditions under study. Because of the intricacy and diversity of the precise makeup, this remains a mystery. In-depth investigations into aspects such as purity, surface area, surface chemistry, size-dependent aggregation/agglomeration (understanding the biological impact of sizedependent aggregation of silver NPs), and biodistribution are additionally required. As the amount of metallic NPs produced via green synthesis is subsequently lower than the normal chemical method of production, this can lead to an increase in cost. Challenges lie in finding more different microalgal species required for better synthesis with respect to quality and quantity of metallic NPs [63]. Though there was huge research on algae but the world is waiting for patents associated formation of NPs from the algae.

6.8 Conclusion and Future Perspectives

This chapter explores the biological process of synthesizing metallic NPs using algae. It highlights the innovative role that various algal species play in reducing metal ions into nanoscale particles, offering a sustainable and environmentally friendly substitute for traditional synthesis techniques. The discussion encompasses the mechanisms involved, the advantages of utilizing algae for this purpose, and the potential applications of these biogenic metallic NPs in various fields. A significant focus is placed on AgNPs and AuNPs due to their distinctive properties, which have garnered considerable scientific and technological interest. The chapter demonstrates how NPs of different sizes and shapes can be produced using various types of algae. The varied applications of metallic NPs in nanomedicine are explored in detail. These tiny particles, with their own physical and chemical properties, play an important role in fields such as targeted medication administration, imaging, and therapy, improving the efficacy of medical therapies. While this developing field has not yet reached a level of maturity that would allow it to be scaled up for commercial applications, algal-mediated biosynthesis of NPs faces several challenges, including slow reaction times, low yields, poor morphological characteristics, and the choice of algal strain. The synthesis of NPs through algal processes is an innovative approach characterized by rapid production, cost efficiency, and high effectiveness. This method leverages the unique biochemical properties of algae, which facilitate the creation of targeted nanomaterials. As a result, nanotechnologists can develop a wide range of desirable nanomaterials while utilizing environmentally friendly and clean energy processes. This advancement not only enhances the scalability and accessibility of nanomaterial production but also contributes to sustainability in the field of nanotechnology, making it an attractive alternative to traditional synthesis methods. To ensure the safe application of eco-friendly NPs in medicine, it is essential to conduct more in-depth research and exploration in this area. Comprehensive studies are necessary to understand their potential benefits, risks, and mechanisms, paving the way for their effective integration into medical practices.

References

1 Chapman RL. Algae: the world's most important "plants"-an introduction. Mitig Adapt Strateg Glob Chang. 2013;18:5–12. doi:10.1007/s11027-010-9255-9.

2 Field CB, Behrenfeld MJ, Randerson JT, Falkowski P. Primary production of the biosphere: integrating terrestrial and oceanic components. Science. 1998;281(5374):237–240. doi:10.1126/science.281.5374.237.

3 Stevenson J. Ecological assessments with algae: a review and synthesis. J Phycol. 2014;50:437–461. doi:10.1111/jpy.12189.

4 Beardall J, Stojkovic S, Larsen S. Living in a high CO_2 world: impacts of global climate change on marine phytoplankton. Plant Ecol Divers. 2009;2(2):191–205. doi:10.1080/17550870903271363.

5 Lardon L, Hélias A, Sialve B, Steyer JP, Bernard O. Life-cycle assessment of biodiesel production from microalgae. Environ Sci Technol. 2009;43(17):6476–6482. doi:10.1021/es900705j.

6 Arora K, Kumar P, Bose D, Li X, Kulshresth S. Potential applications of algae in biochemical and bioenergy sector. 3 Biotech. 2021;11:296. doi:10.1007/s13205-021-02825-5.

7 Sahoo D, Seckbach J. The Algae World. 2015. Springer, Dordrecht.

8 Chaudhary R, Nawaz K, Khan AK, Hano C, Abbasi BH, Anjum S. An overview of the algae-mediated biosynthesis of nanoparticles and their biomedical applications. Biomolecules. 2020;10(11):1–35.

9 Ganesh Kumar V, Dinesh Gokavarapu S, Rajeswari A, Stalin Dhas T, Karthick V, Kapadia Z, et al. Facile green synthesis of gold nanoparticles using leaf extract of antidiabetic potent *Cassia auriculata*. Colloids Surf B: Biointerfaces. 2011;87(1):159–163. doi:10.1016/j.colsurfb.2011.05.016.

10 Azizi S, Ahmad MB, Namvar F, Mohamad R. Green biosynthesis and characterization of zinc oxide nanoparticles using brown marine macroalga *Sargassum muticum* aqueous extract. Mater Lett. 2014;116:275–277. doi:10.1016/j.matlet.2013.11.038.

11 Patel V, Berthold D, Puranik P, Gantar M. Screening of cyanobacteria and microalgae for their ability to synthesize silver nanoparticles with antibacterial activity. Biotechnol Rep. 2015;5(1):112–119. doi:10.1016/j.btre.2014.12.001.

12 Sahoo PC, Kausar F, Lee JH, Han JI. Facile fabrication of silver nanoparticle embedded $CaCO_3$ microspheres via microalgae-templated CO_2 biomineralization: application in antimicrobial paint development. RSC Adv. 2014;4(61):32562–32569. doi:10.1039/C4RA03623A.

13 Prasad R, Pandey R, Barman I. Engineering tailored nanoparticles with microbes: quo vadis? Wiley Interdiscip Rev Nanomed Nanobiotechnol. 2016;8(2):316–330. doi:10.1002/wnan.1363.

14 Sharma A, Sharma S, Sharma K, Chetri SPK, Vashishtha A, Singh P, et al. Algae as crucial organisms in advancing nanotechnology: a systematic review. J Appl Phycol. 2016;28(3):1759–1774. doi:10.1007/s10811-015-0715-1.

15 Senapati S, Syed A, Moeez S, Kumar A, Ahmad A. Intracellular synthesis of gold nanoparticles using alga *Tetraselmis kochinensis*. Mater Lett. 2012;79:116–118. doi:10.1016/j.matlet.2012.04.009.

16 Vijayan SR, Santhiyagu P, Singamuthu M, Kumari Ahila N, Jayaraman R, Ethiraj K. Synthesis and characterization of silver and gold nanoparticles using aqueous extract of seaweed, *Turbinaria conoides*, and their antimicrofouling activity. Sci World J. 2014;1:938272. doi:10.1155/2014/938272.

17 Oza G, Pandey S, Mewada A, Kalita G, Sharon M, Phata J, et al. Facile biosynthesis of gold nanoparticles exploiting optimum pH and temperature of fresh water algae *Chlorella pyrenoidusa*. Pelagia Res Libr. 2012;3(3):1405–1412.

18 Namvar F, Azizi S, Ahmad MB, Shameli K, Mohamad R, Mahdavi M, et al. Green synthesis and characterization of gold nanoparticles using the marine macroalgae *Sargassum muticum*. Res Chem Intermed. 2015;41(8):5723–5730.

19 Kalabegishvili T, Kirkesali EI, Rcheulishvili AN, Ginturi EN, Murusidze IG, Pataraya DT, et al. Synthesis of gold nanoparticles by some strains of Arthrobacter genera. J Mater Sci Eng A. 2012;2(2):164–173.

20 Bhattacharya P, Swarnakar S, Ghosh S, Majumdar S, Banerjee S. Disinfection of drinking water via algae mediated green synthesized copper oxide nanoparticles and its toxicity evaluation. J Environ Chem Eng. 2019;7(1):102867. doi:10.1016/j.jece.2018.102867.

21 Wang L, Hu C, Shao L. The antimicrobial activity of nanoparticles: present situation and prospects for the future. Int J Nanomedicine. 2017;12:1227–1249.

22 Rajeshkumar S, Kannan C, Annadurai G. Synthesis and characterization of antimicrobial silver nanoparticles using marine brown seaweed *Padina tetrastromatica*. Drug Invent Today. 2012;4(10):511–513.

23 Aboelfetoh EF, El-Shenody RA, Ghobara MM. Eco-friendly synthesis of silver nanoparticles using green algae (*Caulerpa serrulata*): reaction optimization, catalytic and antibacterial activities. Environ Monit Assess. 2017;189(7):349. doi:10.1007/s10661-017-6033-0.

24 Sinha SN, Paul D, Halder N, Sengupta D, Patra SK. Green synthesis of silver nanoparticles using fresh water green alga *Pithophora oedogonia* (Mont.) Wittrock and evaluation of their antibacterial activity. Appl Nanosci. 2015;5(6):703–709.

25 Suganya KSU, Govindaraju K, Ganesh Kumar VG, Stalin Dhas T, Karthick V, Singaravelu G, et al. Blue green alga mediated synthesis of gold nanoparticles and its antibacterial efficacy against Gram positive organisms. Mater Sci Eng C. 2015;47:351–356. doi:10.1016/j.msec.2014.11.043.

26 Borase HP, Patil CD, Suryawanshi RK, Koli SH, Mohite BV, Benelli G, et al. Mechanistic approach for fabrication of gold nanoparticles by Nitzschia diatom and their antibacterial activity. Bioprocess Biosyst Eng. 2017;40(10):1437–1446.

27 Rajathi FAA, Parthiban C, Kumar VG, Anantharaman P. Biosynthesis of antibacterial gold nanoparticles using brown alga, *Stoechospermum marginatum* (kützing). Spectrochim Acta A Mol Biomol Spectrosc. 2012;99:166–173. doi:10.1016/j.saa.2012.08.081.

28 Omomowo IO, Adenigba VO, Ogunsona SB, Adeyinka GC, Oluyide OO, Adedayo AA, et al. Antimicrobial and antioxidant activities of algal-mediated silver and gold nanoparticles. IOP Conf Ser Mater Sci Eng. 2020;805(1).

29 Mallmann EJJ, Cunha FA, Castro BNMF, Maciel AM, Menezes EA, Fechine PBA. Atividade antifúngica de nanopartículas de prata obtidas por síntese verde. Rev Inst Med Trop Sao Paulo. 2015;57(2):165–167.

30 Rajeshkumar S, Malarkodi C, Paulkumar K, Vanaja M, Gnanajobitha G, Annadurai G. Algae mediated green fabrication of silver nanoparticles and examination of its antifungal activity against clinical pathogens. Int J Met. 2014;2014:1–8.

31 Dhavale R, Jadhav S, Sibi G. Microalgae mediated silver nanoparticles (Ag-NPs) synthesis and their biological activities. J Crit Rev. 2020;7(2):15–20.

32 El-Sheekh MM, El-Kassas HY. Algal production of nano-silver and gold: their antimicrobial and cytotoxic activities: a review. J Genet Eng Biotechnol. 2016;14(2):299–310. doi:10.1016/j.jgeb.2016.09.008.

33 Romero D, Aguilar C, Losick R, Kolter R. Amyloid fibers provide structural integrity to *Bacillus subtilis* biofilms. Proc Natl Acad Sci USA. 2010;107(5):2230–2234.

34 Bixler GD, Bhushan B. Review article: biofouling: lessons from nature. Philos Trans R Soc A Math Phys Eng Sci. 1967;2012(370):2381–2417.

35 Kumar P, Senthamil SS, Prabha AL, Selvaraj M, Rani LM, Suganthi P, et al. Antibacterial activity and in-vitro cytotoxicity assay against brine shrimp using silver nanoparticles synthesized from *Sargassum ilicifolium*. Dig J Nanomat Biostruct. 2012;7(4):1447–1455.

36 Krishnan M, Sivanandham V, Hans-Uwe D, Murugaiah SG, Seeni P, Gopalan S, et al. Antifouling assessments on biogenic nanoparticles: a field study from polluted offshore platform. Mar Pollut Bull. 2015;101(2):816–825. doi:10.1016/j.marpolbul.2015.08.033.

37 Balaraman P, Balasubramanian B, Kaliannan D, Durai M, Kamyab H, Park S, et al. Phyco-synthesis of silver nanoparticles mediated from marine algae *Sargassum myriocystum* and its potential biological and environmental applications. Waste Biomass Valori. 2020;11(10):5255–5271. doi:10.1007/s12649-020-01083-5.

38 Amutha V, Deepak P, Kamaraj C, Balasubramani G, Aiswarya D, Arul D, et al. Mosquito-larvicidal potential of metal and oxide nanoparticles synthesized from aqueous extract of the seagrass, *Cymodocea serrulata*. J Clust Sci. 2019;30(3):797–812. doi:10.1007/s10876-019-01542-7.

39 Hasan MR, Alotaibi BS, Althafar ZM, Mujamammi AH, Jameela J. An update on the therapeutic anticancer potential of *Ocimum sanctum* L.: "Elixir of life.". Molecules. 2023;28(3):1–28. doi:10.3390/molecules28031193.

40 PKJ P. Multi-functional silver nanoparticles for drug delivery: a review. Int J Cur Res Rev. 2017;9(8):1–5.

41 Gopu M, Kumar P, Selvankumar T, Senthilkumar B, Sudhakar C, Govarthanan M, et al. Green biomimetic silver nanoparticles utilizing the red algae *Amphiroa rigida* and its potent antibacterial, cytotoxicity and larvicidal efficiency. Bioprocess Biosyst Eng. 2021;44(2):217–223. doi:10.1007/s00449-020-02426-1.

42 Babu B, Palanisamy S, Vinosha M, Anjali R, Kumar P, Pandi B, et al. Bioengineered gold nanoparticles from marine seaweed *Acanthophora spicifera* for pharmaceutical uses : antioxidant, antibacterial, and anticancer activities. Bioprocess Biosyst Eng. 2020;43(12):2231–2242. doi:10.1007/s00449-020-02408-3.

43 Acharya D, Satapathy S, Somu P, Parida UK, Mishra G. Apoptotic effect and anticancer activity of biosynthesized silver nanoparticles from marine algae*Chaetomorpha linum*extract against human colon cancer cell HCT-116. Biol Trace Elem Res. 2021;199(5):1812–1822.

44 Kumar P, Govindaraju M, Senthamilselvi S, Premkumar K. Photocatalytic degradation of methyl orange dye using silver (Ag) nanoparticles synthesized from *Ulva lactuca*. Colloids Surf B: Biointerfaces. 2013;103:658–661. doi:10.1016/j.colsurfb.2012.11.022.

45 Husain S, Afreen S, Hemlata YD, Afzal B, Fatma T. Cyanobacteria as a bioreactor for synthesis of silver nanoparticles-an effect of different reaction conditions on the size of nanoparticles and their dye decolorization ability. J Microbiol Methods. 2019;162:77–82. doi:10.1016/j.mimet.2019.05.011.

46 Ramakrishna M, Rajesh Babu D, Gengan RM, Chandra S, Nageswara RG. Green synthesis of gold nanoparticles using marine algae and evaluation of their catalytic activity. J Nanostructure Chem. 2016;6(1):1–13.

47 Borah D, Das N, Das N, Bhattacharjee A, Sarmah P, Ghosh K, et al. Alga-mediated facile green synthesis of silver nanoparticles: photophysical, catalytic and antibacterial activity. Appl Organomet Chem. 2020;34(5):1–10. doi:10.1002/aoc.5597.

48 Zhang Y, Gao G, Qian Q, Cui D. Chloroplasts-mediated biosynthesis of nanoscale Au-Ag alloy for 2-butanone assay based on electrochemical sensor. Nanoscale Res Lett. 2012;7:1–8.

49 Elgamouz A, Idriss H, Nassab C, Bihi A, Bajou K, Hasan K, et al. Green synthesis, characterization, antimicrobial, anti-cancer, and optimization of colorimetric sensing of hydrogen peroxide of algae extract capped silver nanoparticles. Nanomaterials. 2020;10(9):1–19.

50 Kuppusamy P, Mashitah MY, Maniam GP, Govindan N. Biosynthesized gold nanoparticle developed as a tool for detection of HCG hormone in pregnant women urine sample. Asian Pac J Trop Dis. 2014;4(3):237. doi:10.1016/S2222-1808(14)60538-7.

51 Sharma D, Kanchi S, Bisetty K. Biogenic synthesis of nanoparticles: a review. Arab J Chem. 2019;12(8):3576–3600. doi:10.1016/j.arabjc.2015.11.002.

52 Shalaby E. Algae as promising organisms for environment and health. Plant Signal Behav. 2011;6(9):1338–1350.

53 Zhuang D, He N, Khoo KS, Ng EP, Chew KW, Ling TC. Application progress of bioactive compounds in microalgae on pharmaceutical and cosmetics. Chemosphere. 2022;291:132932.

54 Delasoie J, Zobi F. Natural diatom biosilica as microshuttles in drug delivery systems. Pharmaceutics. 2019;11(10):537.

55 Aw MS, Simovic S, Yu Y, Addai-Mensah J, Losic D. Porous silica microshells from diatoms as biocarrier for drug delivery applications. Powder Technol. 2012;223:52–58. doi:10.1016/j.powtec.2011.04.023.

56 Zhang H, Shahbazi MA, Mäkilä EM, da Silva TH, Reis RL, Salonen JJ, et al. Diatom silica microparticles for sustained release and permeation enhancement following oral delivery of prednisone and mesalamine. Biomaterials. 2013;34(36):9210–9219.

57 Khalid M, Khalid N, Ahmed I, Hanif R, Ismail M, Janjua HA. Comparative studies of three novel freshwater microalgae strains for synthesis of silver nanoparticles: insights of characterization, antibacterial, cytotoxicity and antiviral activities. J Appl Phycol. 2017;29:1851–1863.

58 Chokshi K, Pancha I, Ghosh T, Paliwal C, Maurya R, Ghosh A, Mishra S. Green synthesis, characterization and antioxidant potential of silver nanoparticles biosynthesized from de-oiled biomass of thermotolerant oleaginous microalgae *Acutodesmus dimorphus*. RSC Adv. 2016;6(76):72269–72274.

59 Aldayel MF, Al Kuwayti MA, El Semary NA. Investigating the production of antimicrobial nanoparticles by *Chlorella vulgaris* and the link to its loss of viability. Microorganisms. 2022;10(1):145.

60 Sudha SS, Rajamanickam K, Rengaramanujam J. Microalgae mediated synthesis of silver nanoparticles and their antibacterial activity against pathogenic bacteria. Indian J Exp Biol. 2013;51(5):393–399.

61 Sathasivam K, Yanmaz E, Vijayakumar N. Green synthesis approach of nanoparticle $CuFe_2O_4$ with the assistance of chlorella green microalgae biomass extract. Biomass Convers Biorefinery. 2024;23:1–8.

62 Wong SW, Zhou GJ, Kwok KW, Djurišić AB, Han J, Lee JS, Leung KM. in vivo toxicities of nine engineered nano metal oxides to the marine diatom *Skeletonema costatum* and rotifer *Brachionus koreanus*. Mar Pollut Bull. 2020;153:110973.

63 Ashique S, Upadhyay A, Hussain A, Bag S, Chaterjee D, Rihan M, et al. Green biogenic silver nanoparticles, therapeutic uses, recent advances, risk assessment, challenges, and future perspectives. J Drug Deliv Sci Technol. 2022;77:103876.

7

Biological Synthesis of Nanoparticles by Actinomycetes and Biogenic Sustainable Nanotherapeutics

Pratap Kalita[1], *Abhinab Goswami*[2], *Malay K. Das*[3], *and Lalduhsanga Pachuau*[4]

[1]*Department of Pharmaceutics, Pratiksha Institute of Pharmaceutical Sciences, Guwahati, Assam, India*
[2]*Faculty of Pharmaceutical Sciences, Assam Down Town University, Guwahati, Assam, India*
[3]*Drug Delivery Research Laboratory, Department of Pharmaceutical Sciences, Dibrugarh University, Dibrugarh, Assam, India*
[4]*Department of Pharmaceutical Science, Assam University, Silchar, Assam, India*

7.1 Introduction

The applications of nanotechnology in disease theragnostics or nanomedicine have rapidly expanded recently. The application of nanotechnology in medicine creates opportunities for managing, controlling, and fabricating structures and devices at the nanoscale for biomedical purposes, such as drug delivery and therapeutics. In comparison to their larger counterparts, nanoparticulate forms of the same material display fascinating and exceptional characteristics due to surface and quantum effects, which enable nanomaterials to demonstrate improved or unique mechanical, electrical, thermal, optical, catalytic, and other physicochemical properties [1, 2]. By refining synthetic pathways and incorporating appropriate functional groups, one can continually fine-tune the shape, size, and surface characteristics of nanomaterials, thereby broadening their areas of application.

Various top-down approaches, such as mechanical milling, etc., or bottom-up approaches, including sol–gel processes, nanoprecipitation, chemical reduction, and others, have been applied in the synthesis of nanoparticles [3]. In Figure 7.1, the different approaches to synthesizing nanoparticles are provided. Due to the negative environmental impact and the low yield of traditional methods for synthesizing nanoparticles, there has been a demand in recent years for more eco-friendly biogenic nanoparticles that utilize sustainable approaches

Biogenic Synthesis of Nanoparticles and Sustainable Nanotherapeutics,
First Edition. Edited by Malay K. Das.

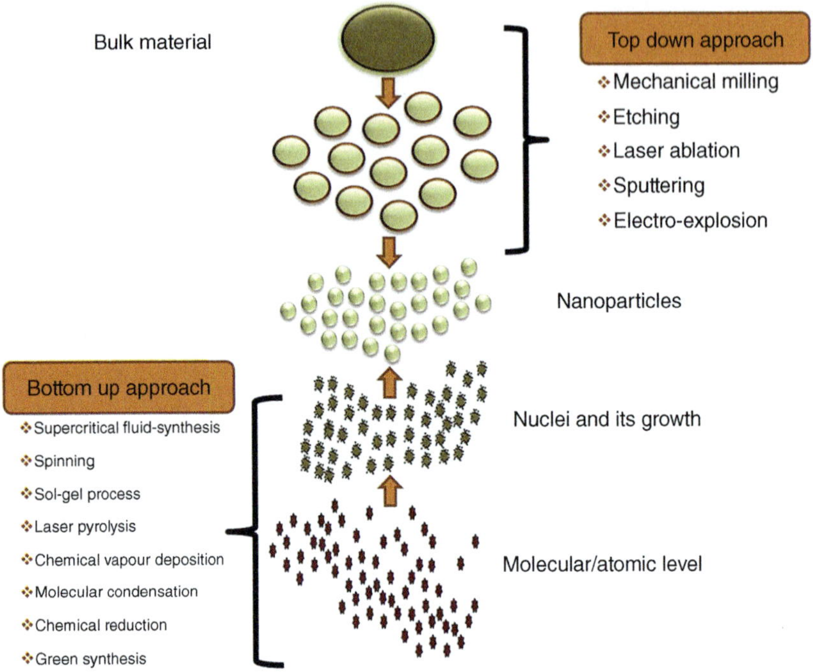

Figure 7.1 Synthesis of nanoparticles through various approaches (*Source:* [3] / with permission of Elsevier).

by combining biological and nanotechnological techniques [4, 5]. Biological synthesis of nanoparticles represents a bottom-up process where nanoparticles are generated by reduction or oxidation of metals, primarily driven by various enzymes secreted by microorganisms and various plant metabolites [6]. Both uni- and multicellular organisms, including various microbes and plants, have been successfully used in the reduction and stabilization of biologically synthesized nanoparticles [4, 7]. Phytochemicals, including flavonoids, terpenes, alkaloids, and polyphenolics from plant extracts and enzymes from microorganisms, played important roles in the reduction of metal ions in producing metallic nanoparticles, including gold, silver, copper, selenium, etc [8–10].

Actinomycetes are Gram-positive bacteria that form spores and thrive in aerobic environments, classified under the order actinomycetales. Actinomycetes exhibit traits of both bacteria and fungi, as evident from their name, which originates from the Greek terms *aktis* (aray) and *mykes* (fungus), but have enough unique characteristics to classify them within the kingdom of bacteria [11]. Actinomycetes are divided into seven families, with over 500 species of actinomycetes identified to date, mainly isolated from soil [7]. These microorganisms are well-known for being a limitless source of antibiotics and are considered a crucial provider of products that hold substantial medical, industrial, and economic importance [4, 12]. Studies in recent years have shown that actinomycetes hold great potential in nanotechnology applications, serving as effective agents for reduction and stabilization for eco-friendly synthesis of metallic nanoparticles as a result of their excellent stability and polydispersity [5–7, 13, 14]. Pathways for biosynthesis of metallic nanoparticles by actinomycetes have been reported to be either extracellular or intracellular pathways, with the extracellular pathway being the more commonly reported [15]. According to reports, biosynthesized nanoparticles generally exhibit enhanced bioactivity, such as antimicrobial property than those produced through conventional methods, which is attributed to the synergistic action of proteins that play a crucial role in capping and stabilizing the synthesized nanoparticles [15]. Some of the different genera of actinomycetes reportedly used to synthesize metallic nanoparticles include *Streptomyces*, *Rhodococcus*, *Thermomonospora*, and *Nocardia*, of which *Streptomyces* is the most used and promising genus [6, 13, 16].

This chapter will provide an insight into recent developments in the biosynthesis of metallic nanoparticles by actinomycetes. Different approaches to the characterization of the biosynthesized nanoparticles and their applications as sustainable nanotherapeutics will also be discussed. Some of the plant extracts and microorganisms used in the synthesis of several metallic nanoparticles are given in Table 7.1.

7.2 Biosynthesis of Metallic Nanoparticles by Actinomycetes

7.2.1 Silver Nanoparticles

There are various techniques to produce silver nanoparticles using actinomycetes, and the procedures are quite straightforward. In summary, silver nanoparticles derived from actinomycetes were prepared by mixing actinomycetes (culture filtrate) with $AgNO_3$ solution in a clean Erlenmeyer flask, followed by incubation at $25 \pm 2\,°C$ in a rotary shaker for 24 hours. Simultaneously, the transition of the solution from colorless to dark brown was monitored continuously during the process of synthesis to verify whether silver ions were reduced within the reaction mixture. Once the reduction of silver ions was confirmed, the silver nanoparticle pellets derived from actinomycetes were collected through ultracentrifugation and then dried at 60 °C in a hot air oven for about six to eight hours. The resulting dried pellets of actinomycetes-based silver nanoparticles were subjected to further analysis [29].

Table 7.1 Some plants and microorganisms-based synthesis of metallic nanoparticles.

Plants/microbes	Nanoparticles	Size and shape	References
Camellia sinensis (green tea) extract	Ag	77.4 nm, spherical	[17]
Myristica fragrans seed extract	TiO_2	2 μm to 100 nm, spherical	[18]
Glycosmis pentaphylla fruit extract	Ag	17 nm, spherical	[19]
Ananas comosus aqueous leaf extract	Au	18.85 nm, spherical	[20]
Eriobotrya japonica leaves extract	ZnO	13 nm, irregular platelets' shape	[21]
Phragmites australis (common reed) leaf extract	Au	18 nm, spherical	[22]
Kniphofia schemperi root extract	TiO_2	~10 nm, spherical	[23]
Chlorella vulgaris	Pd	15 nm, irregular	[24]
Sargassum vulgare	Ag	10 nm, spherical	[25]
Cell-free aqueous extract of Cyanobacterial (*Microchaete* NCCU-342)	Ag	40–100 nm, spherical	[26]
Spirulina platensis alga extract	Pd	10–20 nm, spherical	[27]
Tricoderma citrinoviride extract	TiO_2	10–400 nm, irregular	[28]

In previous research, silver nanoparticles were synthesized by mixing a specific quantity of actinomycete biomass extracts with an equal amount of $AgNO_3$ solution and then incubating the mixture at 30 °C for 48 hours in the dark while being shaken continuously. The biosynthesis of the actinomycete-derived silver nanoparticles was confirmed through visual observation, noting the color switch from pale yellow to dark brown and ultimately black, which was further assessed by determining the ultraviolet–visible (UV–Vis) absorbance of the mixture. After this, the mixture was centrifuged, and the resulting pellets were re-suspended in ultrapure water, followed by another centrifugation step. The pellets were collected and rinsed with ultrapure water several times (two to three) to eliminate any impurities, and the dried actinomycete-derived silver nanoparticles were collected following vacuum oven drying at 50 °C for 24 hours [30, 31].

7.2.2 Copper Oxide Nanoparticles

Numerous studies have documented the synthesis of copper oxide nanoparticles utilizing biological agents such as bacteria. Zhao et al. [32] synthesized copper oxide nanoparticles using marine actinomycetes by combining an aqueous copper sulfate solution with cell-free supernatant and heating the mixture at 100 °C for 15 minutes on a hot plate while stirring continuously. Simultaneously, the transition in color from blue to reddish brown was monitored throughout the process to verify the reduction of copper ions in the mixture. After this, centrifugation of the mixture was performed at 10,000 rpm for 15 minutes to obtain the copper oxide nanoparticle pellets, which were then washed multiple times with distilled water, followed by another centrifugation at the same rpm for 10 minutes after washing with ethanol to eliminate any excess water molecules from the solution of copper oxide nanoparticles. The copper oxide nanoparticles were then dried at 100 °C, and the resulting dried pellets of the actinomycetes-derived copper oxide nanoparticles were put through various analyses.

Actinomycetes have also been used in the synthesis of copper oxide nanoparticles. The process of biosynthesizing these nanoparticles involved mixing 50 ml of the bacterial species in an Erlenmeyer flask, and then centrifuged at 10,000 rpm, after which the supernatant was utilized for preparing the copper oxide nanoparticles. Specifically, copper sulfate was combined with the supernatant from the bacterial isolate in Erlenmeyer flasks, then incubated between 30 °C and 32 °C while being monitored for color changes. As time went on, the color of the reaction mixture changed from a soft blue to green, signaling the formation of biogenic copper oxide nanoparticles. This transformation was confirmed through UV–visible spectroscopy analysis, which was performed over a wavelength range of 200–800 nm [33].

7.2.3 Gold Nanoparticles

Gold nanoparticles garner considerable attention from researchers because of their distinctive properties, including optical, catalytic, biomedical, non-toxic, and biocompatible features. Gold nanoparticles were prepared by mixing cell-free bacterial supernatant with twice its volume of Gold (III) chloride trihydrate ($HAuCl_4.3H_2O$) in distilled water, followed by incubation at 37 °C in the dark for 24–96 hours. The incubated mixture subsequently displayed a color change from red to deep purple, signifying the reduction of gold chloride and the generation of biogenic gold nanoparticles [34].

7.2.4 Selenium Nanoparticles

Biogenic selenium nanoparticles were synthesized by supplementing specific bacterial strains with Se^{4+} ions for a period of five days, after which the mixture underwent centrifugation at 10,000 rpm for a specified time. After pulverizing the collected cell packs in a mortar and pestle, the resulting mixtures were subjected to ultrasonication and repeatedly rinsed with a buffer solution and deionized water, followed by sequential centrifugation. The collected pellets were resuspended in deionized water, creating a suspension containing selenium nanoparticles and cell debris. This suspension was then purified by partitioning the selenium nanoparticles between *n*-octyl alcohol and water, followed by centrifugation. The purified nanoparticles were stored at 4 °C until the biogenic selenium nanoparticles settled at the bottom of the tubes. The precipitated selenium nanoparticles were washed sequentially with chloroform, ethanol, and distilled water. Finally, the purified biogenic selenium nanoparticles were resuspended in deionized water [35, 36].

7.2.5 Zinc Oxide Nanoparticles

In a standard procedure, a few milliliters of biological agents, such as actinomycetes, were combined in an Erlenmeyer flask containing $ZnSO_4$, with the pH adjusted to 8 using sodium hydroxide solution. The solution thus resulted was incubated in a shaker with constant agitation, then cooled with ice, followed by incubation at 40 °C. To obtain the nanoparticle pellets, centrifugation was performed, followed by rinsing with deionized water to effectively remove any impurities. This ensures a higher quality final product. The collected precipitates from the actinomycetes-based zinc oxide nanoparticles were efficiently dried overnight in a hot air oven set to 80 °C, ensuring optimal removal of moisture for subsequent use [37, 38]. In an alternative approach, zinc oxide nanoparticles derived from actinobacteria were synthesized by introducing cultured actinobacterial isolates into a vessel containing a zinc sulfate solution, adjusting the pH of the reaction mixture to 7, and subsequently shaking it on a rotary shaker. Afterward, the mixture was centrifuged under dark conditions, and the yield of actinobacterial-based zinc oxide nanoparticles was ascertained through visual observation, noting the transition from color to a cloudy or milky appearance, which was then used for subsequent studies [39].

7.2.6 Others

Biogenic platinum nanoparticles were also prepared using the bacterial culture. Initially, the bacterial sample was combined with chloroplatinic acid hydrate (H_2PtCl_6), and the mixture was incubated at 29 °C for a duration of one to two weeks in the dark while being subjected to constant agitation on a rotary shaker. The dark environment was maintained to prevent any photochemical reversal throughout the experiment. Biogenic nanoparticles were confirmed by a color change from yellow to greenish brown, indicating the reduction of Pt^{4+} to Pt^0 and the formation of platinum nanoparticles from H_2PtCl_6 [40].

7.3 Characterization of Actinomycetes-Assisted Nanoparticles

7.3.1 Morphological Characterization

Most actinomycetes-assisted synthesis of metallic nanoparticles yields well-dispersed spherical particles. Scanning electron microscope (SEM) studies on silver nanoparticles synthesized using *Streptomyces diastaticus* produced irregular yet spherical shapes and clustered under the observation. The prepared biological silver nanoparticle exhibits a size range of 23–36 nm with an average particle size of 28.5 nm [29]. The copper oxide nanoparticles produced by actinomycetes were observed through SEM and transmission electron microscopy (TEM). They exhibited irregular, small, spherical shapes and were uniformly distributed, with particle sizes ranging from 10 to 30 nm [32]. TEM studies confirm the monodispersed spherical shape and average particle size range from 8.48 to 9.67 nm for the synthesized silver nanoparticles using rare actinomycetes, *Glutamicibacter nicotianae* SNPRA1 and *Leucobacter aridicollis* SNPRA2 [30]. Actinomycetes (*Streptomyces* sp)-mediated silver nanoparticles produced spherical shapes without aggregate formations and smaller diameters of less than 30 nm [41]. The surface topology observed through TEM indicated that the zinc oxide nanoparticles produced with the assistance of actinomycetes exhibited a mixed morphology, characterized by both spherical and hexagonal shapes. The particle size ranged from 0 to 90 nm, with minimal agglomeration [37]. In a separate study, it was found that the zinc oxide nanoparticles synthesized by actinomycetes were spherical in shape, with an average particle size of 21.72 nm and no evidence of aggregation [39].

The TEM characterization of the biosynthesized gold nanoparticles also produced well-dispersed spherical shape nanoparticles with a size varying between 5.42 and 13.34 nm. The studies conducted using energy dispersive x-ray analysis (EDX) offer insights into the elemental mapping of the synthesized nanoparticles. Additionally, the EDX analysis of the biologically synthesized gold nanoparticles reveals their distribution and composition [34]. The TEM-selected area electron diffraction analysis revealed a pattern consisting of individual spots (dots) from the crystalline nanomaterial, affirming the presence of well-defined, bright, circular rings of X-ray diffraction. Each of these diffraction rings corresponds to particular lattice planes of (111), (200), (220), and (311). Together, these specific lattice planes represented by the diffraction patterns indicate that the biosynthesized gold nanoparticles exhibit a face-centered cubic structure [34].

The TEM studies confirmed the spherical particle size of the biogenic copper oxide nanoparticles, and the average particle size was 1.72–13.49 nm. The EDX and elementary mapping also confirmed the purity of the prepared biogenic copper oxide nanoparticles by generating the signals based on the copper element [33].

Another study was conducted on platinum nanoparticles, and the prepared biogenic platinum nanoparticles showed spherical, non-agglomerated, monodisperse particles with a size ranging between 2.44 and 29.57 nm, and the mean particle size was 10.83 nm [40]. The SEM, TEM, and EDX analyses demonstrated that the synthesized biogenic selenium nanoparticles were uniformly distributed and had a spherical morphology, with particle sizes ranging from 48.2 to 135.7 nm and an average size of 73.8 nm [35]. Another study also showed that the fabricated biogenic selenium nanoparticles produced were spherical, with an average 50–70 nm diameter. EDX analysis confirmed that selenium is the principal component of the selenium nanoparticles [36].

7.3.2 Thermal Properties

Actinomycetes-mediated copper oxide nanoparticles displayed thermal stability under thermogravimetric analysis, with decreased weight loss in two stages. The initial phase of thermal degradation occurred below 100 °C, and the next phase occurred at 233 °C [32].

7.3.3 X-Ray Diffraction Properties

Silver nanoparticles synthesized with *S. diastaticus* resulted in a crystalline structure, as demonstrated by XRD analysis. The recorded peaks at 2θ values of 38.11, 46.21, 67.62, and 76.87 correspond to Bragg's reflections of the planes 111, 200, 220, and 311, respectively, validating the face-centered cubic crystal arrangement of silver [29]. The copper oxide nanoparticles derived from actinomycetes exhibited crystalline peaks at 2θ angles of 32.9, 35.1, 38.4, 48.3, 53.8, 58.5, 61.7, and 66.3, which correspond to various diffraction planes [32]. A recent study conducted by Abd El-Ghany et al. [30] indicated that silver nanoparticles synthesized through actinomycetes exhibited diffraction plane values that confirm the face-centered cubic morphology of the biogenic silver nanoparticles. The zinc oxide nanoparticles derived from actinomycetes showed notable diffraction peaks, which confirmed that the nanoparticles possess a hexagonal morphology and a crystalline arrangement [37, 38]. In a different XRD analysis, zinc oxide nanoparticles produced through biosynthesis displayed a polycrystalline diffractogram, with diffraction peaks indicating a distinct hexagonal crystalline structure. The particles contained no secondary phase impurities and exhibited high phase purity [39].

The XRD spectrum of the biosynthesized gold nanoparticles revealed the diffraction peaks at 2θ values of 17.21°, 38.66°, 44.75°, 64.83°, and 78.14°, while scanned under a range of 0–80°, the observed diffraction peaks indicated that the crystals exhibited a preferred growth orientation in the plane (111) [34]. The XRD spectrum of the synthesized biogenic copper oxide nanoparticles displayed peaks at

2θ values of 35°, 38°, 48°, 53°, 58°, 65°, 67°, and 72°. These peaks are associated with the (111), (202), (020), (202), and (113) reflection planes, suggesting that the particles exhibit a face-centered cubic structure [33].

The X-ray diffraction analysis of the biogenic platinum nanoparticles clearly shows distinct peaks at 2θ values of 40.51° and 45.39°. These peaks correspond to the (111) and (200) reflection planes, definitively confirming the crystalline structure of the biogenic platinum nanoparticles, which exhibit a cubic crystal structure [40]. The XRD analysis of biogenic selenium nanoparticles revealed three strong peaks at 2θ values of 23.78, 29.8, and 43.9, which are associated with the (100), (101), and (102) reflection planes, suggesting the face-centered cubic morphology of the red elemental selenium [35].

7.3.4 Fourier Transform Infrared Spectroscopy

The common functional groups found in the produced nanoparticles can be analyzed through Fourier transform infrared spectroscopy (FTIR), which serves as a robust technique for assessing functional groups. FTIR is employed to pinpoint the specific functional groups of biomolecules, such as proteins, polysaccharides, and lipids, attached to the nanoparticle surfaces fabricated from biological sources, thereby disclosing the capping agents that stabilize the nanoparticles and offering understanding of the biogenesis mechanism. This technique has been utilized to recognize the capping agents involved in synthesizing biogenic nanoparticles, aiding in the comprehension of their stability and the biosynthesis processes.

The recent studies revealed that synthesized silver nanoparticles using *S. diastaticus* produced the spectrum band confirming C—H stretching of phenol, C—H stretching of alkanes, P—H phosphine, amide (C=O stretch) group, C—N stretching, and C—O stretching of carboxylic acids [29]. Absorption bands confirm the presence of alcohol groups associated with O—H bonds, along with the stretching of C—H bonds in alkanes, N—H bonds in amines, and C=C stretching in cyclic conjugated alkenes. Furthermore, the stretching of C—N bonds in amines and the bending of C—H bonds in alkanes are unequivocally observed [32, 42].

Actinomycetes mediated zinc oxide nanoparticles displayed characteristic FTIR spectra with absorption peaks that indicate O—H bond vibrations of the moisture attached to the surface of ZnO [37, 38]. Another study on biosynthesized zinc oxide nanoparticles produced several vibration peaks corresponding to the various functional groups, such as OH group, C—H stretching of methylene in proteins, carbonyl and ethylene groups, protein amine bands, C—N stretching, and zinc oxide stretching that confirm the preparation of zinc oxide structure [39]. The FTIR spectra of the biosynthesized gold nanoparticles showed fifteen absorption peaks assigned to the presence of aromatic, alcoholic, and phenolic O—H bonds in the compounds, including bands attributed to the presence of the amide group [34].

FTIR spectra of biogenic platinum nanoparticles revealed the appearance of the absorption peaks corresponding to heterocyclic compounds like proteins and the peaks assigned to alkyl halides [40]. FTIR analysis of biogenic selenium nanoparticles showed various absorption peaks related to peptides, amides, and aldehydes acting as capping and reducing agents in the nanoparticle suspension [36].

7.3.5 UV Visible Spectrophotometer

Silver nanoparticles biosynthesis using actinomycete (such as S *diastaticus*) was demonstrated by the presence of a surface plasmon resonance band at around 424 nm, which is responsible for color change from colorless to dark brown, confirming silver ions getting reduced in the reaction mixture [29]. Another study by Abd El-Ghany et al. [30] confirmed that the absorption peaks of the actinomycetes-mediated silver nanoparticle showed at 405–416 nm due to surface plasmon resonance. The absorption peak confirmed the characteristics of the prepared actinomycetes-based zinc oxide nanoparticles at 372 nm under UV–visible spectroscopy analysis, and this peak is assigned to the intrinsic bandgap absorption of zinc oxide nanoparticles [37]. Another study revealed that the biosynthesized zinc oxide nanoparticles exhibit an absorption peak at 350 nm under the UV–vis absorption spectra and confirmed the synthesis of the nanoparticles [38, 39].

The biosynthesized gold nanoparticles exhibited a maximum absorption peak at 540 nm when analyzed under the wavelength range of 300–800 nm using UV–vis spectroscopy, confirming their successful synthesis [34]. Similarly, the biogenic copper oxide nanoparticles were examined with a UV–visible spectrophotometer, revealing an absorption peak at 550 nm in the 200–800 nm range, which indicates the presence of these nanoparticles [33]. The reduction of biogenic platinum nanoparticles from $Pt4^{+}$ to Pt^{0} was evidenced by a color change from yellow to greenish-brown, further confirming the generation of biogenic platinum nanoparticles. This was supported by an absorption peak observed at 362 nm during UV–Vis spectroscopy analysis [40]. Lastly, the UV spectral analysis of biogenic selenium nanoparticles showed a characteristic peak at 290 nm, confirming their formation [36].

7.3.6 Zeta Potential and DLS Studies

Biosynthesized silver nanoparticles using *S. diastaticus* produced a zeta potential of 11 mV, and polydispersity index and particle size were recorded at 0.346 and 193.4 nm, respectively [29]. Another study by Zwar et al. [41] in 2023 stated that the fabricated actinomycetes-based silver nanoparticles showed a mean particle size around 100 nm and zeta potential with a value greater than ±30 mV, which confirms the stability of the engineered nanoparticles, and the nanoparticles also produced a polydispersity index between 0 and 0.3, which confirms the uniform size distribution within the formulation. The biosynthesized zinc oxide nanoparticles showed a narrow size distribution under DLS studies. Also, they produced an average particle size of 22.4 nm (21.72–22.4 nm) with a polydispersity index of 0.55, and studies revealed that biosynthesized zinc oxide nanoparticles were round crystalline [39]. The zeta potential of the biosynthesized gold nanoparticles exhibits a value of −24.8 mV, confirming good stability, no aggregation, and monodispersion within the colloidal nanoparticle suspension [34]. Another study stated that the prepared biogenic zinc oxide nanoparticles showed the stability of the colloidal nanoparticles with a zeta potential value ranging between 21.8 and −23.2 mv, and the observed high zeta potential negative value (−20 to 25 mv) indicates more stability due to the repulsion behavior each other for an extended period and thus make them more stable [38].

7.4 Application of Biogenic Nanoparticles

7.4.1 Antimicrobial

The rapid rise in antibiotic-resistant pathogens has compelled researchers to explore newer compounds to tackle the problem. However, searching for a new compound to combat antibiotic resistant pathogens is tedious. Traditionally, metals such as gold, silver, iron, copper, etc., have been used since they were immemorable for their antimicrobial activity [43]. Thus, using an existing technique with simple modification can provide adequate time to control the problem. This prompted researchers to explore and utilize the antimicrobial potential of nanoscale metallic/metal oxide particles as an alternative to traditional antimicrobial agents [44–47]. Compared to bulky metallic counterparts, the nanoparticles possess a high surface-to-volume ratio, which facilitates their ease of interaction with the microbial cell membranes. The metallic/metal oxide nanoparticles may exert different mechanisms to show their antimicrobial action. Although several mechanisms have been proposed to explain the antimicrobial action of metallic/metal oxide nanoparticles, an accurate mechanism of action is yet to be justified [48].

The antibacterial properties of biogenic metallic and metal oxide nanoparticles are thought to arise from their interaction with bacterial cell membranes. This interaction can lead to the formation of pits or gaps in the membrane, ultimately resulting in its fragmentation [49]. Additionally, these nanoparticles can interact with sulfur, thiol, and phosphorus groups found in proteins that are part of the hereditary material. This disruption can interfere with various metabolic processes, including protein synthesis, DNA replication, and the mitochondrial respiratory chain, ultimately leading to cell death [50]. Moreover, nanoparticles may produce an antibacterial effect by increasing the production of reactive oxygen species within microorganisms. These reactive species can damage the integrity of the cell wall by oxidizing phospholipids, causing the internal breakdown of proteins, DNA, and RNA [43]. It is also important to note that the mechanism of action of biogenic nanoparticles may differ between Gram-positive and Gram-negative bacteria [51].

Rajivgandhi et al. [38] biosynthesized zinc oxide nanoparticles using actinomycetes and checked their potential antibacterial effect against multi-drug-resistant *K. pneumoniae*. The biosynthesized zinc oxide nanoparticles displayed a concentration-dependent antibacterial effect, with the highest inhibition zone obtained for 250 μg/ml concentration. Furthermore, the biosynthesized zinc oxide nanoparticles displayed 88% biofilm degradation potential at 200 μg/ml of the nanoparticles concentration. The authors suggested that the likely mode of action is the disruption of the bacterial surface by positively charged nanoparticles, followed by essential cell component leakage. Sholkamy et al. [37] fabricated double-functional zinc oxide nanoparticles from *Saccharopolyspora hirsute* and explored the antimicrobial and anticancer activities. The biogenic zinc oxide nanoparticles were investigated for their antibacterial potential against two nosocomial pathogens, *S. aureus* and *P. aeruginosa*. The synthesized zinc oxide nanoparticles displayed a significantly higher inhibition zone for *S. aureus* than *P. aeruginosa* (14.33 ± 2.52 mm versus 12.67 ± 2.08 mm). The findings demonstrate that *S. aureus* shows a higher tendency to interact with biogenic nanoparticles because of its peptidoglycan layer, which allows for easier penetration of the nanoparticles and their subsequent antibacterial impact. The authors propose that the antibacterial action noted can be linked to the synergistic relationship between the filtrate of *S. hirsute* and zinc oxide nanoparticles. Biogenic zinc oxide nanoparticles prepared by using *S. plicatus* displayed antibacterial activity against plant pathogenic bacteria *Erwinia amylovora* [39].

Zhao et al. [32] fabricated copper oxide nanoparticles through marine endophytic actinomycetes. The copper oxide biogenic nanoparticles displayed significant antibacterial properties at 750 μg/ml concentration with 24 and 28 mm inhibition zones against *E. coli* and *P. mirabilis*. Additionally, the copper nanoparticles displayed 92% and 94% alteration against *E. coli* and *P. mirabilis* biofilm at 1000 μg/ml. In a separate investigation, copper oxide nanoparticles synthesized using *Streptomyces* MHM38 displayed significant antibacterial activity against *Enterococcus faecalis*, Gram-positive bacteria and *S. typhimurium, P. aeruginosa*, and *E. coli* (three Gram-negative bacteria). The zone of inhibition values observed were 20.0 ± 0.3 mm, 18.0 ± 0.4, 20.0 ± 0.5, and 18.0 ± 0.5 mm, respectively, for *E. faecalis*, *S. typhimurium*, *P. aeruginosa*, and *E. coli* [33].

Alam et al. [42] biosynthesized ultrafine silver nanoparticles with the help of *R. rhodochrous* and *Streptomyces* sp. extracts and evaluated their therapeutic activities. The nanoparticles fabricated using both bacterial species were checked for antibacterial activity by measuring the zone of inhibition against *E. coli* and *B. subtilis*. At a 10 μg/ml concentration level, the nanoparticles fabricated using *R. rhodochrous* exhibit a higher inhibitory zone against *P. aeruginosa* than *B. subtilis*. At a similar concentration level, the nanoparticles prepared by using *Streptomyces* sp. exhibited a higher inhibitory zone for *B. subtilis* than *E. coli*. Methicillin-resistant *S. aureus* (MRSA) is considered one of the deadliest pathogenic bacteria responsible for eliciting various types of infection in humans [52, 53]. As this pathogenic bacterium is resistant to multiple drugs, it is necessary to develop novel therapeutics to combat MRSA. Biogenic silver nanoparticles possess several activities, such as anticancer, antimicrobial, anti-inflammatory, etc., and are potent against various bacterial pathogens, including MRSA [54]. Considering the efficacy of biogenic silver nanoparticles, Sayed et al. [55] checked the antibacterial activity of biogenic silver nanoparticles against MRSA. The results obtained from this study indicated the exceptional antibacterial property of silver nanoparticles prepared by using *S. virginiae*. Furthermore, it was observed that 84% of the accessed isolates were prone to less than 16 μg/ml concentration of silver nanoparticles. At the same time, only 3.4% of the MRSA isolates were impervious to the silver nanoparticles with a minimum concentration of inhibition at 64 μg/ml.

Silver nanoparticles synthesized by using *S. aizuneusis* displayed an antimicrobial activity against *S. aureus*, *E. coli*, *P. aeruginosa*, *Listeria monoctogense*, *S. paratyphi*, and *B. subtilis* [56]. Shaaban et al. [57] synthesized silver nanoparticles from the cell filtrate of *S. enissocaesilis* and tested their antibacterial and antibiofilm effects against *S. aureus*, *P. aeruginosa*, *S. typhi*, and *E. coli*. The nanoparticles at 50 mg/ml for 24 hours were able to reduce the biofilms of *P. aeruginosa*, *S. typhi*, *E. coli*, and *S. aureus* by 10.7%, 34.6%, 34.75%, and 39.08%, respectively [56]. Biosynthesis of silver nanoparticles by using *S. spiralis* and *S. rochei* displayed significant antimicrobial activity against *P. aeruginosa*, *E. coli*, *S. enterica*, *S. aureus*, and *B. subtilis* [58].

Maktoof and Hateet [40] synthesized platinum nanoparticles by using *S. fradiae* and checked their antibacterial potential against five species of pathogenic bacteria, including four Gram-negative bacteria (*Proteus mirabilis, P. aeruginosa, Acinetobacter baumannii*, and *E. coli*), and one Gram-positive bacteria (*S. aureus*). The zone of inhibition values displayed after receiving biosynthesized nanoparticles were 10.33 mm, 15.2 mm, 16.4 mm, and 10.2 mm, respectively, for *E. coli*, *P. mirabilis*, *S. aureus*, and *A. baumannii*. However, no zone of inhibition was observed for *P. aeruginosa*.

Biogenic nanoparticles also exhibit antifungal activity, similar to their antibacterial effects. Fungi have well-defined structures, with their cell membranes made up of phospholipids and their cell walls consisting of β-1,3-D-glucan, β-1,6-D-glucan, proteins, lipids, mannoproteins, and polysaccharides. These nanoparticles exert their fungicidal effects by interacting with the fungal cell wall, allowing metal ions to diffuse through the cell membrane. Once inside, these metal ions can inhibit β-glucan synthase or *N*-acetylglucosamine, both of which are crucial components of the fungal cell wall. This inhibition disrupts the integrity of the cell wall and membrane. Furthermore, biogenic nanoparticles can induce oxidative stress, ultimately impacting DNA, RNA, and proteins, which may lead to cell death [43, 59].

Silver nanoparticles produced using silver-resistant and rare actinomycetes, *G. nicotianae* and *L. aridicollis*, demonstrated a notable antifungal action against the mycotoxigenic fungi *Aspergillus flavus* and *A. ochraceus*. The nanoparticles effectively inhibited the germination of conidia from both targeted mycotoxigenic fungi. The biosynthesized silver nanoparticles effectively and completely halted the production of total aflatoxins and ochratoxin A at concentrations below 8 μg/ml. The study result indicated that the biogenic silver nanoparticles could disrupt the membrane permeability and integrity, which generated an enhancement in DNA and protein leakage and eventually exhibited antifungal activity [30]. In another study, Kalaba *et al* [39] synthesized zinc oxide nanoparticles mediated by *S. plicatus* and checked the antifungal properties of the prepared nanoparticles against plant pathogenic fungi. The studied pathogenic fungi were *A. flavus*, *A. niger*, *Fusarium oxysporum*, *F. moniliform,* and *Alternaria alternata*, with minimum inhibitor concentration at the value between 15.6 and 500 μg/ml. Extracellular silver nanoparticles from *S. aizuneusis* displayed antifungal activities against *Sclerotium rolfsii*, *Rhizoctonia solani,* and *F. solani* [56].

7.4.2 Anti-Inflammatory Activity

Inflammation serves as a critical component of the body's defense mechanism. It enables the immune system to identify and eliminate harmful foreign stimuli while also facilitating the healing process [60]. Upon injury, infection, and stress, the immune system recruits macrophages and killer cells, cytokines (interleukin-1 [IL-1], IL-1β, and tumor necrosis factor alpha) to the desired site and initiates the inflammation process [61]. Inflammation has been classified into two primary subtypes depending on the duration as acute and chronic. Acute inflammation occurs as the body's reaction to injury, infection, or harmful substances. It has a rapid onset and escalates quickly, with symptoms typically persisting for a few days. In contrast, chronic inflammation is a prolonged response that can endure for several months or even years. The duration and impact of chronic inflammation are influenced by the underlying cause of the injury and the body's capacity for healing and recovery [62]. If left unchecked, chronic inflammation can lead to internal scarring, tissue death, and damage the previously healthy cell's

DNA, which eventually causes damage to cells, tissues, and organs. Even though steroidal as well as nonsteroidal anti-inflammatory drugs can control inflammation, the side effects of these anti-inflammatory drugs elicit an adverse effect [43]. This compels a search for safer and more effective alternatives to conventional anti-inflammatory drugs. Biogenic nanoparticles, synthesized from the identified candidates, have garnered considerable attention due to their unique physical and chemical properties [61].

Elsilk et al. [63] biosynthesized silver nanoparticles using the novel marine actinobacterium *S. rochei* MS-37 and evaluated their biomedical applications. The anti-inflammatory potential of the developed nanoparticles was checked by protein denaturation assay. The results of the test reflect a higher anti-inflammatory potential of the developed nanoparticles as it was received with a lower IC_{50} value than diclofenac sodium (189.44 ± 5.52 μg/ml versus 215.5 ± 4.90 μg/ml). The authors indicate that the observed anti-inflammatory properties of biogenic nanoparticles can be attributed to the synergistic interactions among the bioactive agents present on the surface of silver nanoparticles. This interaction ultimately enhances both the dispersibility and bioavailability of the nanoparticles. In another study, Rajeswaran et al. [64] reported the anti-inflammatory activity of zinc oxide nanoparticles produced using actinobacterium exopolysaccharide. The biosynthesized silver nanoparticles produced through marine actinobacterium strain *S. catenulae* M2 displayed a concentration-dependent anti-inflammatory activity, while the diclofenac sodium displayed an inhibition percentage of 84.3 [65].

7.4.3 Anticancer Activity

Cancer is a multifactorial disease with a complex pathophysiology consisting of factors of both environmental and genetic origin [66]. Knowing cancer's genetic, molecular, and cellular basis can offer new targets and strategies for effective anticancer therapy. The standard drugs used for chemotherapy display various limitations, such as non-specific toxicity, unfavorable physicochemical properties, poor bioavailability, etc. Due to biocompatibility, ease of preparation, surface functionalization, and high surface-to-volume ratio, biogenic nanoparticles can be projected as efficient anticancer therapeutics [67]. Although the mechanism of biogenic nanoparticles is still unclear, it is speculated that these nanoparticles can selectively obstruct the respiratory chain in the mitochondria, further escalating the level of reactive oxygen species inside the cancerous cell and leading to DNA impairment [43].

Copper oxide nanoparticles prepared using marine endophytic actinomycetes displayed concentration-dependent cytotoxicity against A-549 lung cancer cells with 54% cell death in optical density value attained at 500 μg/ml concentration

[32]. Sholkamy et al. [37] evaluated *S. hirsute*-mediated biosynthesized zinc oxide for anticancer potential against MCF-7 and Hep-2 cancer cell lines. The zinc oxide nanoparticles exhibited cytotoxic properties against the two-studied cancer cells depending on the concentration of nanoparticles. At a similar tested concentration of 100 μg/ml, the zinc oxide nanoparticles displayed higher cytotoxicity against Hep-2 cells than MCF-7 cells (cell viability value 35.2% versus 60.5%). Similarly, the IC_{50} values of the nanoparticles obtained in Hep-2 and MCF-7 cells were 73.01 and 112.74 μg/ml, respectively. These results indicated a greater susceptibility of the Hep-2 cells toward zinc oxide nanoparticles than MCF-7 cells. Furthermore, the authors suggested the potential synergistic interaction between the surface-attached secondary metabolites and zinc oxide nanoparticle core. In another study, silver nanoparticles biosynthesized by using *Rhodococcus rhodochrous* and *Streptomyces* sp. displayed a concentration-dependent cytotoxic effect on Hep-2 cells [42]. Sayed et al. [55] checked the cytotoxic effect of biogenic silver nanoparticles on MDA-MB-231 and A-549 cells. The obtained results indicate a profound cytotoxicity of the synthesized nanoparticles towards the two-studied cancer cells. The possible explanation for the observed cytotoxic effect of the nanoparticles can occur due to the activation of lactate dehydrogenase, caspase-3, and reactive oxygen species generation, which initiates apoptosis and, consequently, cell death. Gold nanoparticles synthesized with a cell-free supernatant of *S. albogriselus* displayed a cytotoxic effect on Hep-G2 cancer cells with an IC_{50} value of 22.13 ± 1.3 μg/ml. Furthermore, the combined effect of biogenic gold nanoparticles with doxorubicin displayed an even lower IC_{50} value of 3.52 ± 0.1 μg/ml, indicating the suitability of using biogenic gold nanoparticles together with standard anticancer drugs to effectively kill cancer cells [34]. Silver nanoparticles synthesized through the cell-free extract of *S. enissocaesilis* demonstrated sensitivity specific to cancer *in vitro* against MCF-7 cell lines (human breast cancer cells) and Caco-2 cell lines (colon cancer cells), with IC_{50} values recorded at 0.160 mg/ml and 0.156 mg/ml, respectively [57].

Although biogenic nanoparticles synthesized using actinomycetes displayed promising anticancer activity in *in vitro* investigations performed on various cancer cells. There is still a lack of data from studies using suitable *in vivo* animal models. Therefore, it is unequivocally critical to conduct *in vivo* animal studies and identify the possible mechanisms of biogenic metallic nanoparticles that will aid in their projection as potential anticancer therapeutic. Table 7.2 reveals the applications of actinomycetes-assisted biogenic nanoparticles.

7.4.4 Larvicidal and Nematocidal Activity

Vector-borne diseases are ailments caused by bacteria, parasites, and viruses transmitted through various vectors, including mosquitoes, ticks, sandflies, flies,

Table 7.2 Applications of actinomycetes-assisted biogenic nanoparticles.

Sl. No.	Biogenic nanoparticles	Microbes used for synthesis	Application	Test subjects	References
1	Zinc oxide nanoparticles	Actinomycetes	Antibacterial	*Klebsiella pneumoniae*	[38]
2	Copper oxide nanoparticles	Marine endophytic actinomycetes	Antibacterial/ anticancer	*Escherichia coli* and *Proteus mirabilis*/ A549 lung cancer cells	[32]
3	Silver nanoparticles	*Streptomyces diastaticus*	Larvicidal	*Culex quinquefasciatus*, *Anopheles stephensi*, *Aedes aegypti*, *Spodoptera litura*, and *Plutella xylostella*.	[29]
4	Zinc oxide	*Saccharopolyspora hirsuta*	Antibacterial/ anticancer	*Pseudomonas aeruginosa* and *Staphylococcus aureus*/Hep-2 cells	[37]
5	Silver nanoparticles	*Rhodococcus rhodochrous* (MOSEL-ME29) and *Streptomyces* sp.	Antibacterial/ anticancer	*Escherichia coli, Bacillus subtilis, Staphylococcus aureus*, and *Pseudomonas aeruginosa*/HepG2 cancer cells	[42]
6	Silver nanoparticles	*Streptomyces*	Antibacterial/ anticancer	Methicillin-resistant *Staphylococcus aureus*/breast cancer, non-small cell lung cancer cells	[55]
7	Platinum nanoparticles	*Streptomyces fradiae*	Antibacterial	*Proteus mirabilis*, *Pseudomonas aeruginosa*, *Acinetobacter baumannii, Escherichia coli* and *Staphylococcus aureus*	[40]
8	Copper oxide nanoparticles	*Streptomyces* MHM38	Antibacterial/ antifungal/ antioxidant	*Enterococcus faecalis*, *Salmonella typhimurium*, *Pseudomonas aeruginosa*, *Escherichia coli*/*Rhizoctonia solani*, *Fusarium solani*, and *Aspergillus niger*/ paracetamol-induced oxidative stress in rats	[33]

(Continued)

Table 7.2 (Continued)

Sl. No.	Biogenic nanoparticles	Microbes used for synthesis	Application	Test subjects	References
9	Gold nanoparticles	*Streptomyces albogriseolus*	Anticancer	HeP-G2 human cancer cells	[34]
10	Silver nanoparticles	*Streptomyces enissocaesilis*	Antibacterial/anticancer	*Staphylococcus aureus*, *Pseudomonas aeruginosa*, *Salmonella typhi*, and *Escherichia coli*/Caco-2 and MCF7 cells	[57]
11	Silver nanoparticles	*Glutamicibacter nicotianae* and *Leucobacter aridicollis*	Antifungal	*Aspergillus flavus* and *Aspergillus ochraceus*	[30]
12	Silver nanoparticles	*Streptomyces aizuneusis*	Antibacterial/Antifungal/Larvicidal/Anticancer	*Staphylococcus aureus*, *Escherichia coli*, *Pseudomonas aeruginosa*, *Listeria monocytogense*, *Salmonella paratyphi*, *Bacillus subtilis*/*Sclerotium rolfsii*, *Rhizoctonia solani*, *F. solani*/ *Drosophila melanogaster* Larvae/A549 lung cancer cells	[56]
13	Zinc oxide nanoparticles	*Streptomyces plicatus*	Antibacterial/antifungal/nematocidal	*Erwinia amylovora*/*Aspergillus flavus*, *Aspergillus niger*, *Fusarium oxysporum*, *Fusarium moniliform*, and *Alternaria alternata*/*Meloidogyne incognita*	[39]
14	Silver nanoparticles	*Streptomyces spiralis* and *Streptomyces rochei*	Antibacterial	*Pseudomonas aeruginosa*, *Escherichia coli*, *Salmonella enterica*, *Staphylococcus aureus*, *Bacillus subtilis*	[58]

Table 7.2 (Continued)

Sl. No.	Biogenic nanoparticles	Microbes used for synthesis	Application	Test subjects	References
15	Palladium Nanoparticles and Palladium-reduced graphene oxide nanocomposite	*Streptomyces maritimus*	Antimicrobial/ antioxidant	*Enterococcus* sp., *Staphylococcus aureus*, *Klebsiella* pneumonia, and *Escherichia coli*/ DPPH radicals	[68]
16	Silver Nanoparticles	*Nocardiopsis dassonvillei*	Antibacterial/ antioxidant/ anticancer	*Pseudomonas aeruginosa*/DPPH and OH^- radicals/$CaCo_2$ cells	[65]
17	Collagen nanoparticles	*Streptomyces xinghaiensis*	Anticancer	MCF-7, HeP-G2 and HCT116 cells	[69]
18	Silver nanoparticles	*Streptomyces rochei* MS-37	Antioxidant/ anti-inflammatory/ anticancer	Nitric oxide radicals/Bovine albumin/CAL 27	[63]

and fleas. It is important to note that some of the deadliest transmitted diseases, such as chikungunya, dengue fever, lymphatic filariasis, malaria, Japanese encephalitis, yellow fever, and West Nile fever, are transmitted by mosquitoes (*Aedes aegypti*, *A. albopictus*, *Culex quiquefasciatus*, and *Anopheles stephensi*) [70]. In order to control mosquito larvae and adults, various strategies have been developed. Nonetheless, the complete eradication of mosquitoes is not yet possible. Traditionally, synthetic larvicides such as dichlorodiphenyltrichloroethane, along with organochlorines, pyrethroids, pyrethrins, carbamates, and organophosphates, as well as cyclodiene organochlorines and phenylpyrazoles, have shown potential as effective methods for managing mosquito populations [43]. However, the uncontrolled use of synthetic insecticides resulted in mosquitoes developing resistance [71]. Furthermore, insecticides are toxic and display deleterious effects on non-target organisms, human beings, and the environment. These compelled a search for a safer and more effective alternative to synthetic insecticides. Among them, biogenic metallic nanoparticles can be considered a promising candidate.

The biogenic nanoparticles display an antagonistic effect against various instar larvae of mosquitoes. Biogenic nanoparticles exhibit their effects by infiltrating the cavity of the exoskeleton, thereafter, engaging with sulfur/phosphorus found in proteins/DNA, which results in the swift denaturation of enzymes and organelles [72]. Kalpana et al. [29] successfully synthesized silver nanoparticles utilizing

S. diastaticus. They subsequently assessed the biological effectiveness of these nanoparticles against human vector mosquitoes and agricultural insect pests, demonstrating their potential application in pest management. The organisms studied for the efficacy of the synthesized nanoparticles were *Culex quinquefasciatus*, *A. stephensi*, *A. aegypti*, *Spodoptera litura*, and *Plutella xylostella*. The results indicated that the silver nanoparticles were more toxic to *A. stephensi*, *C. quinquefasciatus*, and *A. aegypti*, and were received Lethal Concentration$_{50}$ (LC$_{50}$)values of 15.70, 11.10, and 17.53 μg/ml, respectively. Antifeedant study results show that the *S. litura* (88.73%) and *P. xylostella* (92.36%) were more prone to silver nanoparticles at 50 μg/ml concentration. Significant larvicidal activity was recorded for *S. litura* and *P. xylostella* larvae at doses greater than 50 μg/ml, achieving 80.67% and 84.05%, respectively. Furthermore, *C. quinquefasciatus*, *A. aegypti*, *A. stephensi*, *S. litura*, and *P. xylostella* when treated with silver nanoparticle treatment displayed morphogenetic variations. Upon checking the influence of silver nanoparticles on levels of superoxide dismutase (SOD), CAT, and glutathione peroxidase (GPx) , it was revealed that the activity of each enzyme was significantly augmented at higher silver nanoparticle concentrations. The silver nanoparticles also impeded the detoxifying enzymes, viz. glutathione S-tranferase (GST) and acetylcholinesterase (AChE) in the studied organisms. The findings from the histological studies indicate notable structural differences in the midgut epithelial cells between the larvae that received treatment and those that did not. Figure 7.2 displays the applications of actinomycetes-assisted biogenic nanoparticles.

The prevalence of plant pathogens is significantly growing and seriously threatens economic and social stability worldwide. Plant pathogens such as nematodes are significant elements that attenuate crop growth and development [73]. Due to excessive agricultural farming, the number of parasites is increasing alarmingly,

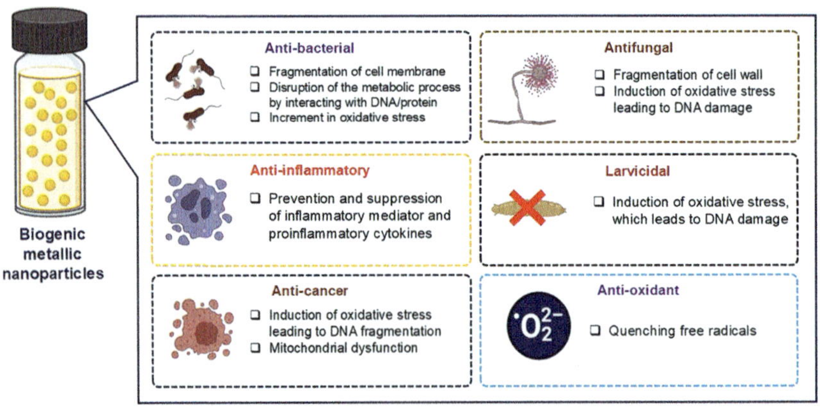

Figure 7.2 Applications of actinomycetes-assisted biogenic nanoparticles.

significantly affecting agricultural yield [74]. Moreover, managing parasites is complex and involves increasing environmental concerns. This necessitates a continuous search for new strategies to control parasites. Utilizing artificial biocidal chemicals to combat pathogenic microorganisms can pose significant threats to both the environment and human health [39]. In this context, biogenic nanoparticles can significantly control the parasite due to their non-toxic nature and effectiveness. Kalaba et al. [39] checked the nematocidal activity of zinc oxide nanoparticles synthesized using *S. plicatus*, which was evaluated against *Meloidogyne incognita*. Post 24-hour exposure to zinc oxide nanoparticles, the amount of living *M. incognita* J_2s reduced significantly, with 96.9% mortality attained after 72 hours. Furthermore, the zinc oxide nanoparticles lead to the deformation of the nematode cuticle.

7.4.5 Antioxidant Activity

Free radicals produced by reactive oxygen species are mainly responsible for numerous human diseases, including cancer, atherosclerosis, and neurodegenerative conditions. Due to their electronic configuration, metallic nanoparticles can quench free radicals by accepting or donating electrons [75]. Actinomycetes-assisted biogenic metallic nanoparticles displayed antioxidant activity with a significant IC_{50} value [33, 68, 69]. The antioxidant activity of the biogenic nanoparticle is usually dictated by the synthesis method, capping agent, and reduction of the biological material used for synthesis [43]. Bukhari et al. [33] investigated the impact of copper oxide nanoparticles generated via *Streptomyces* sp. MHM38 on paracetamol-induced oxidative stress in male albino Sprague–Dawley rats. In comparison to the group of animals that only received paracetamol, the group treated with nanoparticles exhibited a significant reduction in the levels of alanine aminotransferase, aspartate aminotransferase, alkaline phosphatase, lactate dehydrogenase, total and direct bilirubin, urea, and creatinine. Additionally, the levels of SOD and glutathione were significantly elevated in the nanoparticle-treated animals, while the levels of nitric oxide and malondialdehyde were reduced.

7.5 Conclusion

Due to significant environmental concerns and low yield associated with traditional nanoparticle synthesis methods, there has been an increasing interest in eco-friendly biogenic nanoparticles that employ sustainable approaches by blending biological techniques with nanotechnology. Various phytochemicals, including alkaloids, terpenoids, flavonoids, and polyphenolics, have been investigated for their roles in the creation of biogenic nanoparticles. Research carried out over the last decade indicates that metallic nanoparticles produced through microorganisms demonstrate strong biological activities and allow for controlled morphology. Actinomycetes, which are Gram-positive, spore-forming aerobic bacteria, have recently gained attention in the biogenic synthesis of metallic nanoparticles. A diverse array of metallic nanoparticles, including gold, silver, zinc oxide, platinum, palladium, and copper oxide nanoparticles, has been successfully synthesized using actinomycetes. The physicochemical properties and biological activities of these nanoparticles suggest that they represent promising methods for developing sustainable, highly bioactive nanoparticles that are environmentally friendly.

References

1 Joudeh N, Linke D. Nanoparticle classification, physicochemical properties, characterization, and applications: a comprehensive review for biologists. J Nanobiotechnol. 2022;20:262.

2 Roduner E. Size matters: why nanomaterials are different. Chem Soc Rev. 2006;35:583–592.

3 Khanna P, Kaur A, Goyal D. Algae-based metallic nanoparticles: synthesis, characterization and applications. J Microbiol Methods. 2019;163:105656.

4 Nabila MI, Kannabiran K. Biosynthesis, characterization and antibacterial activity of copper oxide nanoparticles (CuO NPs) from actinomycetes. Biocatal Agric Biotechnol. 2018;15:56–62.

5 Sanjivkumar M, Vaishnavi R, Neelakannan M, Kannan D, Silambarasan T, Immanuel G. Investigation on characterization and biomedical properties of silver nanoparticles synthesized by an actinobacterium *Streptomyces olivaceus* (MSU3). Biocatal Agric Biotechnol. 2019;17:151–159.

6 Golinska P, Wypij M, Ingle AP, Gupta I, Dahm H, Rai M. Biogenic synthesis of metal nanoparticles from actinomycetes: biomedical applications and cytotoxicity. Appl Microbiol Biotechnol. 2014;98:8083–8097.

7 Venkateswaran SP, Palaniswamy VK, Vishvanand R, Periakaruppan R. Actinomycetes-assisted nanoparticles: synthesis and applications. In: Abd-Elsalam KA, Periakaruppan R, Rajeshkumar S (eds). Agri-Waste and Microbes for Production of Sustainable Nanomaterials. 2021. Elsevier, p. 375–395.

8 Bhati M. Biogenic synthesis of metallic nanoparticles: principles and applications. Mater Today Proc. 2023;81:882–887.

9 Kumar A, Shah SR, Jayeoye TJ, Kumar A, Parihar A, Prajapati B, et al. Biogenic metallic nanoparticles: biomedical, analytical, food preservation,

and applications in other consumable products. Front Nanotechnol. 2023;5: 1175149.

10 Das RK, Pachapur VL, Lonappan L, et al. Biological synthesis of metallic nanoparticles: plants, animals and microbial aspects. Nanotechnol Environ Eng. 2017;2:18.

11 Bhatti AA, Haq S, Bhat RA. Actinomycetes benefaction role in soil and plant health. Microb Pathog. 2017;111:458–467.

12 Mast Y, Stegmann E. Actinomycetes: the antibiotics producers. Antibiotics (Basel). 2019;8(3):105.

13 Ranjitha VR, Ravishankar Rai V. Actinomycetes mediated synthesis of gold nanoparticles from the culture supernatant of *Streptomyces griseoruber* with special reference to catalytic activity. 3 Biotech. 2017;7:299.

14 Manivasagan P, Venkatesan J, Sivakumar K, Kim S. Actinobacteria mediated synthesis of nanoparticles and their biological properties: a review. Crit Rev Microbiol. 2016;42(2):209–221.

15 Shah M, Fawcett D, Sharma S, Tripathy SK, Poinern GEJ. Green synthesis of metallic nanoparticles *via* biological entities. Materials. 2015;8:7278–7308.

16 Pallavi SS, Rudayni HA, Bepari A, Niazi SK, Nayak S. Green synthesis of Silver nanoparticles using *Streptomyces hirsutus* strain SNPGA-8 and their characterization, antimicrobial activity, and anticancer activity against human lung carcinoma cell line A549. Saudi J Biol Sci. 2022;29:228–238.

17 Prema P, Veeramanikandan V, Rameshkumar K, Gatashch MK, Hatamlch AA, Balasubramani R, Balaji P. Statistical optimization of silver nanoparticle synthesis by green tea extract and its efficacy on colorimetric detection of mercury from industrial waste water. Environ Res. 2022;204:111915.

18 Sagadevan S, Anita Lett J, Vennila S, Varun Prasath P, Saravanan Kaliaraj G, Fatimah I, et al. Photocatalytic activity and antibacterial efficacy of titanium dioxide nanoparticles mediated by *Myristica fragrans* seed extract. Chem Phys Lett. 2021;771:138527.

19 Dutta T, Chowdhury SK, Ghosh NN, Chattopadhyay AP, Das M, Mandal V. Green synthesis of antimicrobial silver nanoparticles using fruit extract of *Glycosmis pentaphylla* and its theoretical explanations. J Mol Struct. 2022;1247:131361.

20 Olajire AA, Mohammed AA. Bio-directed synthesis of gold nanoparticles using *Ananas comosus* aqueous leaf extract and their photocatalytic activity for LDPE degradation. Adv Powder Technol. 2021;32:600–610.

21 Nazir A, Akbar A, Baghdadi HB, ur Rehman S, al-Abbad E, Fatima M, et al. Zinc oxide nanoparticles fabrication using *Eriobotrya japonica* leaves extract: photocatalytic performance and antibacterial activity evaluation. Arab J Chem. 2021;14:103251.

22 El-Borady OM, Fawzy M, Hosny M. Antioxidant, anticancer and enhanced photocatalytic potentials of gold nanoparticles biosynthesized by common reed leaf extract. Appl Nanosci. 2023;13:3149–3160.

23 Bekele ET, Zereffa EA, Gultom NS, Kuo D, Gonfa BA, Sabir FK. Biotemplated synthesis of titanium oxide nanoparticles in the presence of root extract of *Kniphofia schemperi* and its application for dye sensitized solar cells. Int J Photoenergy. 2021;2021:6648325.

24 Arsiya F, Sayadi MH, Sobhani S. Green synthesis of palladium nanoparticles using *Chlorella vulgaris*. Mater Lett. 2021;186:113–115.

25 Govindaraju K, Krishnamoorthy K, Alsagaby SA, Singaravelu G, Premanathan M. Green synthesis of silver nanoparticles for selective toxicity towards cancer cell. IET Nanobiotechnol. 2015;9:325–330.

26 Husain S, Afreen S, Hemlata DY, Afzal B, Fatma T. Cyanobacteria as a bioreactor for synthesis of silver nanoparticles-an effect of different reaction conditions on the size of nanoparticles and their dye decolorization ability. J Microbiol Methods. 2019;162:77–82.

27 Sayadi MH, Salmani N, Heidari A, Rezaei MR. Bio-synthesis of palladium nanoparticle using *Spirulina platensis* alga extract and its application as adsorbent. Surf Interfaces. 2018;10:136–143.

28 Arya S, Sonawane H, Math S, Tambade P, Chaskar M, Shinde D. Biogenic titanium nanoparticles (TiONPs) from *Tricoderma citrinoviride* extract: synthesis,characterization and antibacterial activityagainst extremely drug-resistant *Pseudomonas aeruginosa*. Int Nano Lett. 2021;11:35–42.

29 Kalpana M, Raguvaran K, Manimegalai T, Kalaivani S, Devapriya P, Maheswaran R. Actinobacteria mediated synthesis of silver nanoparticles using *Streptomyces diastaticus* and their biological efficacy against human vector mosquitoes and agricultural insect pests. J Nat Pestic Res. 2024;10:100091.

30 Abd El-Ghany MN, Hamdi SA, Korany SM, Elbaz RM, Emam AN, Farahat MG. Biogenic silver nanoparticles produced by soil rare actinomycetes and their significant effect on *Aspergillus*-derived mycotoxins. Microorganisms. 2023;11(4):1006.

31 Ghabban H, Alnomasy SF, Almohammed H, Al Idriss OM, Rabea S, Eltahir Y. Antibacterial, cytotoxic, and cellular mechanisms of green synthesized silver nanoparticles against some cariogenic bacteria (*Streptococcus mutans* and *Actinomyces viscosus*). J Nanomater. 2022;2022:9721736.

32 Zhao H, Maruthupandy M, Al-mekhlafi FA, Chackaravarthi G, Ramachandran G, Chelliah CK. Biological synthesis of copper oxide nanoparticles using marine endophytic actinomycetes and evaluation of biofilm producing bacteria and A549 lung cancer cells. J King Saud Univ Sci. 2022;34(3):101866.

33 Bukhari SI, Hamed MM, Al-Agamy MH, Gazwi HS, Radwan HH, Youssif AM. Biosynthesis of copper oxide nanoparticles using *Streptomyces* MHM38 and its biological applications. J Nanomater. 2021;(1):6693302.

34 El-Naggar NEA, El-Sawah AA, Elmansy MF, et al. Process optimization for gold nanoparticles biosynthesis by *Streptomyces albogriseolus* using artificial neural network, characterization and antitumor activities. Sci Rep. 2024;14(1):4581.

35 Ameri A, Shakibaie M, Ameri A, Faramarzi MA, Amir-Heidari B, Forootanfar H. Photocatalytic decolorization of bromothymol blue using biogenic selenium nanoparticles synthesized by terrestrial actinomycete *Streptomyces griseobrunneus* strain FSHH12. Desalin Water Treat. 2016;57(45):21552–21563.

36 Sumithra D, Bharathi S, Kaviyarasan P, Suresh G. Biofabrication of selenium nanoparticles using marine *Streptomyces* sp. and assessment of its antibacterial, antibiofilm, antioxidant, and *in vivo* cytotoxic potential. Geomicrobiol J. 2023;40(5):485–492.

37 Sholkamy EN, Abdelhamid MA, Khalifa HO, Ki MR, Pack SP. Bioinspired synthesis and characterization of dual-function zinc oxide nanoparticles from *Saccharopolyspora hirsuta*: exploring antimicrobial and anticancer activities. Biomimetics. 2024;9(8):456.

38 Rajivgandhi G, Gnanamangai BM, Prabha TH, et al. Biosynthesized zinc oxide nanoparticles (ZnO NPs) using actinomycetes enhance the antibacterial efficacy against *K. Pneumoniae*. J King Saud Univ Sci. 2022;34(1):101731.

39 Kalaba MH, Moghannem SA, El-Hawary AS, Radwan AA, Sharaf MH, Shaban AS. Green synthesized ZnO nanoparticles mediated by *Streptomyces plicatus*: characterizations, antimicrobial and nematicidal activities and cytogenetic effects. Plants. 2021;10(9):1760.

40 Maktoof ZM, Hateet RR. Biosynthesis of platinum nanoparticles by *Streptomyces fradiae* and evaluate antibacterial activity. J Nanostruct. 2024;14(2):548–560.

41 Zwar IP, Trotta CDV, Ziotti AB, Lima Neto M, Araújo WL, de Melo IS, et al. Biosynthesis of silver nanoparticles using actinomycetes, phytotoxicity on rice seeds, and potential application in the biocontrol of phytopathogens. J Basic Microbiol. 2023;63(1):64–74.

42 Alam A, Tanveer F, Khalil AT, Zohra T, Khamlich S, Alam MM, et al. Silver nanoparticles biosynthesized from secondary metabolite producing marine actinobacteria and evaluation of their biomedical potential. Antonie Van Leeuwenhoek. 2021;114:1497–1516.

43 Patil S, Chandrasekaran R. Biogenic nanoparticles: a comprehensive perspective in synthesis, characterization, application and its challenges. J Genet Eng Biotechnol. 2020;18(1):67.

44 Nisar P, Ali N, Rahman L, Ali M, Shinwari ZK. Antimicrobial activities of biologically synthesized metal nanoparticles: an insight into the mechanism of action. J Biol Inorg Chem. 2019;24:929–941.

45 Hazarika M, Borah D, Bora P, Silva AR, Das P. Biogenic synthesis of palladium nanoparticles and their applications as catalyst and antimicrobial agent. PLoS One. 2017;12(9):e0184936.

46 Girman A. Alternative mechanisms of action of metallic nanoparticles to mitigate the global spread of antibiotic-resistant bacteria. Cell Surf.;10:100112.

47 Ingle PU, Biswas JK, Mondal M, Rai MK, Kumar PS, Gade AK. Assessment of in vitro antimicrobial efficacy of biologically synthesized metal nanoparticles against pathogenic bacteria. Chemosphere. 2022;291:132676.

48 Wang L, Hu C, Shao L. The antimicrobial activity of nanoparticles: present situation and prospects for the future. Int J Nanomedicine. 2017;12:1227–1249.

49 Slavin YN, Asnis J, Hńfeli UO, Bach H. Metal nanoparticles: understanding the mechanisms behind antibacterial activity. J Nanobiotechnol. 2017;15:1–20.

50 Rajkuberan C, Prabukumar S, Sathishkumar G, Wilson A, Ravindran K, Sivaramakrishnan S. Facile synthesis of silver nanoparticles using *Euphorbia antiquorum* L. latex extract and evaluation of their biomedical perspectives as anticancer agents. J Saudi Chem Soc. 2017;21(8):911–919.

51 Chandrasekaran R, Seetharaman P, Krishnan M, Gnanasekar S, Sivaperumal S. *Carica papaya* (Papaya) latex: a new paradigm to combat against dengue and filariasis vectors *Aedes aegypti* and *Culex quinquefasciatus* (Diptera: Culicidae). 3 Biotech. 2018;8:1–10.

52 Lade H, Kim JS. Molecular determinants of β-lactam resistance in methicillin-resistant *Staphylococcus aureus* (MRSA): an updated review. Antibiotics. 2023;12(9):1362.

53 Minter DJ, Appa A, Chambers HF, Doernberg SB. Contemporary management of Staphylococcus aureus bacteremia—controversies in clinical practice. Clin Infect Dis. 2023;77(11):e57–e68.

54 Burlec AF, Corciova A, Boev M, Batir-Marin D, Mircea C, Cioanca O, et al. Current overview of metal nanoparticles' synthesis, characterization, and biomedical applications, with a focus on silver and gold nanoparticles. Pharmaceuticals. 2023;16(10):1410.

55 Sayed FM, El-Rouby MN, Kamel Z, Farahat MG. Cytotoxicity and antibacterial activity of actinomycetes-mediated biogenic silver nanoparticles against methicillin-resistant *Staphylococcus aureus* (MRSA). Egypt J Med Microbiol. 2024;33(4):21–28.

56 Abd-Elhady HM, Ashor MA, Hazem A, Saleh FM, Selim S, El Nahhas N, et al. Biosynthesis and characterization of extracellular silver nanoparticles from *Streptomyces aizuneusis*: antimicrobial, anti larval, and anticancer activities. Molecules. 2022;27(1):212.

57 Shaaban MT, Mohamed BS, Zayed M, El-Sabbagh SM. Antibacterial, antibiofilm, and anticancer activity of silver-nanoparticles synthesized from the cell-filtrate of *Streptomyces enissocaesilis*. BMC Biotechnol. 2024;24(1):8.

58 Mabrouk M, Elkhooly TA, Amer SK. Actinomycete strain type determines the monodispersity and antibacterial properties of biogenically synthesized silver nanoparticles. J Genet Eng Biotechnol. 2021;19:1–12.

59 Arciniegas-Grijalba PA, Patino-Portela MC, Mosquera-Sanchez LP, Guerrero-Vargas JA, Rodríguez-Paez JE. ZnO nanoparticles (ZnO-NPs) and their antifungal activity against coffee fungus Erythricium salmonicolor. Appl Nanosci. 2017;7:225–241.

60 Agarwal H, Nakara A, Shanmugam VK. Anti-inflammatory mechanism of various metal and metal oxide nanoparticles synthesized using plant extracts: a review. Biomed Pharmacother. 2019;109:2561–2572.

61 Rehman A, John P, Bhatti A. Biogenic selenium nanoparticles: potential solution to oxidative stress mediated inflammation in rheumatoid arthritis and associated complications. Nanomaterials. 2021;11(8):2005.

62 Pahwa R, Goyal A, Jialal I. Chronic inflammation. 2018. Available from: https://www.ncbi.nlm.nih.gov/books/NBK493173/.

63 Elsilk SE, Khalil MA, Aboshady TA, Alsalmi FA, Ali SS. *Streptomyces rochei* MS-37 as a novel marine actinobacterium for green biosynthesis of silver nanoparticles and their biomedical applications. Molecules. 2022;27(21):7296.

64 Rajeswaran S, Veeraiyan B, Saravanan V, Manickam E, Danaraj J. Biogenic facile green synthesis of actinobacterium exopolysaccharide-fabricated zinc oxide nanoparticles for the diverse biomedical applications. Biomass Convers Biorefinery. 2025;15:787–801.

65 Khalil MA, El-Shanshoury AERR, Alghamdi MA, Sun J, Ali SS. *Streptomyces catenulae* as a novel marine actinobacterium mediated silver nanoparticles: characterization, biological activities, and proposed mechanism of antibacterial action. Front Microbiol. 2022;13:833154.

66 Kah G, Chandran R, Abrahamse H. Biogenic silver nanoparticles for targeted cancer therapy and enhancing photodynamic therapy. Cells. 2023;12(15):2012.

67 Todaria M, Maity D, Awasthi R. Biogenic metallic nanoparticles as game-changers in targeted cancer therapy: recent innovations and prospects. Futur J Pharm Sci. 2024;10(1):25.

68 Nithyalakshmi M, Siddharthan N, Lokesh E, Wadaan MA, Dixit S, Balagurunathan R. Fabrication of palladium nanoparticles and rGO-Pd nanocomposite by Streptomyces maritimus: antimicrobial, antioxidant, and scalable applications. Waste Biomass Valori. 2025;16:3449–3465. doi:10.1007/s12649-024-02882-w.

69 El-Sawah AA, El-Naggar NEA, Eldegla HE, Soliman HM. Green synthesis of collagen nanoparticles by *Streptomyces xinghaiensis* NEAA-1, statistical optimization, characterization, and evaluation of their anticancer potential. Sci Rep. 2024;14(1):3283.

70 Muller R, Reuss F, Kendrovski V, Montag D. Vector-borne diseases. In: Marselle M, Stadler J, Korn H, Irvine K, Bonn A (eds). Biodiversity and Health in the Face of Climate Change. 2019. Springer, Cham, p. 67–90.

71 Wanjala CL, Mbugi JP, Ototo E, Gesuge M, Afrane YA, Atieli HE, et al. Pyrethroid and DDT resistance and organophosphate susceptibility among *Anopheles* spp. mosquitoes, western Kenya. Emerg Infect Dis. 2015;21:2178–2181.

72 Benelli G. Plant-mediated biosynthesis of nanoparticles as an emerging tool against mosquitoes of medical and veterinary importance: a review. Parasitol Res. 2016;115(1):23–34.

73 Al-Ani LKT, Furtado EL. The effect of incompatible plant pathogens on the host plant. In: Sharma V, Salwan R, Al-Ani LKT (eds). Molecular Aspects of Plant Beneficial Microbes in Agriculture. 2020. Academic Press, Elsevier, p. 47–57.

74 Borruso L, Salomone-Stagni M, Polsinelli I, Schmitt AO, Benini S. Conservation of *Erwinia amylovora* pathogenicity-relevant genes among *Erwinia* genomes. Arch Microbiol. 2017;199:1335–1344.

75 Chung IM, Abdul Rahuman A, Marimuthu S, Vishnu Kirthi A, Anbarasan K, Padmini P, Rajakumar G. Green synthesis of copper nanoparticles using *Eclipta prostrata* leaves extract and their antioxidant and cytotoxic activities. Exp Ther Med. 2017;14(1):18–24.

8

Biological Synthesis of Nanoparticles by Yeast and Biogenic Sustainable Nanotherapeutics

Balisa Mosisa Ejeta[1], Malay K. Das[1], F. F. Bekere[2,3], and O. T. Ayana[4]

[1] *Drug Delivery Research Laboratory, Department of Pharmaceutical Sciences, Dibrugarh University, Dibrugarh, Assam, India*
[2] *Department of Materials Science and Engineering, Adama Science and Technology University, Adama, Ethiopia*
[3] *Department of Advanced Materials Science and Engineering Center of Excellence, Adama Science and Technology University, Adama, Ethiopia*
[4] *Sunshine College, Nekemte, Oromia, Ethiopia*

8.1 Introduction

8.1.1 Overview of Nanoparticles in Medicine

A nanoparticle represents the essential element in the creation of a nanostructure and is significantly smaller than the realm of everyday items that are explained by Newton's laws of motion, yet larger than an atom or a basic molecule that are controlled by quantum mechanics. Typically, the dimensions of a nanoparticle fall within the spectrum of 1–100 nm (International Organization for Standardization [ISO]). Nevertheless, the perspective from which nanoparticles are regarded and their definitions are greatly influenced by their particular applications [1].

Nanoparticles (NPs) have surfaced as an instrument in contemporary medicine, presenting new therapeutic uses due to their distinct physicochemical characteristics. These particles feature a high surface-area-to-volume ratio, which facilitates improved interaction with biological systems at the molecular and cellular tiers. Their diminutive size permits them to cross biological barriers, such as cell membranes and the blood–brain barrier, rendering them especially effective for drug delivery and targeted therapies [1].

NPs can be fabricated from various materials, including metals (e.g., gold, silver), polymers, lipids, and proteins, with each serving specific biomedical purposes. In the realm of cancer treatment, nanoparticles are crafted to transport chemotherapeutic agents directly to tumor cells, thus minimizing

Biogenic Synthesis of Nanoparticles and Sustainable Nanotherapeutics,
First Edition. Edited by Malay K. Das.

systemic toxicity and enhancing drug effectiveness. NPs also have a crucial function in diagnostic imaging, gene delivery, wound healing, and antimicrobial therapies [2].

The utilization of nanoparticles in disease therapy allows for controlled drug release, enhanced bioavailability, and targeted action, offering significant advantages over conventional treatment methods. Moreover, the development of biogenic nanoparticles using environmentally sustainable methods, such as those synthesized by microorganisms or plant extracts, has further expanded their therapeutic potential by guaranteeing biocompatibility within biological systems and reducing the environmental impact of nanoparticle production [2].

Biological amalgamation of NPs by yeast is an ecological method by which yeast cells act as biological factories to reduce metal ions into nanoparticles through enzymatic processes [3]. Yeast-mediated synthesis involves intracellular or extracellular mechanisms where enzymes like reductases and biomolecules such as proteins facilitate the conversion of metal salts into stable nanoparticles. This process is considered a sustainable alternative to chemical synthesis as it avoids the use of noxious chemicals and energy-intensive methods. The nanoparticles produced by yeast are often well-defined in size, shape, and surface characteristics, making them valuable for biomedical applications such as drug delivery, diagnostics, and antimicrobial agents [4].

Biogenic sustainable nanotherapeutics refer to therapeutic agents created using nanoparticles synthesized from biological sources, such as plants, microorganisms, or enzymes, through environmentally responsible methods. These nanotherapeutics combine the philosophies of green chemistry and nanotechnology to develop biocompatible, nontoxic, and eco-friendly treatment solutions. The sustainable aspect lies in their biogenic production process, which minimizes environmental impact and promotes the use of renewable resources. Biogenic nanotherapeutics hold promise in a wide array of medical fields, including cancer treatment, targeted drug delivery, and antimicrobial therapies, by leveraging the unique properties of biogenically synthesized nanoparticles [5].

Biogenic production of nanoparticles offers some benefits over conventional chemical and physical methods. This eco-friendly approach line up with the principles of green chemistry, making it a more sustainable and safer alternative.

The key reasons why biogenic synthesis is considered superior to conventional chemical and physical methods are

1) **Eco-friendliness:** Biogenic synthesis avoids the utilization of harmful chemicals and high energy consumption, which are typical in chemical and physical methods. Traditional synthesis often involves toxic solvents, stabilizing agents, and energy-intensive processes like high-temperature reduction, which can

have detrimental environmental impacts. In contrast, biological synthesis uses natural reducing agents, making the process greener and more sustainable [6].

2) **Biocompatibility:** NPs synthesized through biogenic methods are often capped with biomolecules such as proteins, lipids, and enzymes, which enhance their biocompatibility and reduce toxicity. This makes biogenic nanoparticles more suitable for biomedical applications, including drug delivery, imaging, and therapeutics, as they are less likely to induce adverse immune responses [7].
3) **Simplicity and scalability:** Biogenic synthesis typically occurs at ambient conditions, requiring neither high pressures nor extreme temperatures. The simplicity of the process makes it easier to scale up for large-scale production, especially when compared to the complex equipment and energy demands of physical and chemical synthesis techniques [8].
4) **Controlled size and shape:** Biogenic methods often provide better control over the shape, size, and methodology of the nanoparticles due to the involvement of biological macromolecules acting as capping agents. This precise control is essential in therapeutic applications where size and shape dictate the nanoparticles' functionality and targeting ability [9].
5) **Reduced toxicity:** The use of nontoxic biological extracts reduces the risk of residual toxicity, which is a significant concern in nanoparticles synthesized chemically. Biogenic synthesis ensures that the nanoparticles produced are safer for biological applications, particularly *in vivo*, where cytotoxicity could impede their therapeutic potential [10].
6) **Sustainability and resource efficiency:** Biogenic synthesis relies on renewable biological materials, such as plant extracts and microorganisms, reducing the dependency on nonrenewable and costly chemicals. This makes it a resource-efficient method, particularly important for sustainable nanotechnology applications [11].

8.1.2 Role of Yeast in Nanoparticle Synthesis

Yeast, one of the eukaryotic microorganisms, possesses several distinctive biological attributes that make it an effective and environmentally friendly platform for the synthesis of nanoparticles. Its ability to act as a bioreactor is attributed to its metabolic and enzymatic properties, which are particularly useful for producing nanoparticles with specific characteristics (Figure 8.1).

The key attributes that make yeast valuable agents in nanoparticle synthesis are addressed below:

1) **Intracellular and Extracellular Synthesis**
   Yeast cells can mediate both an intracellular and extracellular amalgamation of nanoparticles. In intracellular fabrication, the metal ions are reduced inside

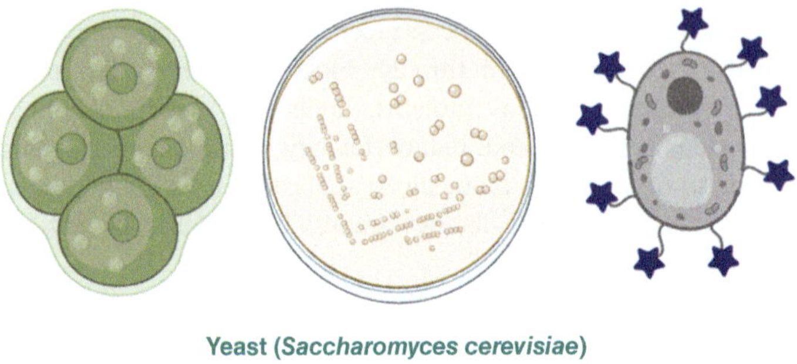

Figure 8.1 Yeast (Saccharomyces cerevisiae).

the yeast cells through the action of enzymes and metabolites, leading to the formation of nanoparticles. In extracellular synthesis, yeast secretes reducing agents into the surrounding environment, facilitating nanoparticle formation outside the cell. This dual capability allows for flexibility in the synthesis process and yields nanoparticles with varied sizes and morphologies [12].

2) **Enzymatic Reduction of Metal Ions**
   Yeast cells contain enzymes such as reductases and oxidoreductases that play a crucial role in reducing metal ions into their nanoparticle forms. These enzymes reduce metal ions like gold (Au^{3+}) and silver (Ag^{+}) to their corresponding zero-valent states (Au^{0}, Ag^{0}), facilitating the synthesis of metallic nanoparticles [13, 14] (Figure 8.2). The enzymatic process allows for precise

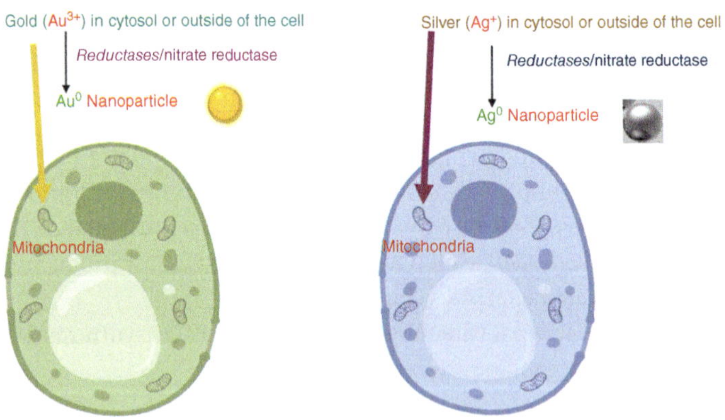

Figure 8.2 Bioreduction mechanism of gold and silver ions into their respective neutral nanoparticles in yeast cells.

control over the reduction rate, which in turn influences nanoparticle size and shape [15].

3) **Secretion of Stabilizing Biomolecules**
Yeast secretes proteins, peptides, and polysaccharides that act as capping agents during nanoparticle synthesis. These biomolecules stabilize the nanoparticles by preventing agglomeration, thus ensuring that they remain uniformly dispersed in solution. This natural capping mechanism helps in producing biocompatible nanoparticles that are ideal for biomedical applications, as they reduce cytotoxicity and enhance stability [4].
4) **Tolerance to Metal Toxicity**
Yeast exhibits a high degree of tolerance to metal ions, allowing it to thrive in environments with elevated metal concentrations. This metal tolerance enables yeast to survive and actively reduce metal ions during nanoparticle synthesis, even under conditions that would be toxic to other microorganisms. As a result, yeast is particularly suited for synthesizing a wide range of metal nanoparticles, including silver, gold, and zinc oxide (Ghosh et al. 2012).
5) **Controlled Size and Morphology**
Yeast-mediated synthesis often results in nanoparticles with controlled shape, size, and morphology. This is due to the influence of yeast's cellular machinery, including enzymes and metabolites, which regulate the nucleation and growth of nanoparticles. Additionally, factors such as yeast strain, culture conditions, pH, and temperature can be optimized to control nanoparticle characteristics, making yeast an adaptable platform for diverse nanoparticle synthesis [16].
6) **Sustainability and Eco-friendliness**
The use of yeast in nanoparticle synthesis aligns with green chemistry principles, as it involves low-energy processes, avoids harmful chemicals, and minimizes environmental impact. Yeast-mediated synthesis typically occurs at room temperature and neutral pH, eliminating the need for toxic solvents or high-energy input. This makes yeast a sustainable alternative to chemical and physical nanoparticle synthesis methods [17].

8.1.3 Advantages of Yeast-Mediated Preparation for Scalability and Sustainability

Yeast-mediated syntheses of nanoparticles are offering numerous significant benefits, particularly in terms of scalability and sustainability. These advantages make it a promising alternative to traditional chemical and physical methods of nanoparticle production.

1) **Eco-friendly and Green Chemistry Approach**
Yeast-mediated synthesis adheres to the philosophies of green chemistry, as it does not require toxic chemicals or high-energy processes. Yeast can

synthesize nanoparticles under ambient conditions (e.g., room temperature and neutral pH), making the process energy-efficient and environmentally benign [17]. This contrasts with conventional methods that often involve harmful solvents, high temperatures, and hazardous reagents.

As a biological system, yeast produces natural capping agents (e.g., proteins and polysaccharides) that stabilize nanoparticles, eliminating the need for synthetic stabilizers. This reduces the generation of hazardous waste and the environmental impact associated with nanoparticle synthesis [4].

2) **Cost-Effective Production**

Yeast-mediated synthesis can be conducted using inexpensive growth media, making it cost-effective for large-scale production. Yeast is a readily available and easy-to-culture organism, allowing for mass cultivation at a low cost. Additionally, the process does not require expensive equipment or chemicals, making it more economically viable compared to conventional synthesis techniques [18].

Yeast cells are highly efficient at reducing metal ions into nanoparticles, minimizing the need for excess reagents. This efficiency contributes to lower operational costs in large-scale production [16].

3) **Scalability**

Yeast-based nanoparticle synthesis is highly scalable, as yeast can be cultured in large bioreactors with ease. Yeast has a short doubling time and can grow to high cell densities, allowing for the manufacturing of nanoparticles on an industrial scale. The scalabilities are further supported by the fact that yeast-mediated synthesis can be performed in batch or continuous bioprocesses, offering flexibility in manufacturing operations [19].

The simple growth conditions required for yeast cultivation, such as standard media and ambient temperatures, make it easier to scale up the process without the need for extensive modifications or infrastructure [20].

4) **Biocompatibility and Safe By-products**

The nanoparticles synthesized by yeast are naturally coated with common biomolecules such as proteins, enzymes, and polysaccharides, enhancing their biocompatibility. This natural biocapping provides a safer alternative to chemically synthesized nanoparticles, which often require additional capping agents that may introduce toxicity [12]. The resulting nanoparticles are ideal for utilization in pharmaceuticals such as drug delivery and diagnostics, due to their reduced cytotoxicity.

Additionally, the by-products of yeast-mediated synthesis are biologically safe and can be easily managed or reused, reducing waste disposal challenges associated with toxic by-products from chemical processes [17].

5) **Sustainable Resource Utilization**
   Yeast is a renewable biological resource, making it a sustainable option for nanoparticle synthesis. Yeast cells can be reused in multiple production cycles, enhancing the overall efficiency of the process. Moreover, the materials required for yeast growth are biodegradable and pose minimal environmental risks [19].
   In contrast to chemical methods that rely on finite, nonrenewable resources, yeast-mediated synthesis supports a more sustainable production model, particularly in industries seeking to minimize their ecological footprint.
6) **Reproducibility and Quality Control**
   Yeast-mediated synthesis offers high reproducibility, which is critical for industrial-scale nanoparticle production. Yeast cultures can be standardized, ensuring consistent nanoparticle characteristics such as size, shape, and morphology across different production batches [16]. This is particularly advantageous for industries that require stringent quality control for medical or electronic applications.

The uniformity and predictability of yeast-mediated nanoparticle synthesis enable the scaling of production without compromising product quality, a common challenge in conventional methods.

So, yeast-mediated synthesis is a sustainable, cost-effective, and scalable technique for fabricating nanoparticles. Its alignment with green chemistry principles, biocompatibility, and ease of large-scale production make it an attractive alternative to both chemical and physical nanoparticle synthesis methods. As industries increasingly prioritize environmentally friendly technologies, yeast-mediated nanoparticle synthesis presents a viable solution for the future of nanotechnology.

8.2 Mechanism of Yeast-Mediated Nanoparticle Synthesis

Yeast-mediated fabrications of inorganic metal nanoparticles involve complex biological pathways that utilize the innate metabolic and enzymatic machinery of yeast cells. These pathways lead to the bio-reduction of metal ions into nanoparticles. This section elaborates on the bioreduction mechanisms that occur within yeast cells and discusses the key enzymes and proteins responsible for facilitating the formation of nanoparticles. Reductases, oxidoreductases, and proteins like glutathione and metallothioneins are central to this process, facilitating the reduction of metal ions into nanoparticles.

8.2.1 Biological Pathways and Enzyme Involvement

In nanoparticle biosynthesis, biological pathways involve the reduction of metal ions into nanoparticles through redox reactions. These processes are guided by the cell's metabolic machinery, involving multiple steps such as electron transport, energy production, and interaction with metal ions. Enzymes secreted by microorganisms like yeast can facilitate the reduction of metal ions into nanoparticles. For instance, in green nanotechnology, enzymes like reductases and oxidases help convert metal salts (e.g., silver nitrate or gold chloride) into metallic nanoparticles. Reductase enzymes donate electrons to metal ions, causing them to reduce into their metallic form, thus forming nanoparticles. These enzymes not only enhance the efficiency of the nanoparticle synthesis but also contribute to controlling the size, shape, and stability of the nanoparticles [21, 22].

Bioreduction refers to the process through which metal ions are reduced from their ionic state to a neutral, metallic form, resulting in nanoparticle formation. In yeast-mediated synthesis, this reduction process is predominantly facilitated by the yeast's metabolic processes, which involve the production of reducing agents, electron donors, and biocatalysts within the cells.

This synthesis of nanoparticles by yeast can be both intracellularly and extracellularly:

Intracellular synthesis: Metal ions are taken up by yeast cells through ion transport mechanisms, where the ions are reduced within the cytoplasm, facilitated by intracellular enzymes and metabolites. Once reduced, nanoparticles may accumulate within the cells.

Extracellular synthesis: Yeast cells secrete enzymes, proteins, and other reducing agents into their environment. These biomolecules react with metal ions existing in the solutions, leading to the extracellular fabrication of nanoparticles. This pathway simplifies the purification of nanoparticles since they are formed outside the cell.

8.2.1.1 Role of Reducing Agents

During yeast metabolism, several reducing agents are produced that play critical roles in nanoparticle synthesis:

The NADH (nicotinamide adenine dinucleotide) and the **NADPH (nicotinamide adenine dinucleotide phosphate)** are essential electron donors produced during cellular respiration and anabolic processes. These molecules transfer electrons to metal ions, reducing them to their zero-valent state, thereby initiating nanoparticle formation [17].

Glutathione, a tripeptide produced by yeast, acts as both a reducing and a capping agent, preventing aggregation of the nanoparticles and stabilization of their growth. Glutathione can bind to metal ions, facilitating their reduction and forming a protective layer around the nanoparticles [23].

8.2.1.2 Key Enzymes and Proteins Facilitating Nanoparticle Formation

Yeast cells contain several enzymes and proteins that catalyze the bioreduction of metal ions and stabilize the synthesized nanoparticles. The primary enzymes involved in yeast-mediated nanoparticle synthesis include reductases, hydrolases, and oxidoreductases. These enzymes act either directly by transferring electrons to metal ions or indirectly by producing metabolites that mediate the reduction process.

8.2.1.3 Reductases

Reductases are critical in the process of reducing metal ions into nanoparticles. These enzymes function by transferring electrons from electron donors such as NADH and NADPH to metal ions, thereby reducing them. Several specific reductases have been implicated in nanoparticle synthesis:

Nitrate reductase: One of the well-documented enzymes in biogenic synthesis, nitrate reductase facilitates the reduction of metal ions such as silver (Ag^+) and gold (Au^{3+}) into their metallic forms (Ag^0 and Au^0). This enzyme catalyzes the conversion of nitrate (NO_3^-) to nitrite (NO_2^-), and in the process, provides electrons that can reduce metal ions [4].

Ferric reductase: Another key enzyme, ferric reductase, reduces Fe^{3+} to Fe^{2+}, aiding in the synthesis of iron-based nanoparticles. The enzyme operates by shuttling electrons from NADPH to ferric ions, a mechanism that can be adapted for reducing other metal ions as well [24].

8.2.1.4 Oxidoreductases

Oxidoreductases are enzymes that facilitate redox reactions, transferring electrons between molecules. In the context of nanoparticle synthesis, these enzymes play a vital role in the reduction of metal ions:

Cytochrome oxidase: Involved in the electron transport chain, cytochrome oxidase is responsible for transferring electrons to oxygen molecules during respiration. This enzyme can also interact with metal ions and contribute to their reduction, particularly in environments where oxygen is limited.

Laccase: This enzyme, typically involved in oxidizing phenols and other aromatic compounds, can reduce metal ions when they act as electron acceptors. Laccase plays a role in the extracellular reduction of metal ions such as gold, leading to the formation of nanoparticles [15].

8.2.1.5 Glutathione and Metallothioneins

In addition to enzymes, proteins like glutathione and metallothioneins contribute to nanoparticle formation by reducing ions of metals and also stabilizing the nanoparticles during synthesis.

Glutathione: Besides acting as a good reducing agent, glutathione binds to metal ions, particularly heavy metals, forming complexes that prevent their oxidation and aggregation. This activity ensures the production of well-dispersed, stable nanoparticles.

Metallothioneins: These are small proteins that are cysteine-rich and bind metal ions, and they can sequester them within the yeast cells. Metallothioneins protect yeast from metal toxicity while simultaneously facilitating nanoparticle formation by concentrating the metal ions in a form that can be reduced by cellular reductants [7].

8.2.2 Stabilization and Capping of Nanoparticles

Once nanoparticles are formed, they are often stabilized by proteins or polysaccharides secreted by yeast. These biomolecules act as capping agents, which prevent the aggregation of nanoparticles and control their growth. The capping proteins create a biocompatible surface on the nanoparticles, improving their stability and preventing oxidation.

Extracellular proteins and polysaccharides secreted by yeast, such as peptidoglycans, bind to the nanoparticle surface, protecting it from aggregation and ensuring that the nanoparticles remain in a dispersed state. This natural capping not only stabilizes the nanoparticles but also enhances their biocompatibility for use in biomedical applications [4].

8.2.3 Types of Metal Nanoparticles Synthesized by Yeast

Yeast has emerged as an efficient and sustainable platform for the production of a variety of metal nanoparticles. This biogenic method leverages the enzymatic and metabolic capabilities of yeast cells to reduce metal ions into their nanoparticulate forms, producing nanoparticles with unique properties. Among the metal nanoparticles synthesized using yeast are gold, silver, zinc oxide, and other metallic nanoparticles like copper, iron, and palladium. The diversity of metal nanoparticles produced is based on the metal ion provided, the strain of yeast used, and the conditions of synthesis. Metal nanoparticles synthesized by yeast have the following advantages compared to its corresponding chemical and physical methods:

1) **Biocompatibility and Low Toxicity**
   Yeast-produced nanoparticles are frequently capped and stabilized by natural biomolecules like proteins, polysaccharides, and enzymes released by the yeast. These natural capping agents improve the biocompatibility of the nanoparticles, rendering them appropriate for biomedical uses. The absence of toxic chemicals in the synthesis process also diminishes the likelihood of harmful

biological reactions, making these nanoparticles more secure for *in vivo* applications [17].

2) **Enhanced Stability**
   The proteins and biomolecules secreted by yeast act as capping agents that prevent agglomeration and oxidation of the nanoparticles. This natural stabilization mechanism results in nanoparticles with better dispersion and long-term stability in various media, including aqueous solutions and biological environments [4].
3) **Controlled Size and Shape**
   Yeast-mediated synthesis permits improved regulation of the size and form of the nanoparticles, which is essential for their use in multiple domains. For instance, spherical nanoparticles are perfect for drug delivery, whereas rod-shaped and hexagonal nanoparticles might have superior optical or electronic characteristics. The capability to adjust the synthesis conditions—like pH, temperature, and metal ion concentration—allows for exact management of nanoparticle morphology [25].
4) **Environmentally Friendly Synthesis**
   Unlike chemical synthesis methods that often require high temperatures, pressures, or toxic solvents, yeast-mediated synthesis is carried out under mild conditions, such as ambient temperature and neutral pH. This makes the process eco-friendly and sustainable, aligning with green chemistry principles [17].
5) **Functionalization and Surface Modifications**
   The surface of nanoparticles produced by yeast can be readily functionalized because of the biological molecules present on their surface. These biomolecules provide reactive functional groups, such as carboxyl or amine groups, that can be conjugated with drugs, targeting ligands, or additional functional molecules for particular biomedical uses like targeted drug delivery or imaging [12].
6) **Tunable Optical and Electronic Properties**
   The optical characteristics (like surface plasmon resonance (SPR) in gold and silver nanoparticles) and electronic characteristics of yeast-synthesized nanoparticles can be modified by changing their size, shape, and composition. This renders them perfect for use in biosensing, diagnostics, and optoelectronics [16].

Therefore, yeast-mediated synthesis provides an eco-friendly and adaptable approach for creating a broad variety of metal nanoparticles, such as gold, silver, zinc oxide, and others. These nanoparticles demonstrate distinctive features such as biocompatibility, stability, and adjustable properties, which make them appropriate for various uses in medicine, electronics, and environmental science.

8.2.4 Effect of Yeast Strain and the Growth Conditions

The influence of yeast strain and growth conditions on nanoparticle synthesis is crucial, as yeast, a type of microorganism, can act as a biological factory for the production of nanoparticles. Different strains of yeast and varying growth conditions can significantly affect the properties of the nanoparticles, such as their size, shape, and stability. This process is integral to green nanotechnology, which utilizes biological systems for nanoparticle synthesis due to their eco-friendly and sustainable nature.

8.2.4.1 Impact of Yeast Strain Variation on Nanoparticle Size, Shape, and Stability

The specific yeast strain employed in the synthesis of nanoparticles has a profound effect on their characteristics. Yeasts are known to produce nanoparticles through biological pathways, such as extracellular or intracellular mechanisms. Different yeast strains possess distinct metabolic pathways and enzymes, which directly influence the nanoparticle synthesis process.

1) **Nanoparticle size:** The yeast strain determines the size of nanoparticles produced. For example, studies have demonstrated that some strains of *Saccharomyces cerevisiae* produce smaller nanoparticles due to their ability to secrete specific enzymes that reduce metal salts to nanoparticles more efficiently [26]. In contrast, other strains may yield larger nanoparticles due to differences in enzymatic activity or the presence of different reducing agents in the cells.
2) **Nanoparticle shape:** Variations in yeast strain can also influence the shape of the nanoparticles. The morphology of nanoparticles, such as spherical, rod-shaped, or triangular, is largely determined by the strain's interaction with the precursor materials. For instance, the strain *Candida albicans* has been found to produce predominantly spherical nanoparticles due to its unique metabolic environment, whereas other yeasts may facilitate the production of more irregularly shaped particles [4].
3) **Nanoparticle stability:** Stability is another critical factor impacted by the yeast strain. The production of capping agents by yeast cells—organic compounds that bind to the surface of nanoparticles—affects stability. Strains that secrete higher amounts of proteins or polysaccharides as capping agents often produce nanoparticles that are more stable in solution, as these biomolecules prevent agglomeration [22]. Yeast strains like *Pichia pastoris* are known for their robust secretion of stabilizing proteins, which enhance nanoparticle stability over extended periods.

8.2.4.2 Optimization of Growth Media, pH, Temperature, and Other Factors

Beyond the yeast strain, various growth conditions, including the composition of growth media, pH, temperature, and aeration, play a pivotal role in optimizing nanoparticle synthesis.

1) **Growth media:** The nutrient composition of the growth media is critical for yeast growth and nanoparticle production. Media enriched with sugars (e.g., glucose, sucrose), proteins (e.g., yeast extract, peptone), and vitamins promote yeast metabolism, which in turn influences the reduction of metal ions and the stabilization of nanoparticles. For instance, nanoparticles synthesized using yeast grown in nutrient-rich media are often smaller and more uniformly distributed due to enhanced metabolic activity [27].
2) **pH:** The pH of the growth medium can significantly impact the synthesis process. Optimal pH levels ensure proper enzymatic function and charge distribution on the yeast cell wall, both of which influence nanoparticle formation. Studies have shown that slightly acidic to neutral pH levels (pH 5–7) are often ideal for nanoparticle production, as these conditions facilitate the reduction of metal salts by yeast cells [25]. Extreme pH levels, either too acidic or too alkaline, may hinder the yeast's metabolic capacity or cause unwanted changes in nanoparticle morphology.
3) **Temperature:** Temperature is another factor that greatly affects nanoparticle synthesis by yeast. Higher temperatures generally accelerate metabolic processes, leading to faster nanoparticle formation. However, excessively high temperatures may denature key enzymes involved in the biosynthesis process. Optimal temperatures for most yeast strains typically range between 25 °C and 35 °C. At these temperatures, the enzymatic activity is maximized, promoting efficient reduction and stabilization of nanoparticles [28].
4) **Aeration and oxygen availability:** The availability of oxygen during yeast cultivation also influences nanoparticle production. Aerobic conditions can enhance the oxidative metabolism of yeast, leading to increased production of reactive oxygen species (ROS), which may facilitate the reduction of metal ions into nanoparticles. However, anaerobic conditions may also promote nanoparticle formation by shifting yeast metabolism toward fermentative pathways, which could yield different nanoparticle characteristics [29].
5) **Other factors:** Additional factors such as agitation speed, incubation time, and the concentration of precursor materials (e.g., metal salts) also play crucial roles. For instance, faster agitation speeds can improve oxygen transfer and the homogeneity of nanoparticle formation, while longer incubation times may result in larger nanoparticles or the formation of agglomerates due to prolonged exposure to reducing agents.

So, the yeast strain and growth conditions are fundamental determinants of nanoparticle characteristics such as size, shape, and stability. By optimizing factors like strain selection, growth media, pH, temperature, and aeration, it is possible to fine-tune the synthesis process to produce nanoparticles with desired properties for various applications, particularly in the field of biomedicine and green nanotechnology.

8.3 Characterization of Nanoparticles Synthesized by Yeast

Characterizing nanoparticles synthesized through biological methods, such as using yeast, is crucial for understanding their properties and optimizing them for specific applications, like drug delivery. Several techniques are employed to assess different aspects of these nanoparticles, including their size, shape, surface characteristics, and stability.

8.3.1 Techniques for Nanoparticle Characterization

1) **Transmission Electron Microscopy (TEM)**
   TEM is used to analyze the morphology (shape and structure) and size of nanoparticles at a high resolution. This technique enables scientists to visualize nanoparticles at the atomic level by transmitting an electron beam through a thin specimen. TEM provides detailed information about the size distribution and internal structure of the nanoparticles [11]. For nanoparticles synthesized by yeast, TEM is particularly useful to confirm their spherical shape and uniformity (Figure 8.3).
2) **Scanning Electron Microscopy (SEM)**
   SEM offers surface morphology analysis by scanning a focused electron beam over the sample surface. SEM produces high-resolution images of the nanoparticle surfaces, providing insights into their topographical features. For yeast-synthesized nanoparticles, SEM can visualize external features and surface texture, though it is generally less detailed than TEM for internal structures [31] (Figure 8.4).
   XRD is used to determine the crystallinity and phase composition of nanoparticles. The technique involves directing X-rays onto a sample and measuring the diffraction patterns to assess the crystalline structure of the nanoparticles. Yeast-synthesized nanoparticles often show distinct XRD patterns that indicate their crystallinity, which can correlate with their stability and functional properties [32].

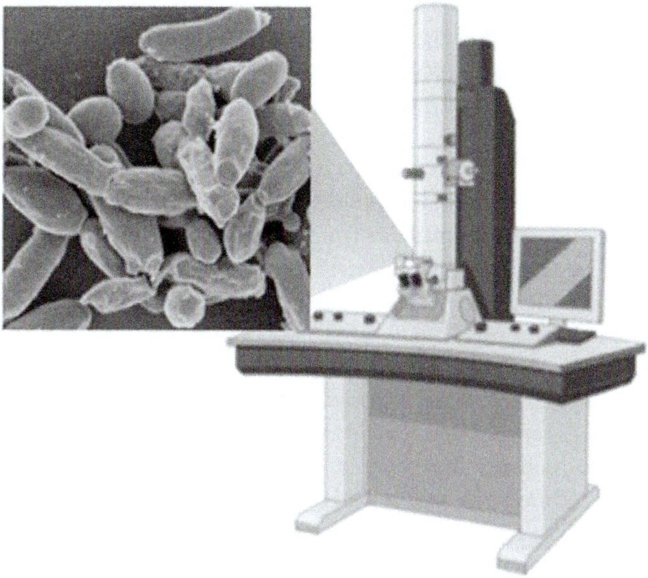

Figure 8.3 Typical hypothetical TEM image of nanoparticles produced by biosynthesis in yeast cells. *Source:* Zhang et al. [30] / with permission of Elsevier.

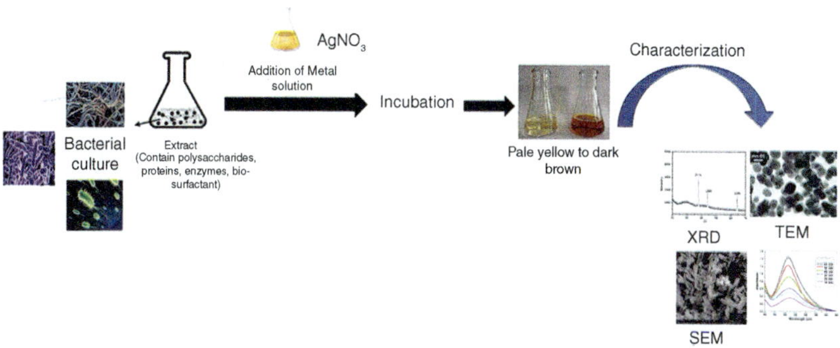

Figure 8.4 Microbial-mediated synthesis of nanoparticles (**X-Ray Diffraction (XRD)**. Bokolia et al. (2024) / with permission of Taylor & Francis.

3) **Fourier Transform Infrared Spectroscopy (FTIR)**
FTIR provides information on the chemical composition and functional groups present on the surface of nanoparticles. In yeast-synthesized nanoparticles, FTIR analysis can help identify organic compounds and biomolecules from the yeast that may be attached to or coating the nanoparticle surfaces, influencing their stability and biological activity [33].
4) **UV–Vis Spectroscopy**
Ultraviolet–visible (UV–Vis) spectroscopy measures the optical properties of nanoparticles, specifically their absorption spectra. This technique is used to monitor the formation and **stability** of nanoparticles by observing characteristic SPR peaks. For metal nanoparticles, such as silver or gold synthesized by yeast, a distinct peak in the UV–Vis spectrum confirms the successful reduction of metal ions to nanoparticles [17].

8.3.1.1 Analysis of Size, Shape, Surface Charge, and Stability

Size: Nanoparticle size is crucial as it affects their functionality in various applications, including drug delivery and catalysis. Techniques like TEM and dynamic light scattering (DLS) are often used to determine the average particle size and size distribution. For yeast-synthesized nanoparticles, the size typically ranges from 10 to 100 nm, depending on the synthesis conditions [32].

Shape: The shape of nanoparticles can affect their surface energy and reactivity. TEM and SEM are key techniques for assessing the shape, which may vary from spherical to more complex geometries. Spherical nanoparticles tend to be more stable, while rod-shaped or other nonspherical particles may have enhanced interactions with biological systems [11].

Surface charge (zeta potential): The surface charge, measured by **zeta potential**, is critical for assessing nanoparticle stability in suspension. A high positive or negative zeta potential indicates good electrostatic stability, as particles repel each other and resist aggregation. NPs synthesized by yeast often have coatings of biomolecules that can stabilize them and influence their zeta potential, making them suitable for biological applications [33].

Stability: Nanoparticle stability is influenced by factors like size, shape, surface charge, and the presence of organic coatings. Stability can be assessed over time using techniques like DLS to measure changes in size distribution or UV–Vis spectroscopy to monitor changes in optical properties. In biological synthesis methods like yeast, the use of organic stabilizers (proteins, lipids) from the yeast cells can provide enhanced stability by preventing aggregation [17].

8.3.2 Structural and Functional Properties of Biogenic Nanoparticles

8.3.2.1 Comparison with Chemically Synthesized Nanoparticles

Biogenic nanoparticles, synthesized through biological processes using organisms such as yeast, bacteria, and plants, differ significantly from chemically synthesized nanoparticles in terms of structural and functional properties.

Size and shape: Biogenic nanoparticles are often more uniform in size due to the controlled biological processes that regulate their formation. In contrast, chemically synthesized nanoparticles may exhibit wider variations in size and shape unless stringent controls are applied [32]. Moreover, biogenic nanoparticles tend to be smaller, which is advantageous for applications in drug delivery, where particle size impacts cell uptake and distribution [34].

Surface chemistry: One of the significant differences lies in the surface chemistry of the nanoparticles. Biogenic nanoparticles are often capped or coated with proteins, enzymes, or other biomolecules from the organisms used in the synthesis process. These natural coatings improve biocompatibility and provide enhanced stability in biological environments, as they reduce aggregation and increase the likelihood of successful integration into biological systems [33]. Chemically synthesized nanoparticles may require additional postsynthesis modification to achieve similar stability and surface properties.

Toxicity: Biogenic nanoparticles generally exhibit lower toxicity than chemically synthesized ones due to their biological origins. The biomolecules associated with the synthesis and surface stabilization reduce the potential for adverse reactions in living organisms, making them more suitable for biomedical applications, such as drug delivery and tissue engineering [17]. Chemically synthesized nanoparticles, especially those with reactive surfaces, may pose higher risks of toxicity or cause unintended immune responses.

Production process: Chemically synthesized nanoparticles typically involve the use of harsh chemicals, high temperatures, and pressure, which can make the process less environmentally friendly. In contrast, biogenic nanoparticle synthesis is often regarded as "green" or eco-friendly, as it relies on biological systems that operate at ambient temperatures and pressures without hazardous reagents [35].

8.3.2.2 Unique Functional Properties of Biogenic Nanoparticles

Biocompatibility and biodegradability: One of the most significant advantages of biogenic nanoparticles is their inherent biocompatibility. Due to their synthesis using biological materials, such nanoparticles exhibit minimal

pathogenic bacteria. The studies reported significant inhibition of bacterial growth due to the nanoparticles' ability to damage bacterial cell membranes and induce ROS production [40].

2) **Fungal pathogens:** Yeast-synthesized nanoparticles have also shown promising results against fungal pathogens such as *C. albicans*. These nanoparticles disrupt the fungal cell wall, leading to cell lysis. Preclinical studies have highlighted the antifungal activity of silver nanoparticles synthesized by yeast against *Candida* species, demonstrating effective inhibition of fungal growth and biofilm formation [41].
3) **Viral pathogens:** Though fewer studies have focused on viral pathogens, some preclinical investigations have explored the antiviral activity of yeast-synthesized nanoparticles, particularly against enveloped viruses such as herpes simplex virus (HSV) and HIV. The nanoparticles have been shown to prevent viral attachment and entry into host cells by interacting with viral surface proteins. In one study, silver nanoparticles synthesized from yeast extracts were reported to exhibit inhibitory effects against HSV-1, reducing viral replication in preclinical models [42].

8.4.2 Cancer Therapy

8.4.2.1 Role of Yeast-Mediated Nanoparticles in Targeted Drug Delivery

Yeast-mediated nanoparticles are increasingly being explored for their role in **targeted drug delivery,** particularly in cancer therapy. Due to the biological capping agents derived from yeast, these nanoparticles demonstrate high biocompatibility, which is crucial for minimizing toxicity to nontarget tissues. Yeast-mediated nanoparticles can be functionalized with ligands or antibodies that specifically recognize cancer cell receptors, enhancing **target specificity** and reducing side effects associated with traditional chemotherapy [17]. For instance, folate-conjugated nanoparticles have been shown to target cancer cells overexpressing folate receptors, facilitating drug delivery directly to tumor sites while sparing healthy cells [32].

8.4.2.2 Studies on Nanoparticle-Assisted Cancer Therapies, Including Preclinical Results

Several preclinical studies have demonstrated the potential of yeast-mediated nanoparticles in cancer therapy. For example, yeast-synthesized gold nanoparticles have been tested for their ability to deliver chemotherapeutic agents like paclitaxel. In these studies, the nanoparticles enhanced drug delivery to tumors and improved drug efficacy while minimizing systemic toxicity. Additionally, the **enhanced permeability and retention effect** seen in tumor tissues allows nanoparticles to accumulate preferentially in tumors, further improving the outcomes of nanoparticle-assisted therapies [11]. Preclinical results have shown that nanoparticle-assisted drug delivery systems can reduce tumor size more effectively than free drugs alone [31].

8.4.3 Wound Healing and Regenerative Medicine

8.4.3.1 Use of Biogenic Nanoparticles in Promoting Wound Healing

Biogenic nanoparticles synthesized by yeast have also been utilized for promoting **wound healing** due to their antimicrobial properties and ability to accelerate cell regeneration. Silver nanoparticles, in particular, have been applied in wound dressings because of their broad-spectrum antimicrobial activity, which prevents wound infections while promoting tissue healing. The nanoparticles interact with bacterial cell walls and membranes, reducing infection rates and promoting a sterile environment conducive to faster healing [39].

8.4.3.2 Preclinical Results on Tissue Regeneration Using Nanomedicine

Preclinical studies on biogenic nanoparticles in regenerative medicine have shown positive outcomes, especially in tissue regeneration. For instance, yeast-mediated gold and silver nanoparticles have been incorporated into scaffolds used for tissue regeneration, where they promote cell proliferation and differentiation. These nanoparticles enhance the bioavailability of growth factors in the wound site, leading to faster tissue repair [34]. In animal models, nanoparticle-embedded wound dressings have shown faster wound closure and reduced scar formation compared to standard treatments [41].

8.4.4 Antioxidant and Anti-Inflammatory Applications

Yeast-synthesized nanoparticles have demonstrated significant **antioxidant** and **anti-inflammatory** properties, which are key in treating conditions associated with oxidative stress and inflammation, such as cardiovascular diseases and chronic inflammatory disorders. Metal nanoparticles, such as silver and gold, synthesized by yeast, can neutralize ROS, reducing oxidative damage to tissues and cells [11]. This property is particularly valuable in mitigating chronic inflammation and preventing the progression of inflammatory diseases.

Studies have shown that biogenic nanoparticles can inhibit pro-inflammatory cytokines, such as tumor necrosis factor-alpha and interleukin-6, which are responsible for promoting inflammatory responses. These nanoparticles modulate immune responses and reduce oxidative stress, making them effective in treating conditions characterized by excessive inflammation, including arthritis and inflammatory bowel disease [33]. Preclinical models have demonstrated that nanoparticle treatments result in lower levels of inflammatory markers and improved clinical outcomes in inflammatory disease models (Table 8.1).

Table 8.1 Summary of some successfully synthesized nanoparticles by yeasts.

Name of nanoparticles	Method of preparation	Potential application	References
Silver nanoparticles (AgNPs)	Yeast cells reduce silver ions ($Ag+$) in aqueous solution to form nanoparticles.	Antibacterial agents, wound healing, biosensors.	[43, 44]
Gold nanoparticles (AuNPs)	Yeast-mediated reduction of gold salts ($AuCl_4$) to gold nanoparticles.	Drug delivery, biosensors, cancer therapy.	[30, 45]
Zinc oxide nanoparticles (ZnONPs)	Yeast cells (e.g., *Candida albicans*) act as reducing agents to synthesize ZnO nanoparticles from zinc salts.	Antifungal activity, UV protection, photocatalysis.	[46]
Copper Nanoparticles (CuNPs)	Yeast cells reduce copper ions (Cu^{2+}) to form copper nanoparticles.	Antimicrobial agents, electronics, sensors.	[47]
Magnesium oxide nanoparticles (MgONPs)	Yeast cells (e.g., *Saccharomyces cerevisiae*) reduce magnesium salts to synthesize MgO nanoparticles.	Antibacterial, anti-inflammatory, biomedical applications.	[48]
Titanium dioxide nanoparticles (TiO_2NPs)	Yeast cells (e.g., *Candida albicans*) reduce titanium precursors to form TiO_2 nanoparticles.	UV blocking, photocatalysis, self-cleaning surfaces.	[49]
Silver sulfide nanoparticles (Ag_2SNPs)	Yeast cells (e.g., *Saccharomyces cerevisiae*) reduce silver and sulfur sources to form Ag_2S nanoparticles.	Antimicrobial, optical devices, drug delivery.	[50]
Magnetite nanoparticles (Fe_3O_4NPs)	Yeast cells synthesize magnetite (Fe_3O_4) nanoparticles from iron salts.	Magnetic resonance imaging (MRI), drug delivery, environmental remediation.	[51]

8.5 Clinical Evidences of Biogenic Nanotherapeutics

8.5.1 Clinical Trials and Case Studies

Biogenic nanoparticles synthesized by microorganisms, such as yeast, are under investigation for their potential therapeutic applications in cancer, infectious diseases, and wound healing. Several clinical trials are exploring their role in targeted drug delivery and antimicrobial therapies.

For example, clinical trials are underway to assess the anticancer properties of nanoparticles loaded with chemotherapeutic agents like doxorubicin or paclitaxel, designed for targeted delivery to cancer cells. In these studies, yeast-synthesized nanoparticles are being evaluated for their ability to enhance drug bioavailability and reduce systemic toxicity [11].

Additionally, the use of silver nanoparticles in wound healing is being tested in clinical settings, particularly for patients with diabetic ulcers, where nanoparticle-based dressings show promise in promoting tissue regeneration and preventing infections [39].

Preclinical studies have demonstrated the potential of biogenic nanoparticles for applications such as cancer therapy, antimicrobial treatments, and tissue regeneration. However, the transition from preclinical success to clinical application has met with several challenges including

Scalability: One significant challenge is the scalability of nanoparticle synthesis for clinical use. While the production of yeast-synthesized nanoparticles is environmentally friendly, scaling up production without compromising the consistency and quality of nanoparticles is a complex process.

Regulatory hurdles: Regulatory approval for biogenic nanoparticles remains a challenge due to the relatively new nature of these materials in medicine. The safety, toxicity, and long-term effects of these nanoparticles need extensive evaluation.

Pharmacokinetics and toxicity: Although preclinical studies show favorable pharmacokinetics and low toxicity, translating these findings into human applications requires more thorough assessment in terms of biodistribution, clearance, and potential accumulation in tissues [33].

8.5.2 Regulatory and Safety Considerations

Regulatory frameworks for nanomedicines, including biogenic nanoparticles, are still evolving. The key regulatory bodies overseeing the approval and commercialization of nanomedicines include the US Food and Drug Administration (FDA) and the European Medicines Agency (EMA). These agencies have established guidelines for the development and testing of nanomedicines, but specific frameworks for biogenic nanoparticles are still in their infancy.

Key regulatory considerations include

Characterization and standardization: Regulatory agencies require comprehensive characterization of nanoparticles, including size, shape, surface charge, and stability, to ensure consistency across batches.

Preclinical and clinical safety: Nanomedicines must undergo extensive safety evaluations in both preclinical and clinical settings. The FDA and EMA assess nanoparticles' **toxicity, biodistribution,** and **immunogenicity** in animal models before approving human trials [52].

Environmental impact: The production of biogenic nanoparticles must also adhere to environmental safety standards, considering their origin from biological sources and the potential for large-scale manufacturing [17].

8.5.3 Assessment of the Safety Profile of Biogenic Nanoparticles

Biogenic nanoparticles, particularly those synthesized using microorganisms like yeast, are considered safer compared to chemically synthesized nanoparticles. This is due to their

Biocompatibility: Yeast-synthesized nanoparticles are capped with biological molecules that improve their interaction with biological systems, reducing immunogenicity and toxicity.

Low toxicity: Preclinical studies have shown that these nanoparticles have low cytotoxicity to normal tissues while being effective against target cells, such as cancer cells or pathogens [31].

Despite these advantages, concerns remain regarding the long-term safety of biogenic nanoparticles, particularly in terms of their biodistribution and clearance from the body. Ongoing clinical trials are essential to better understand the safety profiles of these nanoparticles in human subjects.

8.6 Sustainability and Environmental Impacts

Biogenic nanoparticle synthesis, particularly using yeast, offers a promising alternative to traditional chemical and physical methods of nanoparticle production. These methods align with the principles of sustainability by reducing energy consumption, minimizing hazardous waste, and using eco-friendly, renewable biological materials.

8.6.1 Green Chemistry Principles in Nanoparticle Synthesis

The green synthesis of nanoparticles focuses on using eco-friendly and sustainable methods, which aligns with the 12 principles of green chemistry. Yeast-mediated synthesis employs biological processes to produce nanoparticles without the need for toxic chemicals or high-energy processes, making it an ideal green method.

Key principles applied in yeast-mediated nanoparticle synthesis include

Prevention of waste: Yeast cells act as both the reducing and stabilizing agents during the synthesis, eliminating the need for external chemicals that can generate hazardous byproducts.

Safer solvents and reaction conditions: The synthesis process typically takes place in aqueous environments at ambient temperatures, reducing the need for harmful solvents or high-energy inputs [17].

Use of renewable feedstocks: Yeast, as a biological organism, represents a renewable resource, ensuring the sustainability of the nanoparticle production process. The use of yeast allows for continuous or batch production processes that are scalable and eco-friendly [39].

Green synthesis through yeast not only reduces the environmental footprint but also enhances the biocompatibility of the nanoparticles, making them safer for biomedical applications [32].

8.6.2 Comparison with Chemical and Physical Synthesis Methods

Yeast-mediated nanoparticle synthesis offers several environmental and economic advantages over traditional **chemical** and **physical** methods:

Reduction of toxic chemicals: Conventional methods often rely on toxic reducing agents, stabilizers, and solvents, which pose risks to both the environment and human health. Yeast-based methods eliminate these harmful reagents by utilizing biological pathways that naturally reduce metal ions and cap the resulting nanoparticles.

Energy efficiency: Chemical and physical synthesis methods, such as **chemical reduction** and **laser ablation**, typically require high energy inputs in the form of heat, pressure, or radiation. Yeast-based synthesis occurs at ambient temperature and pressure, significantly reducing energy consumption [11].

Economic viability: Yeast is widely available, inexpensive to culture, and does not require specialized conditions for nanoparticle production. The overall cost of producing nanoparticles using yeast is lower compared to the high energy and material costs associated with physical and chemical methods [17].

One of the most significant advantages of yeast-mediated nanoparticle synthesis is the **reduction of waste:**

Closed-loop systems: Yeast-based nanoparticle production can be integrated into closed-loop bioprocesses where waste products are minimized or recycled, contributing to waste reduction and resource efficiency [34].

Low byproduct toxicity: Since yeast-mediated synthesis avoids harsh chemicals, the waste generated is nontoxic and can often be reused, reducing the need for hazardous waste disposal (Table 8.2).

Table 8.2 Comparison between yeast-mediated synthesis and chemical synthesis methods for producing nanoparticles.

Criteria	Yeast-mediated synthesis	Chemical synthesis
Source of nanoparticles	Biological origin (yeast cells, e.g., *Saccharomyces cerevisiae*, *Candida albicans*)	Chemical precursors (metal salts, reducing agents)
Sustainability	Eco-friendly, uses renewable biological resources (yeast cells)	Often involves toxic chemicals and generates hazardous byproducts
Environmental impact	Low environmental impact, biodegradable, nontoxic	High environmental impact, often requires toxic solvents and reagents
Reaction conditions	Mild conditions (room temperature, neutral pH)	Harsh conditions (high temperature, pressure, acidic or basic media)
Simplicity of process	Relatively simple, one-step bioreduction process	Can be complex, requiring multiple steps and reagents
Size control	Yeast can produce nanoparticles in a controlled range (5–100 nm) depending on conditions	Size control can be difficult without precise tuning of parameters
Shape Control	Yeast cells can help control the shape and morphology (e.g., spherical, rod-like)	Shape can be controlled but may require surfactants or templates
Yield	Moderate yield, typically suitable for small-scale production	High yield, scalable to industrial levels
Cost of production	Generally lower cost, especially for small batches (uses biological agents)	Can be expensive due to the cost of chemicals and reagents
Purity of nanoparticles	High purity as the yeast can selectively reduce metal ions	Purity depends on the precision of the synthesis process
Size distribution	Relatively narrow size distribution due to biological templating	Can have a broader size distribution, requiring additional steps for narrowing size range
Surface functionalization	Yeast cells naturally functionalize the surface of nanoparticles (e.g., proteins, sugars)	Requires additional steps for surface modification, such as coating with polymers or ligands
Applications	Biomedicine (drug delivery, wound healing), biosensors, antimicrobial agents	Electronics, catalysis, environmental cleanup, biomedicine (with proper surface functionalization)

Table 8.2 (Continued)

Criteria	Yeast-mediated synthesis	Chemical synthesis
Reproducibility	Reproducibility can be challenging due to biological variability	High reproducibility with well-controlled chemical processes
Toxicity	Low toxicity, as it utilizes biological systems	High toxicity risk depending on chemicals used (e.g., strong acids, solvents)
Scalability	Limited scalability, mainly suited for laboratory or small-scale production	Highly scalable, suitable for large-scale industrial production

8.7 Future of Sustainable Biogenic Nanomedicine

These nanoparticles contribute to greener healthcare solutions in several ways:

Biocompatibility and reduced toxicity: Biogenic nanoparticles, such as those synthesized by yeast, are often capped with biological molecules, enhancing their compatibility with human tissues and reducing the potential for toxic side effects. This makes them safer for use in drug delivery and therapeutic applications [31].

Reduced environmental impact: The entire lifecycle of biogenic nanoparticles, from production to disposal, has a reduced environmental impact compared to chemically synthesized nanoparticles. The use of renewable resources like yeast, combined with minimal hazardous waste and low energy requirements, makes these nanoparticles a more sustainable option for healthcare.

Circular economy potential: Biogenic nanoparticles offer opportunities for integration into circular economy models, where waste products from other industries (e.g., agriculture or brewing) can be repurposed to produce nanoparticles, reducing waste and promoting resource efficiency [33].

Future trends in sustainable nanomedicine will likely focus on improving the scalability of biogenic nanoparticle synthesis while maintaining the low environmental impact that these methods promise. As regulatory frameworks evolve and clinical trials continue to validate their efficacy, biogenic nanoparticles may become a cornerstone of sustainable medical technology.

8.8 Future Prospects and Challenges

Yeast-mediated nanotherapy has shown great promise in biomedical applications due to its biocompatibility, eco-friendly synthesis process, and potential for producing functional nanoparticles. However, significant challenges remain before this technology can be widely adopted in clinical practice. Future innovations, particularly in genetic engineering and hybrid nanoparticle development, offer exciting opportunities to advance this field.

8.8.1 Innovations and Emerging Trends

One of the most promising areas of research in yeast-mediated nanoparticle synthesis is the application of genetic engineering to optimize the production process. By modifying the metabolic pathways of yeast, researchers can enhance their ability to synthesize specific types of nanoparticles or improve the efficiency of the process.

For example, genetically engineered yeast strains can be designed to express specific enzymes that accelerate the reduction of metal ions to nanoparticles. This approach can not only improve the yield of nanoparticles but also control the size, shape, and surface functionalization of the particles, which are crucial for their biomedical applications [53]. Genetic modifications can also be applied to produce yeast strains that synthesize nanoparticles more quickly or under more favorable environmental conditions, further enhancing the scalability and sustainability of the process.

Moreover, genetic engineering could allow for the functionalization of nanoparticles during synthesis. By introducing genes responsible for specific binding motifs or biologically active molecules, yeast could produce nanoparticles with enhanced targeting capabilities for specific tissues or cancer cells, improving the efficacy of nanoparticle-based drug delivery systems [32].

The development of hybrid nanoparticles, which combine multiple materials or functionalities, represents a significant innovation in yeast-mediated nanotherapy. These hybrid structures can offer multifunctional therapeutic effects, such as combining drug delivery with imaging capabilities for cancer diagnosis and treatment [37].

For example, metal-polymer hybrid nanoparticles synthesized by yeast could be used to deliver a chemotherapeutic agent while simultaneously providing diagnostic imaging through metal components such as gold or silver. These dual-purpose nanoparticles are especially useful in theranostics, where simultaneous diagnosis and treatment are crucial for diseases like cancer. Yeast-mediated synthesis offers a biological platform that can naturally encapsulate drugs and metal components, providing a versatile route to hybrid nanoparticle production.

8.8.2 Challenges and Opportunities

Scalability remains one of the primary challenges for yeast-mediated nanoparticle synthesis. While this method offers eco-friendly production advantages, scaling the process for industrial or clinical use is complex. Factors such as batch-to-batch reproducibility, maintaining nanoparticle consistency, and controlling the size distribution of nanoparticles must be optimized to meet the rigorous standards required for medical applications [39].

Reproducibility is another key issue, particularly in the context of biogenic nanoparticle synthesis, where biological variability can introduce inconsistencies in the nanoparticle properties. Advanced bioprocessing techniques and genetic engineering approaches can help to standardize the process, but further research is needed to develop reliable protocols that ensure consistent nanoparticle production.

In terms of regulatory challenges, the relatively new nature of biogenic nanoparticles in the pharmaceutical industry means that current regulatory frameworks may not fully address their unique properties. Regulatory bodies such as the US FDA and EMA are working to establish guidelines for nanomedicine, but specific frameworks for biogenic nanoparticles, including those synthesized by yeast, are still emerging. These nanoparticles will need to undergo extensive safety evaluations to determine their long-term effects on human health, as well as their environmental impact [11].

Future research will focus on addressing the challenges mentioned above, with particular emphasis on

Improving scalability: Researchers are exploring the development of continuous production processes for yeast-synthesized nanoparticles to improve scalability while maintaining consistency. Innovations in bioreactor design and process control could enable large-scale production of nanoparticles that are suitable for clinical use.

Targeted drug delivery systems: Continued research into the use of targeted drug delivery systems using yeast-synthesized nanoparticles could significantly improve the efficacy of treatments for conditions like cancer, bacterial infections, and inflammatory diseases. The ability to engineer nanoparticles to target specific tissues or cells is a promising area of research [32].

Multimodal therapies: Hybrid nanoparticles, as discussed earlier, will likely become central to multimodal therapeutic strategies. Combining drug delivery, imaging, and diagnostic capabilities within a single nanoparticle system could revolutionize personalized medicine.

Regulatory compliance: Research into the long-term safety and toxicology of biogenic nanoparticles will be crucial for gaining regulatory approval. As biogenic nanoparticles are considered relatively safer due to their biocompatibility, future research will focus on confirming these properties through comprehensive clinical trials [53].

20 Ahmad A, Senapati S, Khan MI, Kumar R, Ramani R, Srinivas V, Sastry M. Intracellular synthesis of gold nanoparticles by a novel alkalotolerant actinomycete, Rhodococcus species. Nanotechnology. 2003;14(7):824–828.

21 Lehninger AL, Nelson DL, Cox MM. Principles of Biochemistry, 7th ed. 2017. W.H. Freeman and Company.

22 Sanghi R, Verma P. A review on biosynthesis of nanoparticles by yeast: green approach for sustainable development. Environ Nanotechnol Monit Manag. 2018;9:111–123.

23 Suresh AK, Pelletier DA, Wang W, Moon JW, Gu B, Mortensen NP, et al. Silver nanocrystallites: biofabrication using *Shewanella oneidensis*, and an evaluation of their comparative toxicity on Gram-negative and Gram-positive bacteria. Environ Sci Technol. 2011;44(13):5210–5215.

24 Wang Y, Wei G, Zhang X, Xu L, Wei D, Sun P. Analysis of ferric reductase gene expression and iron ion concentration during gold nanoparticle biosynthesis by *Rhodopseudomonas capsulata*. Colloids Surf B: Biointerfaces. 2010;80(2):206–211.

25 Ghosh AK, Chapsal BD, Steffey M, Agniswamy J, Wang Y-F, Amano M, et al. Substituent effects on P2-cyclopentyltetrahydrofuranyl urethanes: design, synthesis, and X-ray studies of potent HIV-1 protease inhibitors. Bioorganic & Medicinal Chemistry Letters. 2012;22(6):2308–2311. doi:10.1016/j.bmcl.2012.01.061.

26 Anshup A, Mehta S, Prasad S, Maitra A. Role of yeast in the synthesis of gold nanoparticles. J Nanopart Res. 2020;12(3):1105–1110.

27 Kumar V, Yadav SK, Yadav A. Biological synthesis of nanoparticles by yeast and their applications. J Biomed Nanotechnol. 2016;12(2):234–245.

28 Li X, Xu H, Chen Z, Chen G. Biosynthesis of nanoparticles by microorganisms and their applications. J Nanomater. 2017;2017:7896027.

29 Bhargava S, Gupta M, Verma P. Green synthesis of nanoparticles using yeast: influence of aeration and growth conditions. Biotechnol Adv. 2021;34(4):799–810.

30 Zhang X, Yuanyuan Q, Shen W, et al. Biogenic synthesis of gold nanoparticles by yeast Magnusiomyces ingens LH-F1 for catalytic reduction of nitrophenols. Colloids Surf A Physicochem Eng Asp. 2016;497(2016):280–285. ISSN 0927–7757

31 Bansal V, Ahmad A, Sastry M. Biosynthesis of metal nanoparticles: physico-chemical processes and application. Int J Nanotechnol. 2011;2(5):341–350.

32 Singh G, Bedi PMS, Faruk A. Biomedical Nanomaterials: A Short Insight of Applications. Journal of Drug Delivery & Therapeutics. 2018;8(6):304–309. doi:10.22270/jddt.v8i6.2020.

33 Kora AJ, Rastogi L. Biosynthesis of nanoparticles by microbes and their applications in medicine, agriculture, and environmental remediation: a review. Adv Colloid Interf Sci. 2013;29(5):23–30.

34 Mittal AK, Chisti Y, Banerjee UC. Synthesis of metallic nanoparticles using plants. Biotechnol Adv. 2013;31(2):346–356.

35 Mandal D, Bolander ME, Mukhopadhyay D, Sarkar G, Mukherjee P. The use of microorganisms for the formation of metal nanoparticles and their application. Appl Microbiol Biotechnol. 2006;69(5):485–492.

36 Lara HH, Ayala-Núñez NV, Ixtepan-Turrent L, Rodriguez-Padilla C. Mode of antiviral action of silver nanoparticles against HIV-1. J Nanobiotechnol. 2011;8(1):1–10.

37 Zhang X-F, Shen W, Gurunathan S. Silver Nanoparticle-Mediated Cellular Responses in Various Cell Lines: An in vitro Model. International Journal of Molecular Sciences. 2016;17(10):1603. doi:10.3390/ijms17101603.

38 Dakal TC, Kumar A, Majumdar RS, Yadav V. Mechanistic Basis of Antimicrobial Actions of Silver Nanoparticles. Frontiers in Microbiology. 2016;7:1831. doi:10.3389/fmicb.2016.01831.

39 Durán N, Durán M, de Jesus MB, Seabra AB, Fávaro WJ, Nakazato G. Silver nanoparticles: A new view on mechanistic aspects on antimicrobial activity. Nanomedicine: Nanotechnology, Biology and Medicine. 2016;12(3):789–799. doi:10.1016/j.nano.2015.11.016.

40 Kim JS, Kuk E, Yu KN, Kim JH, Park SJ, Lee HJ, et al. Antimicrobial effects of silver nanoparticles. Nanomed Nanotechnol Biol Med. 2007;3(1):95–101.

41 Monteiro DR, Gorup LF, Takamiya AS, Ruvollo-Filho AC, de Camargo ER, Barbosa DB. The growing importance of materials that prevent microbial adhesion: antimicrobial effect of medical devices containing silver. Int J Antimicrob Agents. 2012;39(2):118–128.

42 Elechiguerra JL, Burt JL, Morones JR, Camacho-Bragado A, Gao X, Lara HH, Yacaman MJ. Interaction of silver nanoparticles with HIV-1. J Nanobiotechnol. 2005;3(1):6.

43 Kapoor V, Saini M, Saini K. Biosynthesis of silver nanoparticles using *Saccharomyces cerevisiae* and their antibacterial activity. J Nanosci Nanotechnol. 2008;8(9):4081–4088. doi:10.1166/jnn.2008.759.

44 Zhang XF, Liu ZG, Shen W, Gurunathan S. Silver nanoparticles: synthesis, characterization, properties, applications, and therapeutic approaches. Int J Mol Sci. 2016;17(9):1534.

45 Araujo AC, Fernandes AR, Pereira MM. Biosynthesis of gold nanoparticles by *Saccharomyces cerevisiae* and their antimicrobial activity. J Nanosci Nanotechnol. 2015;15(3):2292–2297. doi:10.1166/jnn.2015.9254.

46 Muneer M, Shankar M, Prakash M. Biosynthesis of ZnO nanoparticles using *Candida albicans* and their antimicrobial activities. J Mater Sci Mater Electron. 2012;23(6):1266–1270. doi:10.1007/s10854-012-0909-2.

47 Ahamed M, Alhadlaq HA, Khan MA. Biosynthesis and characterization of copper nanoparticles using *Saccharomyces cerevisiae* and its antimicrobial activity. J Nanosci Nanotechnol. 2010;10(8):5117–5122. doi:10.1166/jnn.2010.2429.

48 Galdiero S, Galdiero M, Caserta G. Biosynthesis of magnesium oxide nanoparticles using *Saccharomyces cerevisiae* and their antimicrobial properties. J Nanopart. 2014;2014:101754. doi:10.1155/2014/101754.

49 Ameen F, Al-Farsi S, Al-Gheethi A. Biosynthesis of titanium dioxide nanoparticles using *Candida albicans* and their photocatalytic properties. Environ Sci Pollut Res. 2016;23(15):15296–15302. doi:10.1007/s11356-016-6943-7.

50 Raliya R, Tarafdar JC, Yadav S. Biosynthesis of silver sulfide nanoparticles using *Saccharomyces cerevisiae* and their antimicrobial activity. Int J Nanomedicine. 2015;10:123–132. doi:10.2147/IJN.S89702.

51 De Lima R, Siqueira P, Chaves E. Biosynthesis of magnetite nanoparticles using *Saccharomyces cerevisiae* and their potential biomedical applications. Nanotechnology. 2014;25(15):155 702. doi:10.1088/0957-4484/25/15/155702.

52 Sekhon BS, Kamboj S. Nanotechnology in bio-pharmaceuticals: a review. Int J Nanomedicine. 2010;5:803–812.

53 Bhattacharya D, Gupta RK. Nanotechnology and potential of microorganisms. Crit Rev Biotechnol. 2012;23(3):199–229.

9

Biological Synthesis of Nanoparticles by Viruses and Biogenic Sustainable Nanotherapeutics

Hemanga Hazarika[1], Padmanath Pegu[1], Harshita Krishnatreyya[2], Naba Jyoti Kalita[1], and Ratan Chowdhury[3]

[1] *School of Pharmaceutical Sciences, Girijananda Chowdhury University- Tezpur Campus, Tezpur, Assam, India*
[2] *National Institute of Pharmaceutical Education & Research, Guwahati, Assam, India*
[3] *Rangapara College, Rangapara, Assam, India*

9.1 Introduction

Nanotechnology bridges biological and physical sciences, enabling advancements in nanomedicine and nano-based drug delivery systems. It revolutionizes medicine by targeting specific diseases and contributing to diagnosis, oncology, regenerative medicine, and gene therapy. Biological systems offer significant advantages for nanoparticle (NP) synthesis and are increasingly favored by researchers. The field's growth is driven by both biological and synthetic methods, utilizing materials at the nanoscale (1–100 nm). With diverse applications, nanotechnology holds the potential for a transformative "nano revolution" [1]. Biological systems can act as "biolaboratories" for synthesizing pure metal and metal oxide NPs at the nanoscale using biomimetic approaches [2]. Polymeric NPs, ranging in size from 1 nm to 1000 nm, are composed of natural or synthetic polymers. These NPs can encapsulate drugs or bioactive agents within their core (nanocapsules) or attach them to their surface (nanospheres). Polymers are categorized based on their structure, composition, and origin [3]:

1) Based on Composition:
    a) Organic
        i) Lipid-based: Liposomes
        ii) Polymeric: Polycaprolactone (PCL) Nanocapsules
        iii) Carbon-based: Fullerenes, CNTs

Biogenic Synthesis of Nanoparticles and Sustainable Nanotherapeutics,
First Edition. Edited by Malay K. Das.

are biocompatible, cost-effective, and environment-friendly [13]. Virus-mediated synthesis of nanomaterials has garnered significant interest due to the inherent nanoscale size (20–500 nm) and self-sufficient nature of viruses, making them highly suitable for nanotechnology applications. Plant viruses, in particular, are commonly employed for NP fabrication as they pose no harm to humans. Previous studies have explored virus-mediated synthesis of nanocrystals, highlighting the crucial role of capsid proteins. These proteins provide a reactive surface for binding metal ions, facilitating efficient NP synthesis [14]. Animal viruses are rarely used for NP synthesis as they pose a safety threat to humans. Viruses have specific properties, such as biocompatibility, increased nanoparticulate synthesis, and easy gene manipulation, which make them favorable for biogenic and green synthesis [15]. The cowpea mosaic virus (CPMV), a plant virus non-infectious to mammals, has been utilized for NP synthesis and demonstrates therapeutic efficacy in tumor models, including glioma, melanoma, colon, and ovarian cancers. The tobacco mosaic virus (TMV) is another plant virus that has garnered significant research interest. Notably, cisplatin NPs synthesized via TMV-mediated methods show promise in the treatment of ovarian cancer [16]. Bacteriophages are viruses that infect and replicate within bacterial cells, making them easy to cultivate in laboratory settings using *Escherichia coli* for the synthesis of barium titanate NPs [17].

Virus-like particles (VLPs) can be used for drug encapsulation and conjugation in targeted drug delivery. In addition to nucleic acids, VLPs are flexible and can be used to incorporate and deliver a variety of small and large molecules of both biological and synthetic origin. VLPs of CCMV, brome mosaic virus, red clover necrotic mosaic virus (RCNMV), and *Hibiscus chlorotic* virus have been used to encapsulate drug materials [18]. Encapsulation of CCMV CpG oligodeoxynucleotides (ODNs) for use in cancer immunotherapy has also been reported [19].

9.2 Role of Viruses in NP Synthesis

Viruses were initially defined as viral nanoparticles (VNPs). In recent times, viruses have been used as nanocontainers to encapsulate synthetic nanomaterials. Nowadays, viruses are even used as scaffolds or templates to prepare metal NPs and nanotube structures using mineralization techniques [20]. Liposomes/lipids, natural as well as synthetic polymers, and inorganic NPs constitute traditional nanodrug carrier systems, which may sometimes be toxic or with low drug-delivery efficiency. Viruses and protein cages currently constitute a novel and emerging class of nanocarrier systems. Generally, VLPs are non-infectious as they are the genome-free versions of their VNP counterparts. Natural VNP-based carriers have shown ideal nanomedical approaches in developing smart nanosized cages. VLPs encapsulate drugs using the genetically enabled direct loading *in vivo* assembly technique. After expression of both the drug and capsid protein, *in vitro/in vivo* self-assembly occurs [18]. This technique was used by Douglas et al., who utilized scaffolding proteins for the assembly of P22 procapsids with their helix domains within the C-terminus. Streptavidin genetically fused with scaffolding proteins was used as a molecular linker to entrap the biotin-labeled delivery system. By binding the scaffold proteins to the interior streptavidin, biotinylated green fluorescent protein (GFP) was exposed on the P22 procapsid. The scaffold proteins penetrated, allowing for exterior functionalization whereas the GFP was too large to traverse the viral capsid [21]. VLPs comprising several capsid proteins signify repetitive, multivalent scaffolds which assist the antigen presentation. VLPs have become ideal for epitope delivery and immune vaccination systems and because of such properties and they usually induce more robust immunity than soluble antigens [18].

VLPs act as frameworks for conjugation against infectious diseases. An example is the cucumber mosaic virus (CuMV), which is used as a scaffold for conjugation of the nerve growth factor (NGF) to its exterior surface. These VLPs induce anti-NGF antibodies in mice to reverse pain caused by partial meniscectomy-induced osteoarthritis [22]. NPs are extensively used as platforms for MRI contrast agents. VLPs have been popularly used as the base or support for MRI contrast agents to avoid the restrictions of traditional contrast agents [23, 24]. They have good capsular configuration with a higher loading efficiency, increased intracellular uptake capacity, and easy functionalization on different coat proteins. These are a few reasons for using bacteriophage M13. VLPs were also used as support for contrast agents [25]. CPMV-based VNPs are widely used in biomedical applications because they act as strong frameworks that offer chemical conjugation, genetic modification, etc., and finally, targeted delivery to specific cells and tissues [26].

9.2.1 Mechanism of Virus-Mediated NP Synthesis

9.2.1.1 Overview of Virus Structure

Virus-mediated NP synthesis leverages the unique biological characteristics and structural features of viruses. The structural complexity of viruses, combined with their ability to self-assemble and display specific surface properties, make them effective frameworks for developing nanomaterials with high precision.

9.2.1.2 Viral Capsid Structure and Function

Viruses are primarily composed of a nucleic acid core, which contains viral genetic material, and a protein shell known as the capsid. The capsid is often made up of repeating protein subunits, forming a highly ordered structure that can take on various morphologies, including icosahedral, helical, or more complex shapes, depending on the virus type [27, 28] This capsid not only protects the viral genome but also plays a critical role in NP synthesis by serving as a natural scaffold for NP formation [29]. Some viruses are enveloped, meaning they possess a lipid bilayer derived from the host cell membrane, surrounding the capsid. This envelope is often studded with glycoproteins that facilitate attachment to and entry into host cells. Non-enveloped viruses, on the other hand, lack this lipid layer, which may influence their behavior in NP synthesis applications, as non-enveloped capsids are more stable and can withstand harsher conditions [30]. The nuclear capsid consists of the genome along with nucleic acid-protein complexes. Recently, these viral capsids have gained significant interest as innovative nanomaterials for drug delivery [31, 32]. Bacterial viruses (like phages or bacteriophages) consist of a double-stranded DNA genome, an icosahedral capsid head (to protect the genome), and a tail structure (to deliver the genome to the host cell) [33]. A single or a few structural proteins assemble into a single/double protein shell to form a capsid. Therefore, several protein shells construct in a continuous three-dimensional manner to form the capsid structure. Viral capsids follow two simple patterns for self-assembly. The viral capsid morphology, its size, shape, and orientation are determined by the specific arrangement of protein molecules [32].

9.2.1.3 Viral Self-Assembly and Reassembly

One of the defining features of viruses is their ability to self-assemble, which enables them to form highly ordered structures from individual capsid proteins without the need for external guidance [34]. This self-assembly process is a cornerstone of virus-mediated NP synthesis, as the virus can spontaneously encapsulate various materials or coat its surface with NPs, forming stable nanostructures [35]. The ability to reassemble viral capsids *in vitro* has allowed researchers to modify viruses to create novel NP systems. For example, the CCMV is known for its ability to disassemble and reassemble under varying pH and ionic conditions, allowing it to encapsulate foreign molecules during the reassembly process [36] (Figure 9.1).

Bioconjugation-based approaches have demonstrated the significant effectiveness of bacteriophage MS2 in drug and cellular delivery. Peptides, quantum dots, ricin toxin, siRNA, and doxorubicin were successfully targeted using a VLP assembly co-labeled with capsid proteins. These multifunctional MS2 VLPs selectively targeted and killed cancerous cells while leaving healthy cells unharmed. Another strategy to enhance drug delivery involved encapsulating doxorubicin within the core of MS2 particles. Loading empty MS2 viral capsids resulted in reduced intracellular accumulation of doxorubicin in human hepatocellular carcinoma cells. Compared to free drug formulations, the targeted and drug-loaded viral NPs were able to bypass the P-glycoprotein efflux mechanism, enabling cancer cell destruction at up to 20 times lower drug concentrations than conventional free drugs [37, 38]. Rod-shaped VLPs serve as scaffolds for the conjugation of various small molecules. Doxorubicin is frequently incorporated into the red clover

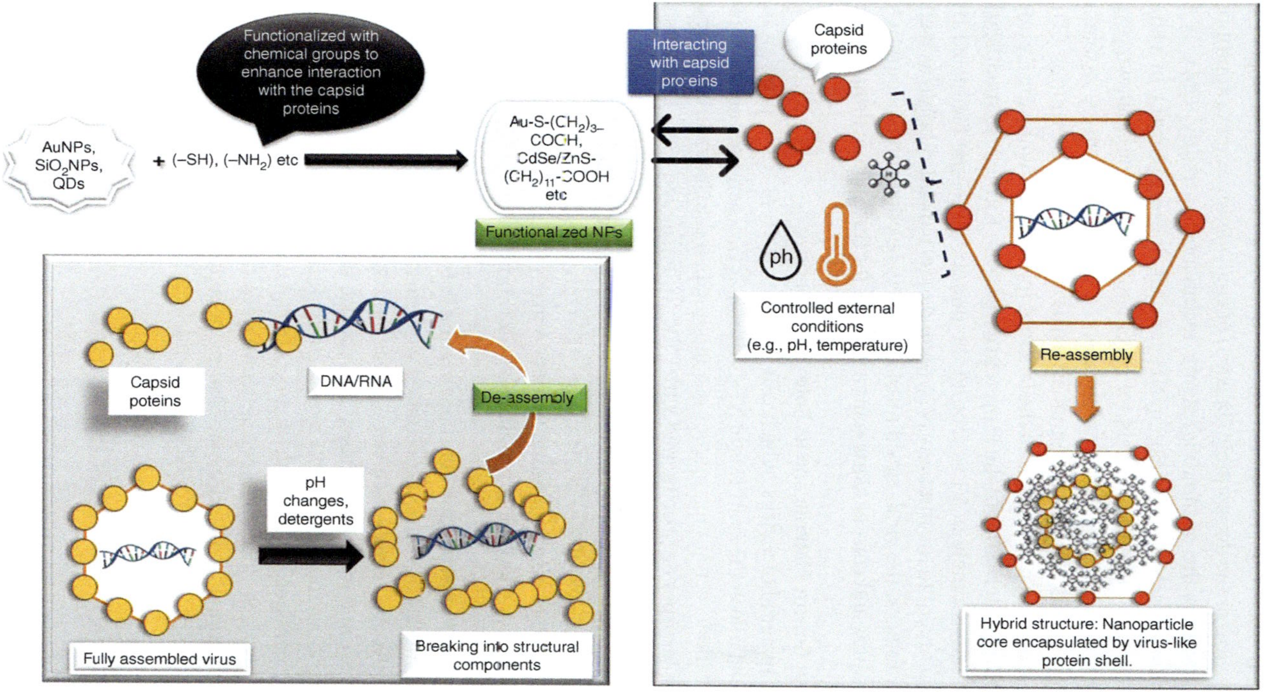

Figure 9.1 Virus self-assembly and reassembly mechanisms in nanoparticle synthesis.

necrotic mosaic virus (RCNMV) by binding it to the viral nucleic acid (VNA) [39]. Doxorubicin has also been loaded into the CuMV using the same technique. The assembly (*in vivo*) or the disassembly (*in vitro*) of viral NPs around the appropriate payload determines the scaffolding/caging technique. Currently, CMV is the plant virus most widely used for caging NPs for drug delivery [40, 41]. In a study, CCMV NPs were explored as carriers for regenerative medicine. Following viral disassembly, the viral RNA was extracted, and CCMV dimers were assembled around heterologous Sindbis mammalian virus RNA. The N-terminus of the CCMV capsid protein was fused with the leucine zipper K-coil, while the E protein was linked to the C-terminus of the E-protein cargo from *E. coli*. This modification enabled the self-assembly of VLPs around the protein cargo and facilitated non-covalent attachment to the dimmers [42]. Owing to its biological origin as well as its other advantages, the viral capsid must definitely be considered a merit when debating the pros and cons of using nanomaterials for therapeutic drug delivery.

9.2.1.4 Viral Genetic Material and Its Role in Synthesis

The genetic material of a virus, which can be either DNA or RNA, plays an indirect yet crucial role in the synthesis of NPs. Although viral genomes are often inert in NP applications, certain techniques use modified viral RNA to direct synthesis processes or incorporate specific molecules for NP formation [30]. This versatility enables the creation of tailored NPs with specific functionalities by encoding instructions for NP assembly or modification. For instance, engineered viruses can display functional peptides on their surface through genetic modification of their genomes, enabling selective binding to metal ions or other substrates, which aids in the formation of NPs with controlled size and composition [43].

9.2.1.5 Surface Properties and Functionalization Potential

The exterior surface of viruses is decorated with various functional groups and, in some cases, glycoproteins, which can serve as binding sites for metal ions or other precursors required for NP synthesis. The surface chemistry of the virus can be modified to control the nature of NP synthesis. By functionalizing the viral surface with specific peptides, it is possible to direct the synthesis of metallic or semiconducting NPs [44]. Surface engineering techniques have enabled viruses like the TMV to be functionalized with metal-binding peptides, facilitating the nucleation and growth of NPs with controlled shapes and sizes [45].

9.3 Metal NP Synthesis by Viruses

The synthesis of metal NPs (MNPs) through biological entities, particularly viruses, has gained attention due to its environmentally benign and sustainable approach. Viruses, being natural NPs, offer a biological platform for the precise and controlled synthesis of MNPs. They present an efficient template for nano-assembly due to their high surface-to-volume ratio and the ability to self-assemble, offering synthesized particles of uniform size and shape.

9.3.1 Mechanism of Virus-Mediated MNP Production

The process of viral-mediated MNP synthesis is a multi-step mechanism that leverages the virus's natural structural and biochemical properties. Certain viruses can mediate the reduction of metal ions into stable metallic NPs on their surface proteins (Figure 9.2). Generally, the virus surface proteins provide binding sites for metal ions, and subsequent biochemical or redox processes lead to the nucleation and growth of MNPs [46–48].

For instance, the bacteriophage T4 has been shown to act as a scaffold for the synthesis of silver nanoparticles (AgNPs). Its capsid proteins contain specific binding domains that facilitate the attachment and reduction of silver ions (Ag^{+}) into silver NPs [49]. Another significant example is the cowpea mosaic virus (CPMV), where its negatively charged coat proteins help attract and bind positively charged metal ions, such as gold (Au3+) and palladium (Pd^{2+}), facilitating their nucleation into NPs [45].

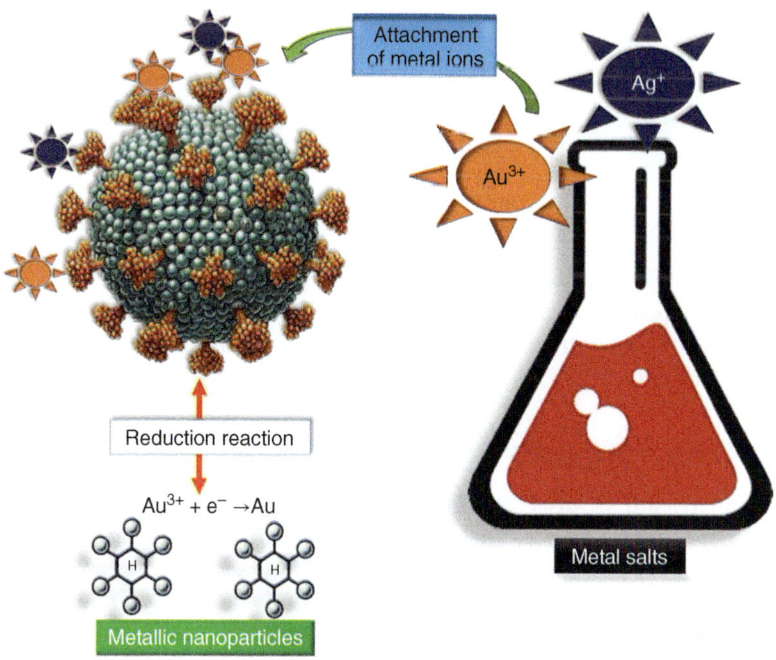

Figure 9.2 Virus-driven nanoparticle fabrication using reduction reaction.

Recent studies indicate that factors such as pH, ionic concentration, and temperature significantly affect the rate and size of NP formation on viral surfaces. These parameters enable precise control over the size and morphology of the NPs [50]. The ability to precisely fine-tune NP synthesis offers promising applications in biomedicine, chemistry, and pharmaceuticals.

9.3.2 Case Studies of MNPs Synthesized Using Viral Templates

Cowpea mosaic virus (CPMV) has been successfully used to synthesize AuNPs due to its robust capsid, which offers ample binding sites for gold ions [51, 52]. These AuNPs have shown potential in biomedical applications such as drug delivery and bio-imaging due to their stable and biocompatible nature [53, 54].

TMV, a plant virus, has been widely studied for synthesizing AgNPs. TMV particles act as templates, aligning silver ions into a one-dimensional assembly along the viral filament. These AgNPs exhibit enhanced antimicrobial properties, showing effectiveness against pathogens such as *E. coli* and *Staphylococcus aureus* [55].

The M13 bacteriophage has been used to template palladium NP formation, leading to monodispersed particles that exhibit high catalytic activity in organic synthesis. Recent work has demonstrated that PdNPs synthesized with M13 templates show enhanced performance in catalytic applications, such as the Suzuki-Miyaura coupling reactions [56].

9.4 Virus-Derived NPs in Nanomedicine and Nanotherapeutics

9.4.1 Viral NPs as Drug Delivery Systems

Viruses offer a highly customizable platform for targeted drug delivery, leveraging their ability to penetrate cellular barriers and release cargo with high precision. Viral NPs (VNPs) have an innate capability to interact with biological systems in a way that makes them suitable for drug delivery applications.

9.4.2 Virus-Like Particles (VLPs) for Targeted Drug Delivery

VLPs are derived from virus capsid proteins but lack the viral genome, making them safe for therapeutic use. These NPs retain the virus's ability to enter cells, allowing them to be used for delivering drugs directly to the target site. A significant advantage of VLPs is their ability to encapsulate a wide variety of therapeutic agents, including small molecules, proteins, and nucleic acids [57].

VLP NPs can be used for the delivery of small molecules in cancer therapy. Drug delivery systems that can provide targeted and intracellular delivery encourage drug accumulation in tumor tissues, and minimalize drug doses, which can improve therapeutic outcomes. VLPs are highly suitable for the targeted delivery of therapeutic agents [19]. In an initial study, bacteriophage MS2 VLPs were used to deliver chemotherapeutic drugs DOX, cisplatin, and 5-FU, to human hepatocellular carcinoma cells (HCC). These VLPs demonstrated high cytotoxicity in HCC with nominal uptake in healthy cells *in vitro* even at very low doses. In fact, several viral NPs and VLP systems have been utilized for DOX delivery [38].

VLPs may be used to deliver small molecules in anti-microbial therapy. Bacterial populations continue to evolve and escape current antibacterial treatments requiring improved antimicrobial approaches [58]. VLPs have been engineered to be used in such cases. Benhar et al. utilized phage display on filamentous fd phages to display specific peptides against *S. aureus* [59]. Recently, Finn et al. demonstrated the conjugation of variants of macrolide antibiotics for treatment against pulmonary macrophages such as *Mycobacterium tuberculosis* and *Legionella pneumophilia* [60].

Viruses are natural and broadly used vehicles for the encapsulation and delivery of therapeutic genes and nucleic acids. Tian et al. utilized the trans-activating transduction (TAT) peptide by conjugating it to the outer surface of the TMV for microRNA delivery [61]. This engineered TMV–TAT system successfully delivered GFP silencing RNA (siRNA) to hepatocellular carcinoma tumors (MHCC97-H/GFP) *in vivo* in Balb/c-nude mice. Apolipoprotein-E-altered RNAi-loaded Qβ bacteriophage VLPs were used successfully in targeting malignant brain tumors by crossing the blood-brain barrier [62].

VLPs can deliver protein and peptides. Polypeptides and protein delivery are challenging as they require different strategies for encapsulation. Black et al. validated the packaging of protein and DNA together within phage T4 [63]. Proteins displayed on the surface of VLPs can be utilized for cell targeting or therapeutic applications. For instance, Francis et al. conjugated anti-epidermal growth factor receptor (EGFR) antibodies to MS2 VLPs to specifically target EGFR expressed on breast cancer cells [64].

Viruses have theranostic applications in nanotechnology. Fluorescent VLPs are used in optical imaging for the specific labeling of cells and tissues and evaluating pharmacokinetic biodistribution and biological interactions [65]. For example, CPMV has been used comprehensively for optical imaging applications. VLPs loaded with magnetic resonance imaging (MRI), positron emission tomography (PET), or computed tomography (CT) contrast agents can be used to provide very accurate imaging of deep tissues. Engineered dodecane tetraacetic acid (DOTA) MS2 bacteriophages were used by Francis et al. to deliver 64Cu for microPET/CT imaging [66]. Qβ-based VLPs engineered with 7-nm gold nanoparticles and loaded internally with DOX molecules were used in photothermal therapy [67].

For example, the Hepatitis B virus (HBV) core protein has been re-engineered to form VLPs that efficiently deliver chemotherapeutic agents to liver cancer cells, thereby reducing systemic toxicity [68]. Another promising example includes Norwalk virus-like particles, which have shown efficient delivery of mRNA vaccines due to their ability to enter antigen-presenting cells (APCs) effectively [69].

9.4.3 Functionalization of VNPs for Enhanced Therapeutic Efficacy

The functionalization of VNPs, by modifying their surface with specific ligands, antibodies, or targeting moieties, can enhance their specificity and therapeutic efficacy. This modification enables these particles to recognize and bind to target cell receptors, providing a high degree of selectivity in delivering therapeutic payloads [70, 71].

For instance, surface-functionalized CPMV VLPs have been designed to target cancer cells expressing specific markers, enhancing uptake and cytotoxicity of the encapsulated drug in tumor cells. Similarly, TMV-based NPs have been modified with polyethylene glycol (PEG) to improve blood circulation time and reduce immune clearance [72, 73]. These advancements highlight the potential of VNPs and VLPs in drug delivery applications, offering a promising alternative for treating various diseases, including cancer, viral infections, and autoimmune disorders.

9.4.4 Functionalization of Viral NPs for Theranostics

Natural polymer-based nanomaterials known as VLPs lack viral genomes but resemble viral structures by using viral coat proteins in order to put them together hierarchically. VLPs have drawn a lot of interest in a variety of medical devices based on nanotechnology diagnostic and therapeutic applications, such as theranostics, imaging, and cancer treatment. VLPs have a consistent structure, are controlled in assembly, and are both biodegradable and biocompatible. They can be chemically or genetically altered, and can include a broad variety of medicinal and diagnostic chemicals. These characteristics have given rise to highly developed multifunctional therapeutic platforms utilizing nanomedicines within the body [24]. Molecular imaging is a rapidly expanding area of study in biomedicine that enables the observation, characterization, and measurement of biological processes *in vivo*. Numerous imaging modalities, such as optical imaging, computed tomography (CT), magnetic resonance imaging (MRI), and positron emission tomography (PET), have been extensively employed to illustrate the cellular and molecular patterns and causes of disease in living individuals. With the least amount of toxicity, ideal molecular imaging methods should deliver good signal-to-noise ratios at the target site rapidly. Because VLPs have a shorter circulation and retention period than manufactured NPs, they might be more beneficial as agents for the dissemination of images and reduce the risk of adverse effects [74]. VLPs are compatible with medications and imaging reagents at the same time in more recent imaging applications; this is called theranostics, and it allows for both diagnosis and treatment. PTT, which uses NIR laser photo absorbers to produce heat for the thermal ablation of cancer cells following laser irradiation, is one of the most promising approaches [75]. PTT offers distinct benefits in cancer treatment when compared to traditional therapeutic modalities, including high specificity, low invasiveness, and accurate spatial-temporal selectivity [76]. Gold NPs smaller than 7 nm were attached to the surface of Qβ-based VLPs in order to tailor them for PTT; up to 500 DOX molecules were injected into the inner cavity [67]. When exposed to a 6-ns laser, DOX was quickly liberated without heating the bulk solution, from the capsid, enabling the *in vitro* death of A549 lung cancer cells and RAW264.7 macrophages with excellent precision. According to live-cell imaging, only the cells within the irradiation region displayed dissociation and morphological changes, indicating highly targeted therapeutic effects. In a different study, photodynamic treatment (PDT) was conducted using Qβ coupled to the derivatives of metalloporphyrin and a CD22 receptor that targets glycan [77]. Dose-responsive phototoxicity and increased cell death in targeted versus nontargeted viral particles were demonstrated *in vitro* using CD22 positive Chinese hamster ovary cells. To further combat estrogen-receptor cells that are positive for breast cancer, PDT has additionally been utilized with P22 VLPs encasing CYP, which transforms the prodrug tamoxifen, and is embellished with the photosensitizer protoporphyrin IX [78]. By loading Gd onto TMV nanorods and coating them with polydopamine (PDA), which improves MRI characteristics and provides PDA contrast while permitting Photoacoustic, MRI and PTT imaging capabilities were added to the PTT therapy for application in theranostics [79]. Commercial contrast agents for MRI have r1 values of 3–5 mM^{-1}/s, but NPs of Gd-TMV without PDA achieved r1 values of up to ~13 mM^{-1}/s. Greater r1 values, up to around 80 mM^{-1}/s, were obtained by applying PDA coating onto the TMV (Gd-TMV-PDA). The reason for this rise in relaxivity could be that the PDA layer improves the magnetic characteristics and Gd centers' interaction with water by trapping bulk water in the TMV structure. *In vitro*, the GdTMV-PDA particles also demonstrated high-efficiency photothermal conversion (28.9%) and robust NIR absorption. Other widely utilized photothermal agents exhibit photothermal conversion efficiencies of 13% (gold nanocages) [80], 21% (gold nanorods) [81], 22% (Cu2-xSe) [82], and 24.4% ($Fe@Fe_3O_4$) [83]. However, the cells + irradiation and cells + controls GdTMV-PDA without irradiation showed negligible cytotoxicity and the NPs demonstrated strong *in vitro* efficacy in the management of 4T1 breast cancer cells and PC-3 prostate cancer. PTT has also made use of mammalian viruses. Methotrexate and iron oxide have been loaded into HBc for MRI and PTT [84], conjugated to an RGD peptide for tumor targeting, copper sulfide encapsulation for PTT only [85], and indocyanine green loading for image-guided PTT [86]. Finally, copper sulfate [87] and gold NPs have been loaded into ferritin-based VLPs for cancer diagnostic purposes.

9.5 Sustainable Aspects of Virus-Mediated Biogenic Production of Nanomaterials

The production of nanomaterials through natural processes and principles, aimed at minimizing the use of hazardous components and reducing waste, is referred to as sustainable synthesis. This approach seeks to mitigate the environmental harm caused by conventional synthesis methods. Natural processes such as biomineralization and self-assembly offer valuable insights into the nanocircular economy (NE), which provides numerous benefits. Firstly, it minimizes the environmental impact of nanotechnology by reducing energy consumption and waste generation. Additionally, it enhances resource efficiency by integrating the principles of recycling and reusability [88]. By imitating natural processes, a variety of nanomaterials that are biocompatible, biodegradable, and less damaging to the environment can be produced.

Naturally occurring NPs' presence in organisms does not mean that they are innocuous; depending on the circumstance, they may have negative impacts. The ability of NPs to accumulate in organisms is directly correlated with their impact. Furthermore, when stabilizers and surfactants are used, generated NPs tend to last longer than naturally occurring NPs, which usually merge over time to create bigger materials. While more complex living beings usually absorb NP through their skin, gastrointestinal tract, or respiratory system, microorganisms take NPs through their cell surfaces. The way that NPs enter the body is directly influenced by the characteristics of live organisms [89].

For instance, cylinder-shaped carbon nanotubes can easily enter the human body through the pulmonary epithelium and cause toxicity, in contrast to other carbon nanotubes, like fullerenes, which do not. Prokaryotes' cell surfaces provide some defense against a range of NPs because they lack mechanisms for moving particles through their cell walls. However, the surfaces of eukaryotic cells may absorb a wide range of environmental particles by phagocytosis or endocytosis, which makes it easier for the potentially dangerous NPs to penetrate and have negative consequences [90]. Once within the cell, the type of NPs will dictate their toxicity and the harm they cause to different cell compartments. For example, carbon nanotubes usually induce toxicity, but silver NPs usually stick to the membrane and alter its permeability and structure. NPs are eventually released from the cell by organelles called lysosomes, but they can potentially be dangerous. Damage to the nucleus occurs when NPs penetrate the nuclear membrane and interfere with the Golgi complex's normal function. The production of ROS is the primary cause of NPs' detrimental effects [89].

It is crucial to consider the possible positive effects of NPs on the environment in addition to their possible negative ones. NPs have several uses that might have a beneficial effect on their surroundings. NPs have a wide range of potential applications in reducing pollution and enhancing environmental health, including supporting energy-efficient industrial processes and helping to clean up soil and water [91].

Soil remediation is one of the most important applications of NPs to stop environmental damage. Zinc nanoparticles (ZnNP) can be used to degrade dyes and medicines efficiently. Using iron NPs to eliminate heavy metals or organochlorine compounds by electronic donation has also been studied [70]. Consequently, FeO NPs have shown a 98.1% efficiency in the adsorption of Cr(VI). Most of these nanomaterials are biodegradable and do not leave any toxic leftovers behind. Because it may be utilized *in situ* and save time and money, the nano-remediation approach holds tremendous promise for decontaminating groundwater. Directly converting a contaminant into a detoxifying agent in water by using NPs is known as an *in situ* remediation. This has led to research on carbon nanotubes. They have a high adsorption capacity because of their porousness, which results in a high surface-volume ratio, making them ideal candidates as nanofilters. As a result, we can eliminate both organic and inorganic pollutants.

Photovoltaic cells have several significant issues, including high production costs and low solar energy absorption efficiency (only 40% of solar energy is converted to electric energy). Because they are simple to develop and effectively store energy, carbon nanotubes have been investigated as a potential substitute material for cell architecture. Moreover, they can tolerate abrupt shocks and temperature fluctuations. As a potential material for solar energy collection, TiO_2 NPs are also becoming increasingly well-liked due to their strong photocatalytic activity and good adsorption. The use of carbon nanotubes in windmills has demonstrated numerous advantages, such as the ability to create lightweight blades that can be stretched to store energy or serve as building blocks for antifreeze coatings that increase windmill lifespan in colder climates. The use of certain NPs that reduce the temperature and depth needed to capture geothermal energy is also recommended [92].

9.6 Challenges and Prospects in Virus-Mediated Biogenic Synthesis of Nanoparticles

While NP-based techniques have advanced at the lab level, comprehensive nanotoxicology studies on their effects on human health and the environment remain scarce. Research is needed on NP interactions with plant nutrients and other substances. Accurate assessments of exposure time, application methods, environmental factors, NP shape, and concentration are crucial. Establishing regulatory frameworks is essential to ensure the safe and responsible use of NPs in agriculture. Such measures will promote sustainable and safe agricultural practices involving nanotechnology [93]. Biogenic NP studies often focus on synthesis, characterization, and *in vitro* evaluation, with limited *in vivo* studies on kinetics and biodistribution, which are crucial for regulatory approval. While some NPs show promise in wound healing, they are mainly applied topically, lacking comprehensive degradation pathway analysis. Simulation studies offer limited insights into these challenges. The green synthesis and biomedical potential of biofabricated NPs are promising but require focused efforts on *in vivo* validation and regulatory criteria. Future research must address these gaps to advance nanomedicine applications [10]. The impact and uptake of NPs depend on their persistence, organism complexity, and entry pathways, which are influenced by stabilizers and organism traits [89]. Carbon nanotubes' shape and cellular mechanisms like phagocytosis in eukaryotes facilitate NP uptake and toxicity, while prokaryotes' cell walls offer limited defense [90]. Challenges in reproducibility and scalability, as well as public concerns limit the use of genetically modified organisms (GMOs) in NP biofabrication. Despite these hurdles, the use of GMOs is anticipated to gain prominence for scalable, efficient, and eco-friendly NP production in the future [93].

One of the primary challenges in microbial NP synthesis lies in the labor-intensive purification processes required to recover intracellular NPs [91]. The development of virus-inspired NPs raises key questions about whether a universal "artificial virus formulation" can suffice with minor modifications or if designs should mimic specific natural virus targets, such as the rabies virus for brain-targeting or HBV for hepatic cells. While tunable formulations may expand applicability, clinical translation depends on realistic *in vitro* and *in vivo* models that mimic physiological conditions, addressing factors like immune recognition and barrier penetration. Challenges include the structural complexity of virus-mimetic NPs, which complicates large-scale production, regulation, and characterization. Encouragingly, research is focusing on safe, FDA-approved materials, but further studies are essential to advance this promising field [94]. Viruses, being naturally immunogenic, have been utilized to develop vaccines and trigger immune responses against pathogens and dysfunctional proteins. Their surface can densely align imaging agents, enabling high-resolution, non-invasive tools for early disease detection and treatment. These advancements have fostered a groundbreaking intersection of nanotechnology and medicine [95]. Moreover, key considerations such as material selection, regulatory compliance, cost-efficiency, and scalability must be thoroughly addressed to achieve widespread adoption. The future of sustainable synthesis appears promising, driven by advancements in advanced materials and the application of circular economy principles to environmental technologies and nanomedicine.

9.7 Conclusion

The synthesis of synthetic NPs is hindered by challenges such as hazardous byproducts, instability, high costs, and environmental concerns. Green synthesis, utilizing sustainable and biodegradable materials, has emerged as a promising alternative, offering benefits like biocompatibility, cost-effectiveness, and environmental sustainability. Metal NPs produced via green methods are particularly appealing due to their unique physicochemical properties and potential applications. This approach is cost-efficient, eco-friendly, and stable, aligning with the goals of sustainable development by employing renewable resources and recyclable processes. Virus-mediated NP synthesis has gained significant attention due to the nanoscale properties of viruses (20–500 nm). Plant viruses, which are non-toxic to humans, are widely utilized in NP fabrication, with their capsid proteins providing reactive binding sites for metal ions. Although animal viruses also show promise in NP synthesis, safety concerns limit their exploration. Viruses offer unique advantages, including high NP yield, biosimilarity, and genetic manipulability for desired properties, making them ideal for green chemistry applications. Yeast-based NP synthesis further enhances green approaches by delivering higher yields, ease of genetic manipulation, enzyme production, and rapid growth on simple nutrients. These biological systems highlight the potential of biogenic NP synthesis in revolutionizing sustainable nanotechnology and advancing nanotherapeutic applications.

References

1 Mughal B, Zaidi SZJ, Zhang X, Hassan SU. Biogenic nanoparticles: synthesis, characterization and applications. Appl Sci [Internet]. 2021;11(6):2598. Available from: https://www.mdpi.com/2076-3417.
2 Kapoor RT, Salvadori MR, Rafatullah M, Siddiqui MR, Khan MA, Alshareef SA. Exploration of microbial factories for synthesis of nanoparticles – a sustainable approach for bioremediation of environmental contaminants. Front Microbiol [Internet]. 2021:12. Available from: https://www.frontiersin.org/articles/10.3389/fmicb.2021.658294/full.
3 Zielińska A, Carreiró F, Oliveira AM, Neves A, Pires B, Venkatesh DN, et al. Polymeric nanoparticles: production, characterization, toxicology and ecotoxicology. Molecules [Internet]. 2020;25(16):3731. Available from: https://www.mdpi.com/1420--3049/25/16/3731.
4 Soni V, Raizada P, Singh P, Cuong HN, Rangabhashiyan S, Saini A, et al. Sustainable and green trends in using plant extracts for the synthesis of biogenic metal nanoparticles toward environmental and pharmaceutical advances:

a review. Environ Res [Internet]. 2021;202:111622. Available from: https://linkinghub.elsevier.com/retrieve/pii/S0013935121009166.

5 Patil S, Chandrasekaran R. Biogenic nanoparticles: a comprehensive perspective in synthesis, characterization, application and its challenges. J Genet Eng Biotechnol [Internet]. 2020;18(1):67. Available from: https://linkinghub.elsevier.com/retrieve/pii/S1687157X23005322.

6 Raj S, Trivedi R, Soni V. Biogenic synthesis of silver nanoparticles, characterization and their applications – a review. Surfaces [Internet]. 2021;5(1):67–90. Available from: https://www.mdpi.com/2571--9637/5/13.

7 Hebbalalu D, Lalley J, Nadagouda MN, Varma RS. Greener techniques for the synthesis of silver nanoparticles using plant extracts, enzymes, bacteria, biodegradable polymers, and microwaves. ACS Sustain Chem Eng [Internet]. 2013;1(7):703–712. Available from: https://pubs.acs.org/doi/10.1021/sc4000362.

8 Begum SJP, Pratibha S, Rawat JM, Venugopal D, Sahu P, Gowda A, et al. Recent advances in green synthesis, characterization, and applications of bioactive metallic nanoparticles. Pharmaceuticals [Internet]. 2022;15(4):455. Available from: https://www.mdpi.com/1424--8247/15/4/455.

9 Puri A, Mohite P, Maitra S, Subramaniyan V, Kumarasamy V, Uti DE, et al. From nature to nanotechnology: the interplay of traditional medicine, green chemistry, and biogenic metallic phytonanoparticles in modern healthcare innovation and sustainability. Biomed Pharmacother [Internet]. 2024;170:116083. Available from: https://linkinghub.elsevier.com/retrieve/pii/S0753332223018814.

10 Kulkarni D, Sherkar R, Shirsathe C, Sonwane R, Varpe N, Shelke S, et al. Biofabrication of nanoparticles: sources, synthesis, and biomedical applications. Front Bioeng Biotechnol [Internet]. 2023:11. Available from: https://www.frontiersin.org/articles/10.3389/fbioe.2023.1159193/full.

11 Mathivanan K, Selva R, Chandirika JU, Govindarajan RK, Srinivasan R, Annadurai G, et al. Biologically synthesized silver nanoparticles against pathogenic bacteria: synthesis, calcination and characterization. Biocatal Agric Biotechnol [Internet]. 2019;22:101373. Available from: https://linkinghub.elsevier.com/retrieve/pii/S1878818119312885.

12 Markus J, Mathiyalagan R, Kim Y-J, Abbai R, Singh P, Ahn S, et al. Intracellular synthesis of gold nanoparticles with antioxidant activity by probiotic *L. kimchicus* DCY51 T isolated from Korean kimchi. Enzym Microb Technol [Internet]. 2016;95:85–93. Available from: https://linkinghub.elsevier.com/retrieve/pii/S0141022916301697.

13 Parveen K, Banse V, Ledwani L. Green synthesis of nanoparticles: their advantages and disadvantages. AIP Conf Proc. 2016;1724:020048. Available from: https://pubs.aip.org/aip/acp/article/580977.

14 Pandit C, Roy A, Ghotekar S, Khusro A, Islam MN, Bin ET, et al. Biological agents for synthesis of nanoparticles and their applications. J King Saud Univ Sci [Internet]. 2022;34(3):101869. Available from: https://linkinghub.elsevier.com/retrieve/pii/S1018364722000507.

15 Mandhata CP, Sahoo CR, Padhy RN. Biomedical applications of biosynthesized gold nanoparticles from cyanobacteria: an overview. Biol Trace Elem Res [Internet]. 2022;200(12):5307–5327. Available from: https://link.springer.com/10.1007/s12011-021-03078-2.

16 Franke CE, Czapar AE, Patel RB, Steinmetz NF. Tobacco mosaic virus-delivered cisplatin restores efficacy in platinum-resistant ovarian cancer cells. Mol Pharm [Internet]. 2018;15(8):2922–2931. Available from: https://pubs.acs.org/doi/10.1021/acs.molpharmaceut.7b00466.

17 Jeong CK, Kim I, Park K-I, Oh MH, Paik H, Hwang G-T, et al. Virus-directed design of a Flexible BaTiO3 nanogenerator. ACS Nano [internet]. 2013;7(12):11016–11025. Available from: https://pubs.acs.org/doi/10.1021/nn404659d.

18 Chung YH, Cai H, Steinmetz NF. Viral nanoparticles for drug delivery, imaging, immunotherapy, and theranostic applications. Adv Drug Deliv Rev [Internet]. 2020;156:214–235. Available from: https://linkinghub.elsevier.com/retrieve/pii/S0169409X20300703.

19 Cai H, Shukla S, Steinmetz NF. The antitumor efficacy of CpG oligonucleotides is improved by encapsulation in plant virus-like particles. Adv Funct Mater [Internet]. 2020;30(15). Available from: https://onlinelibrary.wiley.com/doi/10.1002/adfm.201908743.

20 Pokorski JK, Steinmetz NF. The art of engineering viral nanoparticles. Mol Pharm [Internet]. 2011;8(1):29–43. Available from: https://pubs.acs.org/doi/10.1021/mp100225y.

21 Sharma J, Uchida M, Miettinen HM, Douglas T. Modular interior loading and exterior decoration of a virus-like particle. Nanoscale [Internet]. 2017;9(29):10420–10430. Available from: https://xlink.rsc.org/?DOI=C7NR03018E.

22 von Loga IS, El-Turabi A, Jostins L, Miotla-Zarebska J, Mackay-Alderson J, Zeltins A, et al. Active immunization targeting nerve growth factor attenuates chronic pain behavior in murine osteoarthritis. Ann Rheum Dis [Internet]. 2019;78(5):672–675. Available from: https://ard.bmj.com/lookup/doi/10.1136/annrheumdis-2018-214489.

23 Gao D, Lin X-P, Zhang Z-P, Li W, Men D, Zhang X-E, et al. Intracellular cargo delivery by virus capsid protein-based vehicles: from nano to micro nanomedicine nanotechnology. Biol Med [Internet]. 2016;12(2):365–376. Available from: https://linkinghub.elsevier.com/retrieve/pii/S1549963415005985.

24 Kim KR, Lee AS, Kim SM, Heo HR, Kim CS. Virus-like nanoparticles as a theranostic platform for cancer. Front Bioeng Biotechnol [Internet]. 2023:10. Available from: https://www.frontiersin.org/articles/10.3389/fbioe.2022.1106767/full.

25 Ghosh D, Lee Y, Thomas S, Kohli AG, Yun DS, Belcher AM, et al. M13-templated magnetic nanoparticles for targeted *in vivo* imaging of prostate cancer. Nat Nanotechnol [Internet]. 2012;7(10):677–682. Available from: https://www.nature.com/articles/nnano.2012.146.

26 Wu Z, Chen K, Yildiz I, Dirksen A, Fischer R, Dawson PE, et al. Development of viral nanoparticles for efficient intracellular delivery. Nanoscale [Internet]. 2012;4(11):3567. Available from: https://xlink.rsc.org/?DOI=c2nr30366c.

27 Katzourakis A, Gifford RJ. Endogenous viral elements in animal genomes. Malik HS, editor. PLoS Genet [Internet]. 2010;6(11):e1001191. Available from: https://dx.plos.org/10.1371/journal.pgen.1001191.

28 Villagomez-Bernabe B, Chan SW, Coulter JA, Roseman AM, Currell FJ. Fast ion-beam inactivation of viruses, where radiation track structure meets RNA structural biology. Radiat Res [Internet]. 2022;198(1). Available from: https://bioone.org/journals/radiation-research/volume-198/issue-1/RADE-21-00133.1/Fast-Ion-Beam-Inactivation-of-Viruses-Where-Radiation-Track-Structure/10.1667/RADE-21-00133.1.full.

29 Iglesias SM, Hou C-FD, Reid J, Schauer E, Geier R, Soriaga A, et al. Cryo-EM analysis of pseudomonas phage Pa193 structural components. Commun Biol [Internet]. 2024;7(1):1275. Available from: https://www.nature.com/articles/s42003--024-06985-x.

30 Iravani S, Zolfaghari B. Plant viruses and bacteriophages for eco-friendly synthesis of nanoparticles: recent trends and important challenges. Comments Inorg Chem [Internet]. 2022;42(4):226–248. Available from: https://www.tandfonline.com/doi/full/10.1080/02603594.2021.1993837.

31 Hartzell EJ, Lieser RM, Sullivan MO, Chen W. Modular hepatitis B virus-like particle platform for biosensing and drug delivery. ACS Nano [Internet]. 2020;14(10):12642–12651. Available from: https://pubs.acs.org/doi/10.1021/acsnano.9b08756.

32 Aljabali AA, Hassan SS, Pabari RM, Shahcheraghi SH, Mishra V, Charbe NB, et al. The viral capsid as novel nanomaterials for drug delivery. Future Sci OA [Internet]. 2021;7(9). Available from: https://www.tandfonline.com/doi/full/10.2144/fsoa-2021-0031.

33 White HE, Sherman MB, Brasilès S, Jacquet E, Seavers P, Tavares P, et al. Capsid structure and its stability at the late stages of bacteriophage SPP1 assembly. J Virol [Internet]. 2012;86(12):6768–6777. Available from: https://journals.asm.org/doi/10.1128/JVI.00412-12.

34 Pitek AS, Wang Y, Gulati S, Gao H, Stewart PL, Simon DI, et al. Elongated plant virus-based nanoparticles for enhanced delivery of thrombolytic therapies. Mol Pharm [Internet]. 2017;14(11):3815–3823. Available from: https://pubs.acs.org/doi/10.1021/acs.molpharmaceut.7b00559.

35 Fuertes MA, López Mateos D, Valiente L, Rodríguez Huete A, Valbuena A, Mateu MG. Electrostatic screening, acidic pH and macromolecular crowding increase the self-assembly efficiency of the minute virus of mice capsid *in vitro*. Viruses [Internet]. 2023;15(5):1054. Available from: https://www.mdpi.com/1999--4915/15/5/1054.

36 Garmann RF, Goldfain AM, Tanimoto CR, Beren CE, Vasquez FF, Villarreal DA, et al. Single-particle studies of the effects of RNA–protein interactions on the self-assembly of RNA virus particles. Proc Natl Acad Sci [Internet]. 2022;119(39). Available from: https://pnas.org/doi/full/10.1073/pnas.2206292119.

37 Flynn CE, Lee S-W, Peelle BR, Belcher AM. Viruses as vehicles for growth, organization and assembly of materials11The Golden Jubilee issue – selected topics in materials science and engineering: past, present and future, edited by S. Suresh. Acta Mater [Internet]. 2003;51(19):5867–5880. Available from: https://linkinghub.elsevier.com/retrieve/pii/S1359645403005123.

38 Ashley CE, Carnes EC, Phillips GK, Durfee PN, Buley MD, Lino CA, et al. Cell-specific delivery of diverse cargos by bacteriophage MS2 virus-like particles. ACS Nano [Internet]. 2011;5(7):5729–5745. Available from: https://pubs.acs.org/doi/10.1021/nn201397z.

39 Loo L, Guenther RH, Lommel SA, Franzen S. Encapsidation of nanoparticles by red clover necrotic mosaic virus. J Am Chem Soc [Internet]. 2007; 129 (36): 11111–11117. Available from: https://pubs.acs.org/doi/10.1021/ja071896b.

40 Lockney DM, Guenther RN, Loo L, Overton W, Antonelli R, Clark J, et al. The red clover necrotic mosaic virus capsid as a multifunctional cell targeting plant viral nanoparticle. Bioconjug Chem [Internet]. 2011;22(1):67–73. Available from: https://pubs.acs.org/doi/10.1021/bc100361z.

41 Zeng Q, Wen H, Wen Q, Chen X, Wang Y, Xuan W, et al. Cucumber mosaic virus as drug delivery vehicle for doxorubicin. Biomaterials [Internet]. 2013;34(19):4632–4642. Available from: https://linkinghub.elsevier.com/retrieve/pii/S0142961213003141.

42 Lavelle L, Michel J-P, Gingery M. The disassembly, reassembly and stability of CCMV protein capsids. J Virol Methods [Internet]. 2007;146(1–2): 311–316. Available from: https://linkinghub.elsevier.com/retrieve/pii/S0166093407002868.

43 Lin Y, Schuphan J, Dickmeis C, Buhl EM, Commandeur U, Fischer H. Attachment of ultralow amount of engineered plant viral nanoparticles to mesenchymal stem cells enhances Osteogenesis and mineralization. Adv Healthc Mater [Internet]. 2020;9(21). Available from: https://onlinelibrary.wiley.com/doi/10.1002/adhm.202001245.

44 Zhou K, Wang Q. Nanowires and nanoparticle chains inside tubular viral templates. Methods Mol Biol. 2018;1776:215–227. Available from: http://link.springer.com/10.1007/978-1-4939-7808-3_14.

45 Wu Z, Zhou J, Nkanga CI, Jin Z, He T, Borum RM, et al. One-step supramolecular multifunctional coating on plant virus nanoparticles for bioimaging and therapeutic applications. ACS Appl Mater Interfaces [Internet]. 2022;14(11):13692–13702. Available from: https://pubs.acs.org/doi/10.1021/acsami.1c22690.

46 Khalil AT, Ovais M, Iqbal J, Ali A, Ayaz M, Abbas M, et al. Microbes-mediated synthesis strategies of metal nanoparticles and their potential role in cancer therapeutics. Semin Cancer Biol [Internet]. 2022;86:693–705. Available from: https://linkinghub.elsevier.com/retrieve/pii/S1044579X21001760.

47 Šebesta M, Vojtková H, Cyprichová V, Ingle AP, Urík M, Kolenčík M. Mycosynthesis of metal-containing nanoparticles – fungal metal resistance and mechanisms of synthesis. Int J Mol Sci [Internet]. 2022;23(22):14084. Available from: https://www.mdpi.com/1422--0067/23/22/14084.

48 Nargis J, Khan N, Khan K, Bibi N, Ullah F, Meenhaz A, et al. *Monotropa hypopitys* mediated metal (Ag, Ni and Cu) nanoparticles in microbial inhibition and mercury (ii) ion detection. Pak J Bot [Internet]. 2022;54(2):743–750. Available from: http://pakbs.org/pjbot/paper_details.php?id=11169.

49 Abdelsattar AS, Gouda SM, Hassan YY, Farouk WM, Makky S, Nasr A, et al. *in vitro* bacteriophage-mediated synthesis of silver nanoparticles for antibacterial applications and heavy metal detection. Mater Lett [Internet]. 2022;318:132184. Available from: https://linkinghub.elsevier.com/retrieve/pii/S0167577X22005377.

50 Agudelo W, Montoya Y, Garcia-Garcia A, Bustamante J. Concentration and pH influence on the reducing property of silk fibroin from silk fibrous waste in the green synthesis of silver nanoparticles. Ezzat a, editor. J Nanotechnol [Internet]. 2024;2024(1). Available from: https://onlinelibrary.wiley.com/doi/10.1155/2024/3781764.

51 Durán-Meza AL, Escamilla-Ruiz MI, Segovia-González XF, Villagrana-Escareño MV, Vega-Acosta JR, Ruiz-Garcia J. Encapsidation of different Plasmonic gold nanoparticles by the CCMV CP. Molecules [Internet]. 2020;25(11):2628. Available from: https://www.mdpi.com/1420--3049/25/11/2628.

52 Beiss V, Mao C, Fiering SN, Steinmetz NF. Cowpea mosaic virus outperforms other members of the Secoviridae as in situ vaccine for cancer immunotherapy. Mol Pharm [Internet]. 2022;19(5):1573–1585. Available from: https://pubs.acs.org/doi/10.1021/acs.molpharmaceut.2c00058.

53 Nithin BR, Bhuyar P, Maniam GP, Rahim MHA, Govindan N. Environment friendly approach for plant mediated green biosynthesis of gold nanoparticles and their modern applications in biomedical aspects – an updated report. BioNanoScience [Internet]. 2023;13(4):1517–1540. Available from: https://link.springer.com/10.1007/s12668-023-01134-w.

54 Velázquez-Lam E, Tome-Amat J, Segrelles C, Yuste-Calvo C, Asensio S, Peral J, et al. Antitumor applications of polyphenol-conjugated turnip mosaic virus-derived nanoparticles. Nanomedicine [Internet]. 2022;17(14):999–1012. Available from: https://www.tandfonline.com/doi/full/10.2217/nnm-2022-0067.

55 Ahsan T. Biofabrication of silver nanoparticles from *Pseudomonas fluorescens* to control tobacco mosaic virus. Egypt J Biol Pest Control [Internet]. 2020;30(1):66. Available from: https://cjbpc.springeropen.com/articles/10.1186/s41938-020-00268-3.

56 Karczewska-Golec J, Sadowska K, Golec P, Karczewski J, Węgrzyn G. Engineered M13-derived bacteriophages capable of gold nanoparticle synthesis and Nanogold manipulations. Int J Mol Sci [Internet]. 2024;25(20):11222. Available from: https://www.mdpi.com/1422--0067/25/20/11222.

57 Mejía-Méndez JL, Vazquez-Duhalt R, Hernández LR, Sánchez-Arreola E, Bach H. Virus-like particles: fundamentals and biomedical applications. Int J Mol Sci [Internet]. 2022;23(15):8579. Available from: https://www.mdpi.com/1422--0067/23/15/8579.

58 Ventola CL. The antibiotic resistance crisis: part 1: causes and threats. P T [Internet]. 2015;40(4):277–283. Available from: http://www.ncbi.nlm.nih.gov/pubmed/25859123.

59 Yacoby I, Shamis M, Bar H, Shabat D, Benhar I. Targeting antibacterial agents by using drug-carrying filamentous bacteriophages. Antimicrob Agents Chemother [Internet]. 2006;50(6):2087–2097. Available from: https://journals.asm.org/doi/10.1128/AAC.00169-06.

60 Crooke SN, Schimer J, Raji I, Wu B, Oyelere AK, Finn MG. Lung tissue delivery of virus-like particles mediated by macrolide antibiotics. Mol Pharm [Internet]. 2019;16(7):2947–2955. Available from: https://pubs.acs.org/doi/10.1021/acs.molpharmaceut.9b00180.

61 Tian Y, Zhou M, Shi H, Gao S, Xie G, Zhu M, et al. Integration of cell-penetrating peptides with rod-like bionanoparticles: virus-inspired gene-silencing technology. Nano Lett [Internet]. 2018;18(9):5453–5460. Available from: https://pubs.acs.org/doi/10.1021/acs.nanolett.8b01805.

62 Pang H-H, Huang C-Y, Chou Y-W, Lin C-J, Zhou Z-L, Shiue Y-L, et al. Bioengineering fluorescent virus-like particle/RNAi nanocomplexes act synergistically with temozolomide to eradicate brain tumors. Nanoscale

[Internet]. 2019;11(17):8102–8109. Available from: https://xlink.rsc.org/?DOI=C9NR01247H.

63 Liu Q, Jin C, Wang Y, Fang X, Zhang X, Chen Z, et al. Aptamer-conjugated nanomaterials for specific cancer cell recognition and targeted cancer therapy. NPG Asia Mater [Internet]. 2014;6(4):e95–e95. Available from: https://www.nature.com/articles/am201412.

64 Finbloom JA, Aanei IL, Bernard JM, Klass SH, Elledge SK, Han K, et al. Evaluation of three morphologically distinct virus-like particles as Nanocarriers for convection-enhanced drug delivery to glioblastoma. Nanomaterials [Internet]. 2018;8(12):1007. Available from: https://www.mdpi.com/2079–4991/12 August 1007. doi:10.3390/nano8121007.

65 Bachmann MF, Jennings GT. Vaccine delivery: a matter of size, geometry, kinetics and molecular patterns. Nat Rev Immunol [Internet]. 2010;10(11): 787–796. Available from: https://www.nature.com/articles/nri2868.

66 Farkas ME, Aanei IL, Behrens CR, Tong GJ, Murphy ST, O'Neil JP, et al. PET imaging and biodistribution of chemically modified bacteriophage MS2. Mol Pharm [Internet]. 2013;10(1):69–76. Available from: https://pubs.acs.org/doi/10.1021/mp3003754.

67 Benjamin CE, Chen Z, Kang P, Wilson BA, Li N, Nielsen SO, et al. Site-selective nucleation and size control of gold nanoparticle Photothermal antennae on the pore structures of a virus. J Am Chem Soc [Internet]. 2018;140(49):17226–17233. Available from: https://pubs.acs.org/doi/10.1021/jacs.8b10446.

68 Depta PN, Dosta M, Wenzel W, Kozlowska M, Heinrich S. Hierarchical coarse-grained strategy for macromolecular self-assembly: application to hepatitis B virus-like particles. Int J Mol Sci [Internet]. 2022;23(23):14699. Available from: https://www.mdpi.com/1422--0067/23/23/14699.

69 Ikwuagwu B, Tullman-Ercek D. Virus-like particles for drug delivery: a review of methods and applications. Curr Opin Biotechnol [Internet]. 2022;78:102785. Available from: https://linkinghub.elsevier.com/retrieve/pii/S0958166922001197.

70 Truchado DA, Rincón S, Zurita L, Sánchez F, Ponz F. Isopeptide bonding in planta allows functionalization of elongated flexuous Proteinaceous viral nanoparticles, including non-viable constructs by other means. Viruses [Internet]. 2023;15(2):375. Available from: https://www.mdpi.com/1999-4915/15/2/375.

71 Zanganeh S, Doroudian M, Nowzari ZR, Nasirmoghadas P, Nahand JS, Sepehrpour S, et al. Viral nanoparticles-mediated delivery of therapeutic cargo. Adv Ther [Internet]. 2023;6(10). Available from: https://onlinelibrary.wiley.com/doi/10.1002/adtp.202300082.

72Årataitong K, Huault S, Chotwiwatthanakun C, Jariyapong P, Thongsum O, Chawiwithaya C, et al. Chimeric virus-like particles (VLPs) designed from shrimp nodavirus (MrNV) capsid protein specifically target EGFR-positive human colorectal cancer cells. Sci Rep [Internet]. 2021;11(1):16579. Available from: https://www.nature.com/articles/s41598--021-95891-x.

73 Yur D, Sullivan MO, Chen W. Highly modular hepatitis B virus-like nanocarriers for therapeutic protein encapsulation and targeted delivery to triple negative breast cancer cells. J Mater Chem B [Internet]. 2023;11(18):3985–3993. Available from: https://xlink.rsc.org/?DOI=D3TB00445G.

74 Steinmetz NF. Viral nanoparticles as platforms for next-generation therapeutics and imaging devices. Nanomedicine [Internet]. 2010;6(5):634–641. Available from: http://www.ncbi.nlm.nih.gov/pubmed/20433947.

75 Shanmugam V, Selvakumar S, Yeh C—S. Near-infrared light-responsive nanomaterials in cancer therapeutics. Chem Soc Rev [Internet]. 2014;43(17):6254–6287. Available from: https://xlink.rsc.org/?DOI=C4CS00011K.

76 Alkilany AM, Thompson LB, Boulos SP, Sisco PN, Murphy CJ. Gold nanorods: their potential for photothermal therapeutics and drug delivery, tempered by the complexity of their biological interactions. Adv Drug Deliv Rev [Internet]. 2012;64(2):190–199. Available from: https://linkinghub.elsevier.com/retrieve/pii/S0169409X11000500.

77 Rhee J-K, Baksh M, Nycholat C, Paulson JC, Kitagishi H, Finn MG. Glycan-targeted virus-like nanoparticles for photodynamic therapy. Biomacromolecules [Internet]. 2012;13(8):2333–2338. Available from: https://pubs.acs.org/doi/10.1021/bm300578p.

78 Chauhan K, Hernandez-Meza JM, Rodríguez-Hernández AG, Juarez-Moreno K, Sengar P, Vazquez-Duhalt R. Multifunctionalized biocatalytic P22 nanoreactor for combinatory treatment of ER+ breast cancer. J Nanobiotechnol [Internet]. 2018;16(1):17. Available from: https://jnanobiotechnology.biomedcentral.com/articles/10.1186/s12951-018-0345-2.

79 Hu H, Yang Q, Baroni S, Yang H, Aime S, Steinmetz NF. Polydopamine-decorated tobacco mosaic virus for photoacoustic/magnetic resonance bimodal imaging and photothermal cancer therapy. Nanoscale [Internet]. 2019;11(19):9760–9768. Available from: https://xlink.rsc.org/?DOI=C9NR02065A.

80 Wang Y, Black KCL, Luehmann H, Li W, Zhang Y, Cai X, et al. Comparison study of gold nanohexapods, nanorods, and nanocages for photothermal cancer treatment. ACS Nano [Internet]. 2013;7(3):2068–2077. Available from: https://pubs.acs.org/doi/10.1021/nn304332s.

81 Vankayala R, Huang Y, Kalluru P, Chiang C, Hwang KC. First demonstration of gold nanorods-mediated photodynamic therapeutic destruction of Tumors via near infra-red light activation. Small [Internet]. 2014;10(8):1612–1622. Available from: https://onlinelibrary.wiley.com/doi/10.1002/smll.201302719.

82 Hessel CM, Pattani VP, Rasch M, Panthani MG, Koo B, Tunnell JW, et al. Copper selenide nanocrystals for photothermal therapy. Nano Lett [Internet]. 2011;11(6):2560–2566. Available from: https://pubs.acs.org/doi/10.1021/nl201400z.

83 Zhou Z, Sun Y, Shen J, Wei J, Yu C, Kong B, et al. Iron/iron oxide core/shell nanoparticles for magnetic targeting MRI and near-infrared photothermal therapy. Biomaterials [Internet]. 2014;35(26):7470–7478. Available from: https://linkinghub.elsevier.com/retrieve/pii/S014296121400461X.

84 Zhang Q, Shan W, Ai C, Chen Z, Zhou T, Lv X, et al. Construction of multifunctional Fe 3 O 4 -MTX@HBc nanoparticles for MR imaging and Photothermal therapy/chemotherapy. Nanotheranostics [Internet]. 2018;2(1):87–95. Available from: http://www.ntno.org/v02p0087.htm.

85 Shan W, Chen R, Zhang Q, Zhao J, Chen B, Zhou X, et al. Improved stable Indocyanine green (ICG)-mediated cancer Optotheranostics with naturalized hepatitis B Core particles. Adv Mater [Internet]. 2018;30(28). Available from: https://onlinelibrary.wiley.com/doi/10.1002/adma.201707567.

86 Jia Q, Li D, Zhang Q, Ye S, Xi Z, Wang X, et al. Biomineralization synthesis of HBc-CuS nanoparticles for near-infrared light-guided photothermal therapy. J Mater Sci [Internet]. 2019;54(20):13255–13264. Available from: http://link.springer.com/10.1007/s10853-019-03613-6.

87 Wang Z, Huang P, Jacobson O, Wang Z, Liu Y, Lin L, et al. Biomineralization-inspired synthesis of copper Sulfide–ferritin Nanocages as cancer Theranostics. ACS Nano [Internet]. 2016;10(3):3453–3460. Available from: https://pubs.acs.org/doi/10.1021/acsnano.5b07521.

88 Han J, Zeng Q, Chen K, Yu X, Dai J. Lattice thermal conductivity of monolayer InSe calculated by machine learning potential. Nanomaterials [Internet]. 2023;13(9):1576. Available from: https://www.mdpi.com/2079--4991/13/9/1576.

89 Maharramov AM, Hasanova UA, Suleymanova IA, Osmanova GE, Hajiyeva NE. The engineered nanoparticles in food chain: potential toxicity and effects. SN Appl Sci [Internet]. 2019;1(11):1362. Available from: http://link.springer.com/10.1007/s42452-019-1412-5.

90 Vietti G, Lison D, van den Brule S. Mechanisms of lung fibrosis induced by carbon nanotubes: toward an adverse outcome pathway (AOP). Part Fiber Toxicol [Internet]. 2015;13(1):11. Available from: http://particleandfibretoxicology.biomedcentral.com/articles/10.1186/s12989-016-0123-y.

91 Osman AI, Zhang Y, Farghali M, Rashwan AK, Eltaweil AS, Abd El-Monaem EM, et al. Synthesis of green nanoparticles for energy, biomedical, environmental, agricultural, and food applications: a review. Environ Chem Lett [Internet].

2024;22(2):841–887. Available from: https://link.springer.com/10.1007/s10311-023-01682-3.

92 Martínez G, Merinero M, Pérez-Aranda M, Pérez-Soriano E, Ortiz T, Villamor E, et al. Environmental impact of nanoparticles' application as an emerging technology: a review. Materials (Basel) [Internet]. 2020;14(1):166. Available from: https://www.mdpi.com/1996--1944/14/1/166.

93 Singh NA, Narang J, Garg D, Jain V, Payasi D, Suleman S, et al. Nanoparticles synthesis via microorganisms and their prospective applications in agriculture. Plant Nano Biol [Internet]. 2023;5:100047. Available from: https://linkinghub.elsevier.com/retrieve/pii/S2773111123000244.

94 Maslanka Figueroa S, Fleischmann D, Goepferich A. Biomedical nanoparticle design: what we can learn from viruses. J Control Release [Internet]. 2021;329:552–569. Available from: https://linkinghub.elsevier.com/retrieve/pii/S0168365920305617.

95 Koudelka KJ, Pitek AS, Manchester M, Steinmetz NF. Virus-based nanoparticles as versatile Nanomachines. Annu Rev Virol [Internet]. 2015;2(1):379–401. Available from: https://www.annualreviews.org/doi/10.1146/annurev-virology-100114-055141.

10

Synthesis of Biogenic Nanoparticles Using Onion and Biogenic Nanomedicine

Roohi Kesharwani[1], Shahnawaz Shameem[1], and Dilip K. Patel[2]

[1]*Faculty of Pharmacy, United University, Prayagraj, Uttar Pradesh, India*
[2]*Department of Pharmacy, Government Polytechnic, Jaunpur, Uttar Pradesh, India*

10.1 Introduction

One of the most active research fields in the contemporary material sciences is nanotechnology. A significant area of current research is nanotechnology, which deals with the production of bioinspired nanoparticles (NPs), strategy, and structural manipulation of particles with sizes ranging from around 1 to 100 nm. The field of nanotechnology is advancing daily, influencing every aspect of human existence and generating increasing interest in the biological sciences. In the design, development, and use of chemical processes and products, green chemistry, which seeks to minimize or remove compounds harmful to human health and the environment, has gained increasing attention in recent years [1, 2]. Because NPs have specific chemical, optical, and mechanical capabilities, their application is becoming more and more significant in the twenty-first century. The prospective uses of metal NPs in biomedicine, photonics, optics, antibacterial activity, and catalysis make them significant [3–5]. By customizing materials at the atomic level, nanotechnology is anticipated to create new ways to combat and prevent illness. Indeed, this is a difficult area of study with countless opportunities for the future. One significant symbiotic relationship between nanotechnology and green chemistry is the use of phytochemicals in the creation of NPs. Generally speaking, biosynthesis, also known as green synthesis, uses a variety of plant parts, such as leaves, roots, and fruits [6, 7]. It is crucial to establish nano-nature links between nanotechnology and environmentally friendly areas of the world as the nano revolution progresses. Numerous biomedical applications have made use of NPs of metals such as gold, silver, copper, platinum, iridium, osmium, and palladium [8–10]. To lessen inflammation, nonsteroidal anti-inflammatory

Biogenic Synthesis of Nanoparticles and Sustainable Nanotherapeutics,
First Edition. Edited by Malay K. Das.

medications (NSAIDs) come in a variety of forms. The alternate use of medicinal herbs can help mitigate the adverse effects of these anti-inflammatory NSAIDs. Secondary metabolites from plants have been employed extensively for decades in a variety of medicinal applications, including cancer, neurological illnesses, and cardiovascular ailments. Numerous studies have documented the anti-inflammatory and antioxidant properties of secondary metabolites found in plants, as well as their use in the management of arthritis [11–13]. Continuous and chronic inflammatory responses brought on by the fast production of superoxide radicals and excess phagocyte activation are the cause of cardiovascular illnesses, malignancies, and a number of autoimmune disorders. The plant metabolite content of onions (*Allium cepa* [AC]), a multipurpose vegetable, has recently piqued the interest of researchers due to a number of epidemiological studies showing that eating onions lowers the risk of diabetes, cancer, and a number of neurological illnesses [14, 15].

10.1.1 Onion (*Allium Cepa*)

The perennial herb AC, commonly referred to as onions, has a stem in the underground bulb. AC was one of the first therapeutic herbs to be cultivated worldwide. Onions have numerous health benefits, including anticarcinogenic, antioxidant, hypolipidemic, and hypoglycaemic properties, due to their high phytocompound content. The plant and many sections are rich in organo-sulfur compounds, which are a great source of antioxidants, and are packed with dietary flavonoids and phytochemicals [16, 17]. The sugar content of onions is around 90% water, and they have a considerable amount of dietary fiber. Vegetable-rich diets have been linked to several health benefits, including the prevention of two of the more common and pertinent diseases of our day. Onions are rich in potassium, selenium, and many vitamins (B2, C, and B1). They can treat stomach cancer, cardiovascular diseases, and diabetes mellitus. Keloid prevention and biologically effective hypertrophic scarring are revealed by onion peel [17, 18].

10.1.2 Morphological and Chemical Characteristics of *Onion*

Although some writers refer to onions as Alliaceae, they actually belong to the Liliaceae family. Red, white, or yellow, onions are a perennial crop that can be consumed raw, cooked, pickled, or powdered when still tender. Typically, the plants have tiny flowers and are either purple or white. One or two leafless flower stalks that are 75–180 cm (2.5–6 ft) tall are characteristic of common onions. The fibrous adventitious root system is present. A cluster of progressive leaves gives way to the underground bulb, which is fleshy, cylindrical, and has a sheathy leaf foundation with a parallel venation [15, 17, 18] (Table 10.1).

Table 10.1 Various parts of the onion with pharmacological activity.

Plant part used	Pharmacological category	Biological action
Bulb	Antioxidant	Free radical scavenger [19]
Bulb	Antimicrobial	Effective against *Escherichia coli*, *Staphylococcus aureus*, and *Pseudomonas aeruginosa* [20]
Bulb	Antidiabetic	Act by enhancing insulin secretion [21]
Skin	Anticancer	Inhibiting the proliferation of cancerous cells [22]
Skin	Anti-inflammatory	Inflammatory cytokines suppression [23]
Leaf	Antiviral	Suppress respiratory viruses [24]
Leaf	Cardioprotective	Improves lipid profiles and reduces cholesterol and blood pressure [25]
Seed	Hepatoprotective	Protects hepatic tissues against oxidative stress [26]

Biologically active substances such as flavonoids, phenolics, and anthocyanins are abundant in onions. They are mostly composed of saponins, quercetin, and anthocyanins. Among other flavonoids, quercetin can change a number of metabolic pathways and is crucial in inhibiting lipid peroxidation and scavenging oxygen radicals. Dried onion peels contain important phenolic components, including kaempferol, gallic acid, ferulic acid, protocatechuic acid, and quercetin. With 20 times the amount of quercetin and quercetin glycosides compared to the edible portion, onion peel is regarded as a waste product. Based on *in vitro* antioxidant activities, the previous data indicates that the outer layer of red onions has the strongest antioxidant activities compared to other onion kinds and may be used to treat stress and cancer [27, 28]. This incredible plant is also rich in calcium, magnesium, phosphorus, potassium, sodium, vitamin B6, and folic acid. Onions are pungent because they contain a lot of sulfur compounds that burn the back of the mouth and throat. A useful method for determining the pungency of onions is the analysis of pyruvic acid, which is produced in stoichiometric proportions by thiosulfins. The carbohydrate, lipid, protein, and polyphenol moieties make up the majority of the onion extract and phytochemical profile. Additionally, quercetin and quercetin and β-glucopyranoside contributed to the creation of metal oxide NPs. Strong degree of polymerization (DP) fructans found in onion bulbs can be used to replace fats, which have positive health effects. Compared to other vegetables, onions have a greater soluble/insoluble nutritional fiber ratio (SDF:IDF), which is associated with a number of physiological and metabolic impacts. Methiin, *S*-alk(en)yl-L-cys-teine derivatives, isoalliin, alliin, deoxyalliin, cycloalliin, and *N*-(gamma-glutamyl) are among the 11 sulfur-containing flavor precursors found in onions, according to Yamazaki et al.-*S*-(2-propenyl)-*N*-(gamma-glutamyl), L-cysteine. The compounds include *S*-methyl-L-cysteine and *N*-gamma-glutamyl-*S*-(2-propenyl) sulfoxide of L-cysteine, *N*-(gamma-glutamyl)-*S*-(*E*-1-propenyl)*N*-(gamma-glutamyl), *S*-(2-carboxypropyl) glutathione, and L-cysteine (Glu-PEC)-*S*-(*E*-1-propenyl)-Glu-PECSO (L-cysteine sulfoxide) [14, 28, 29] (Table 10.2).

10.2 Approaches of Nanoparticle Synthesis

NPs and nanocrystals have found many uses in the realm of medicine in contemporary material science. Controlling the size, shape, and surrounding medium of NPs has long been of obvious interest because it is well known that these factors greatly influence the electrical and optical properties of the particles [36, 37]. NPs can be synthesized using a variety of physical, chemical, and biological techniques. Numerous NPs with distinct sizes and shapes are produced using physical synthesis techniques. Nevertheless, the drawbacks of this approach include the generation of significant waste materials, high energy consumption, synthesis time, and achieving thermal stability of NPs [29, 38, 39] (Figure 10.1).

Chemical synthesis techniques are complex and have a variety of consequences, including the production of waste and a large number of hazardous substances. These substances are bad for human health as well as the environment. In order to create NPs, environmentally safe materials such as extracts from various plant parts and microorganisms are used instead of hazardous chemicals. Numerous advantages, environmental friendliness, and compatibility for pharmaceutical and other biological applications are provided by this type of green nanoparticle manufacturing. It has long been known that the biogenic creation of metal NPs inhibits the growth of microorganisms found in industrial and medical processes [15, 41, 42]. Numerous microorganisms have been studied for their ability to create various metal NPs, including silver (Ag), gold, zinc, palladium, magnesium, copper, iron, lead, titanium, and others. These include bacteria, actinomycetes, fungi, and viruses. In the physical world, two methods are primarily employed: the self-assembly method (bottom-up) and the destructive approach (top-down) (Figure 10.2). Although the synthesis principles of these two methods vary, they both ultimately result in NPs with the desired properties. The most popular method for creating silver nanoparticles (AgNPs) via a bottom-up strategy is chemical reduction. Silver ions (Ag^+) are reduced in aqueous or nonaqueous

Table 10.2 Various onion-mediated biosynthesized nanoparticles and their biomedical application [8, 17, 30–35]

Onion part	Type of nanoparticles	Mechanism	Activity
Allium cepa Bulb extract	Gold (Au) NPs	Au^{3+} gets reduced to Au^0 thanks to the onion extract, which also keeps the gold nanoparticles from aggregating. This process may allow the gold nanoparticles to self-assemble into a stable colloidal suspension.	Anticancer
Allium cepa Bulb extract	Silver (Ag) NPs	Flavonoids and phenolic compounds act as reducing agents that transform silver ions (Ag^+) into elemental silver nanoparticles (Ag^0). The extract is rich in natural compounds that facilitate this process.	Antifungal
Allium cepa Bulb extract	Silver (Ag) NPs	As reducing agents, flavonoids and phenolic chemicals convert Ag^+ into elemental silver (Ag^0) nanoparticles. The extract contains natural chemicals that catalyze the process.	Antioxidant
Allium cepa Bulb extract	Gold (Au) NPs	Au^{3+} is reduced to Au^0 by the onion extract, which also stabilizes the AuNPs and stops them from clumping together. The process might entail the gold nanoparticles self-assembling into a stable suspension of colloids.	Anticancer
Allium cepa Peel extract	Zinc oxide (ZnO) NPs	The phytochemicals included in the extract from onion peels, mainly flavonoids, phenols, and organic acids, convert zinc ions (Zn^{2+}) into ZnO nanoparticles. When exposed to light or microbial cells, ZnO NPs can produce reactive oxygen species (ROS), including superoxide, hydroxyl radicals, and hydrogen peroxide. This is the primary cause of their antibacterial action.	Antimicrobial
Allium cepa skin extract	Iron oxide (Fe_3O_4) NPs	Natural components of the extract, such as flavonoids and other organic compounds, facilitate the synthesis of Fe_3O_4 nanoparticles by acting as reducing agents for Fe^{3+} to Fe^{2+}.	Magnetic properties
Allium cepa Bulb extract	Copper oxide (CuO) NPs	CuO nanoparticles are formed when elemental copper (Cu^0) undergoes oxidation after being converted from Cu^{2+} ions by the bioactive compounds in the onion extract acting as reducing agents. In addition to preventing nanoparticle agglomeration, the onion extract's flavonoids and phenols aid in lowering copper ions.	Antifungal

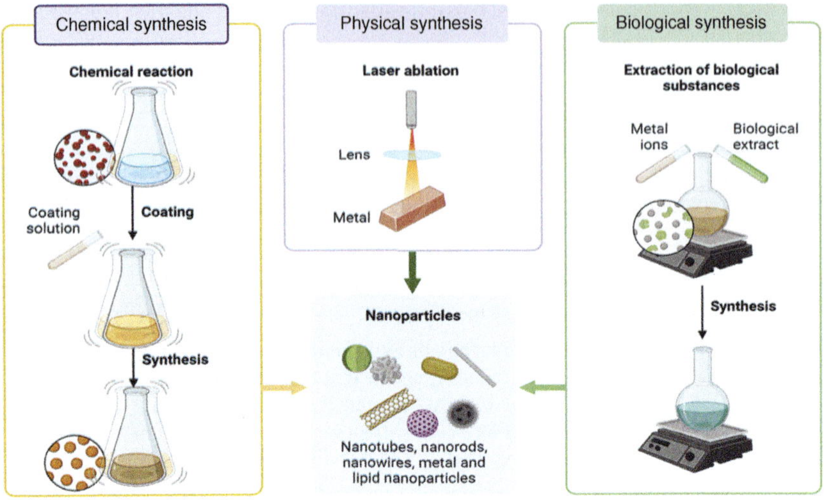

Figure 10.1 Approaches of nanoparticles synthesis [40]. *Source:* Created using BioRender.

solutions using a variety of organic and inorganic reducing agents, including sodium borohydride (NaBH4), sodium citrate, ascorbate, elemental hydrogen, Tollen's reagent, *N*,*N*-dimethyl formamide (DMF), and poly (ethylene glycol) block copolymers [8, 43]. Additionally, capping agents are utilized to stabilize the NP size. The ability to generate a huge number of NPs quickly is one of this method's greatest advantages. The chemicals utilized in these kinds of syntheses are hazardous and produce byproducts that are not environmentally friendly. This could be the cause of the green biosynthesis of NPs, which avoids the use of hazardous chemicals and is therefore proving to be an increasing desire to create environmentally friendly procedures. As a result, the development of green syntheses of NPs is advancing as a major area of nanotechnology; using biological entities such as microorganisms, plant extract, or plant biomass to produce NPs could be an environmentally friendly substitute for chemical and physical methods [44, 45]. In contrast to the top-down method, the bottom-up technique starts at the atomic level by creating molecules. The bottom-up approaches are implemented in a variety of ways, including pyrolysis, spinning, sol–gel, and chemical vapor deposition [9, 43, 46]. When using a top-to-bottom method, bulk materials were broken up into tiny fragments, which resulted in the production of fine NPs. Milling, sputtering, grinding, and photolithographic procedures were used to produce this type of NP. The most common method for creating metallic NPs is evaporation-condensation in a tube furnace at air pressure. In this synthesis, the foundation material is vaporized into a carrier gas inside a boat

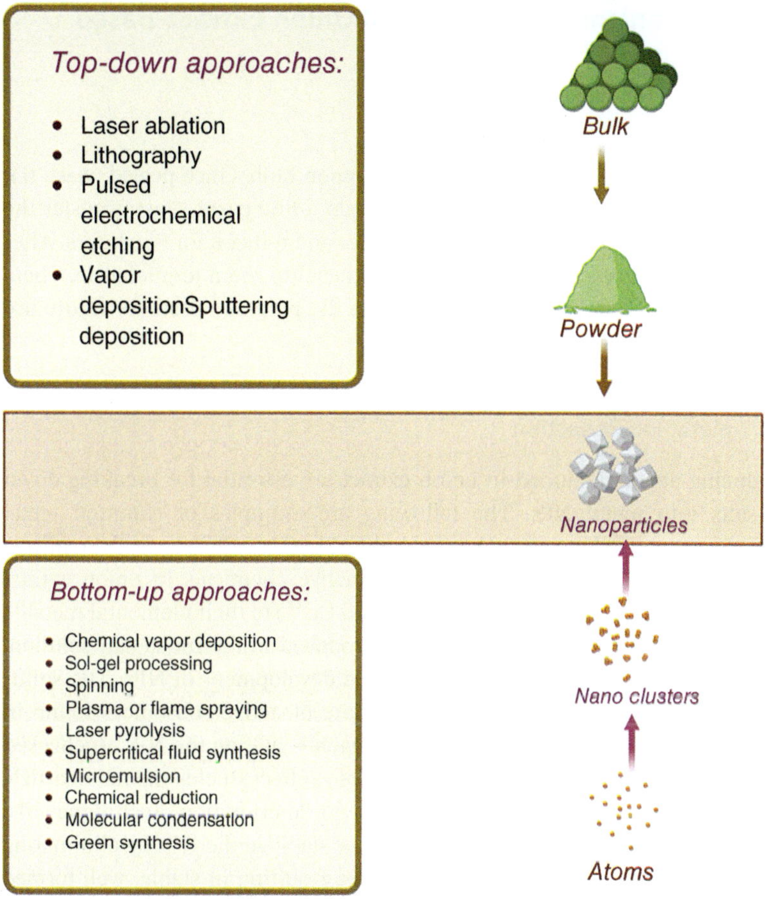

Figure 10.2 Summary of the top-down and bottom-up approaches to green synthesizing nanoparticles [40]. *Source:* Created using BioRender.

that is centered at the furnace. Evaporation/condensation has been used to create Ag, Au, PbS, and fullerene NPs in the past. Every technique has its own setbacks and NP output capability. The top-down strategy is a very practical method that produces a significant amount of NPs. However, top-down approaches have drawbacks, such as surface imperfections in NPs, the possibility of NP damage, and higher nanoparticle dimensions, necessitating the use of mechanical techniques or the addition of acids to reduce the particle size. Complex analysis (thermal decomposition method, mechanical method/ball-milling method, lithographic methods, laser ablation, sputtering) is typically needed for the top-down approach [14, 39, 47–49].

10.3 Mechanism for Synthesis of Onion Extract-Based Nanoparticles

10.3.1 Extraction Method

First, begin by peeling the outer layer of the onion bulb. Once peeled, wash the onion bulb with distilled water and grind it into a fine paste. Next, transfer the paste into a round-bottom flask filled with water and reflux it for two hours. After this time has elapsed, allow the onion paste to cool to room temperature. Then, centrifuge the paste at 5000 rpm. Finally, store the prepared paste for future use [12, 49, 50].

10.3.2 Metal Ion Reduction

The reducing agents included in onion extract are essential for breaking down metal salts into metal NPs. The following are examples of common metal precursors: copper sulfate ($CuSO_4$), gold chloride ($HAuCl_4$), and silver nitrate ($AgNO_3$) for silver semiconductors. The bioactive chemicals in onion extract decrease the metal ions (such as Ag^+, Au^{3+}, and Cu^{2+}) to their elemental metallic form (Ag^0, Au^0, and Cu^0) when they are introduced to the metal salt solution. A color shift frequently follows, signifying the development of NPs. The sulfur containing chemicals and flavonoids, which are bioactive substances found in onion extract, serve as capping and reducing agents, halting the NPs' unchecked growth and aggregation. Additionally, the capping effect shields the nanoparticle from external elements, including oxidation and deterioration. Techniques like centrifugation or dialysis can be used to further stabilize the NPs by eliminating excess onion extract and unreacted metal ions, resulting in stable, well-formed NPs [46].

Additionally, an *in vitro* study showed that onion peel extract inhibits the response to lipopolysaccharide in human colon cancer cells, and that crude onion powder inhibits the activation of nuclear factor kappa-B (NF-κB) in RAW264.7 cells by the receptor activator of nuclear factor kappa-B ligand. It can be inferred from earlier research that fruits and vegetables contain NPs in large quantities. In this study, we postulated that onions may be used to create NPs with anti-inflammatory qualities [12, 46].

Prioritizing the sustainable use of natural resources, the crude extract from the entire onion peel was separated into four fractionates in this work: ethyl acetate (EA), butanol, methanol, and water. The produced fractionates underwent a variety of antioxidant testing. The fractionate of EA had the most antioxidant activity among them. The anti-inflammatory properties of the EA fractionate and the pure component quercetin were examined. The EA fractionate was also used

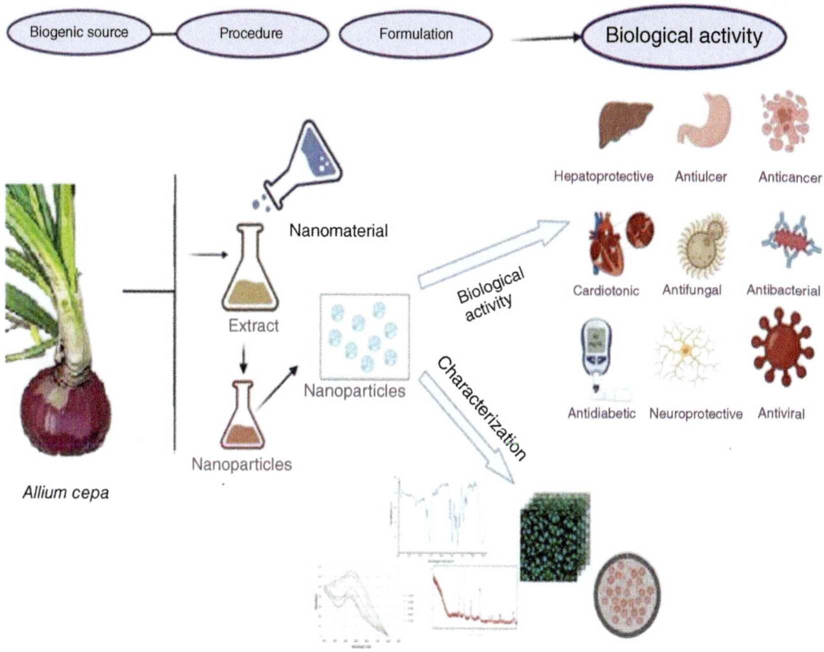

Figure 10.3 Schematic representation of onion-mediated biogenic synthesis of nanoparticles and their application [40]. *Source:* Created using BioRender.

to create gold nano-bioconjugates, which were then tested for antioxidant and anti-inflammatory properties by contrasting them with the crude onion peel drug (OPD) to validate the enhanced anti-inflammatory properties of the alloys [29] (Figure 10.3).

10.4 Synthesis of Different Types Nanoparticle Containing Onion Extract

Using onion (*A. cepa*) extract to create NPs is an eco-friendly and sustainable process that makes use of onions' abundant phytochemical makeup. Without the use of hazardous chemicals, this method transforms metal salts into stable NPs by using the reducing agents and natural antioxidants found in onion extract, such as flavonoids, phenolic acids, and sulfur compounds. To improve the performance of the produced NPs for specific applications and obtain the necessary nanoparticle properties, it is imperative to optimize their size, shape, and stability [34, 51–53].

10.4.1 Silver Nanoparticle

Since AgNPs exhibit strong antibacterial, antioxidant, and catalytic qualities, they have garnered a lot of study interest. Onion (*A. cepa*) extract facilitates the green synthesis of AgNPs by utilizing the rich phytochemical profile of onions, which contains sulfur compounds, phenolic acids, and flavonoids. When creating NPs, these substances function as organic stabilizing and reducing agents, offering a sustainable and environmentally beneficial substitute for traditional chemical processes. A 1:1 mixture of the onion extract and $AgNO_3$ solution was made. At room temperature, stirring the reaction mixture caused the color to shift from pale yellow to brown, signifying the creation of AgNP. To guarantee total reduction, the solution was kept undisturbed for a full day. It has therapeutic uses in anti-inflammatory and wound-healing therapies [17, 27, 52, 54–56].

10.4.2 Gold Nanoparticle

Due to their exceptional optical characteristics, excellent biocompatibility, and potential for use in targeted medication administration and diagnostics, gold nanoparticles, or AuNPs, are well known. Using onion extract's antioxidant qualities to stabilize NPs, onion-mediated production of AuNPs provides a sustainable method. The onion extract was mixed 1:1 with a solution of chloroauric acid ($HAuCl_4$). When the solution was agitated at room temperature, the hue turned ruby red, signifying the production of AuNPs. For full synthesis, the reaction mixture was incubated for 24 hours. Applications comprise targeted delivery of medicinal agents, biosensing, and imaging applications, including drug delivery and cancer therapy [2, 52, 57–60].

10.4.3 ZnO Nanoparticle

Several studies have been conducted on zinc oxide nanoparticles (ZnO NPs) because of their antibacterial, ultraviolet (UV)-protective, and photocatalytic qualities. In order to transform zinc salts into stable NPs, onion extract uses its abundant antioxidant and reducing components to create ZnO NPs in an environmentally friendly and effective manner. Onion extract and zinc acetate solution were combined in a 1:1 ratio. A white precipitate formed as a result of heating the mixture to 60–80 °C while stirring constantly. For additional characterization, the precipitate was gathered, cleaned, and dried. Utilized as antibacterial and antifungal compositions, sunscreens, and UV-protective coatings [10, 28, 41, 61–63].

10.4.4 Cu Nanoparticle

The electrical, catalytic, and antibacterial qualities of copper nanoparticles (Cu NPs) are well recognized. By using onion extract as a green synthesis method, the cost and environmental toxicity are reduced. To the onion extract, a solution of CuSO4 was added. When the solution was agitated, a hue shift that indicated the production of NPs occurred. For analysis, the product was gathered, cleaned, and dried. It serves as a wound dressing and an antimicrobial [51, 64–67].

10.4.5 Key Characteristics and Attributes of Biogenic Nanoparticles

Biogenic NPs show various properties as shown in Figure 10.4.

10.4.6 Benefits Associated with Onion-Mediated Biogenic Nanoparticles

Green synthesis is an environmentally safe, economical, and easily scalable method for large-scale nanoparticle production, offering an improvement over chemical and physical approaches. Additionally, it eliminates the need for hazardous chemicals, high temperatures, pressures, and energy. However, because microbe-mediated synthesis necessitates extremely aseptic conditions and their upkeep, it is not an industrially viable way among the other biological approaches for producing AgNPs. Therefore, because of their ease of improvement, lower biohazard, and simpler cell culture maintenance method, plant extracts may be

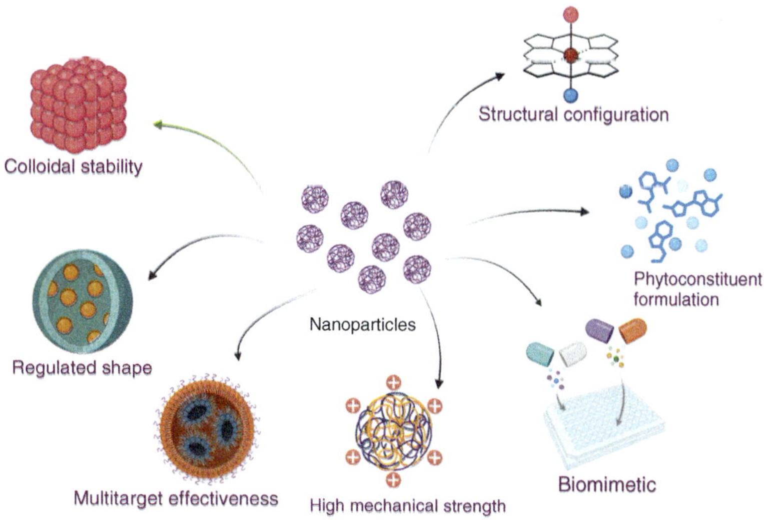

Figure 10.4 Key characteristics and attributes of biogenic nanoparticles [40]. *Source:* Created using BioRender.

Table 10.3 Benefits of onion-mediated biogenic nanoparticle.

Category	Advantage	Details
Eco-friendly	Green-synthesis	Avoids hazardous chemicals and decreases its impact on the environment by using components that are biological.
Cost-effective	Low-cost Manufacturing	Since it uses readily available natural resources, production costs are minimized.
Biocompatible	Safe for biological activity	Minimizes cytotoxicity and is compatible with both human and animal cells.
Stable	Enhanced physical and chemical stability	Biogenic nanoparticles are persistent across an array of environmental conditions.
Multifunctional	Rich in bioactive compounds	Since biomolecules have been incorporated, they can combine medicinal, antibacterial, and catalytic capabilities.
Targeted drug delivery	High drug delivery specificity	Functionalized with biological ligands to ensure effective delivery to the targeted sites.
Scalable	Feasible for large-scale synthesis	Capable of being created utilizing basic biological processes on large scales.
Least toxic	Minimal side effects in therapeutic applications	When compared to chemical approaches, biogenic synthesis minimizes the occurrence of hazardous chemical leftovers.

a better option for this purpose than microorganisms. In Table 10.3, the several benefits of biogenic generated NPs are displayed [2, 14, 68–70].

10.4.7 Contributing Factors to Nanoparticle Biosynthesis

The biosynthesis of NPs using onion (AC) extract is influenced by several critical factors that impact the size, shape, stability, and yield of the NPs. These factors include temperature of reaction, pH of reaction mixture, concentration of onion extract, reaction time, and phytochemical composition of onion extract might affect the stability of biogenic NPs.

10.4.8 Effect of Temperature

The kinetics of nanoparticle production are affected by temperature: For the production of gold and AgNPs with good control over size and shape, room temperature (25–30 °C) is adequate. The reaction rate for ZnO and Cu NPs is increased by using elevated temperatures (60–80 °C), which improves yield and crystallinity. Aggregation and polydispersity can be brought on by high temperatures [41, 42].

10.4.9 Effect of pH

One important aspect affecting the stability and size of NPs is pH. Larger and irregular NPs are the result of slower reduction rates caused by an acidic pH. pH 7–8, a range from neutral to slightly alkaline, encourages ideal reduction and even size distribution. Rapid decrease may result from a highly alkaline pH (>9), although unchecked growth and aggregation can also occur. Although the pH of the process can be maintained by the natural buffering agents found in onion extract, some NPs may require further modification [38, 71].

10.4.10 Effect of Concentration of Onion Extract

Flavonoids, phenolics, and sulfur compounds are examples of reducing and stabilizing agents that are dependent on the amount and concentration of onion extract. Increased onion extract concentration speeds up the reduction process and improves nanoparticle stability by halting aggregation. According to Sharma et al., inadequate onion extract can result in partial reduction and decreased yield [52].

10.4.11 Effect of Reaction Time

The amount of metal ion reduction and nanoparticle formation is dependent on the reaction time. Incomplete reduction may result from shorter reaction times, but secondary growth and aggregation may be caused by reaction times that are too long. The ideal incubation period for full biosynthesis and acceptable stability is typically 24 hours [3, 31].

10.4.12 Effect of Phytochemical Composition of Onion Extract

The high phytochemical content of onion extract is responsible for its efficacy in the creation of NPs: Flavonoids (kaempferol and quercetin) are potent reducing agents. Phenolic acids assist with stability and capping. Compounds of sulfur protect against aggregation and improve reduction. During synthesis, vitamins C and E function as antioxidants.

The phytochemical composition of the extract may be somewhat impacted by seasonal variation and onion variety, which could reduce the effectiveness of synthesis [30, 72, 73].

10.4.13 Application of Onion Mediated Biosynthesized Nanoparticles

10.4.13.1 Anticancer Activity

AgNPs and AuNPs, in particular, are onion-mediated biosynthesized NPs that have demonstrated great promise in cancer treatment because of their capacity to trigger apoptosis (programmed cell death) and prevent the growth of cancer cells. AgNPs cause mitochondrial malfunction and oxidative stress, which damages cell membranes. Apoptotic pathways are activated and DNA breakage occurs when ROS levels rise.

Targeted delivery via AuNPs minimizes damage to healthy tissues by improving selectivity toward cancer cells when functionalized with targeting ligands. For example, AgNPs synthesized using onion extract induced apoptosis in Michigan Cancer Foundation – 7 (MCF-7) breast cancer cell lines, while AuNPs showed cytotoxicity against HepG2 liver cancer cells through reactive oxygen species (ROS)-mediated pathways [15, 74, 75].

10.4.13.2 Antibacterial Activity

Significant antibacterial qualities have been demonstrated by onion-mediated biosynthesized NPs, specifically AgNPs, Cu NPs, and ZnO NPs. Onion extract, which has natural reducing and capping agents such as flavonoids, phenolic acids, and sulfur compounds that strengthen their antibacterial effect, is used to create these NPs. Through their interactions with the bacterial cell wall, AgNPs and Cu NPs cause structural damage. Within the cell membrane, the NPs attach to proteins and DNA that contain phosphorus and sulfur, increasing permeability and allowing cellular contents to seep out. When metal ions from NPs attach to bacterial proteins and enzymes, denaturation and functional loss result. Cell viability and metabolic processes are disrupted. The chemicals quercetin, kaempferol, and sulfur found in onion extract strengthen the antibacterial properties of NPs. These phytochemicals decrease virulence and the development of resistance by blocking bacterial quorum sensing [28, 49].

10.4.13.3 Antioxidant Activity

The inclusion of phenolic chemicals and flavonoids, which function as natural reducing and stabilizing agents during synthesis, gives onion-mediated NPs their strong antioxidant qualities. Quercetin and kaempferol, two phytochemicals found in onion extract, neutralize free radicals and lessen oxidative stress. The expression of antioxidant enzymes such as catalase and superoxide dismutase is increased by NPs, which improves cellular defense. Biomolecule protection involves halting the oxidation of proteins and lipids in biological systems. For example, AgNPs prevent oxidative damage in cell membranes, and ZnO NPs synthesized with onion [69] extract shield cells from UV-induced oxidative stress [12, 21, 76].

10.4.13.4 Antidiabetic Activity

By altering glucose metabolism and improving insulin sensitivity, onion-mediated biosynthesized NPs--particularly ZnO NPs--have demonstrated encouraging antidiabetic benefits. By simulating the effects of insulin on cells, ZnO NPs improve glucose absorption. Cellular glucose absorption is enhanced by glucose transporter-4 activation. NPs increase insulin production by modifying pancreatic β-cell function. Regarding the effects of insulin on cells, ZnO NPs have been shown to improve glucose absorption. α-Amylase and α-glucosidase inhibition reduces postprandial blood sugar rises by blocking the conversion of carbohydrates to glucose. For example, ZnO NPs synthesized with onion extract demonstrated effective glucose regulation in diabetic cell models [69, 77, 78].

10.4.13.5 Anti-Inflammatory Activity

Significant anti-inflammatory qualities of onion-mediated NPs, particularly AgNPs and Cu NPs, aid in lowering oxidative stress and inflammation in tissues. NPs inhibit inflammatory mediators by lowering the expression of cytokines such as tumor necrosis factor alpha, interleukin-6, and interleukin-1-beta. Inflammation is managed by blocking the cyclooxygenase and nitric oxide synthase pathways. In inflammatory tissues, antioxidant qualities lessen oxidative damage. AgNPs have been shown to have anti-inflammatory properties in wound dressings, which speeds up the healing process. Cu NPs in rheumatoid arthritis models decrease inflammation [79]. The high exterior-position-to-interior ratio of NPs makes them useful for blocking inflammatory chemicals like cytokines and enzymes that mediate inflammation, as well as other supplements. Effective anti-inflammatory effects have been described for a number of metal-mediated NPs, including those based on iron oxide, copper, silver, and gold [43].

10.4.13.6 Antiviral

By obstructing viral replication and blocking entry into host cells, onion-mediated NPs have shown antiviral properties against a variety of viruses. AgNPs impede viral replication by interacting with its DNA and RNA. By attaching themselves to viral surface proteins, NPs stop the virus from connecting to host cell receptors. Viral envelopes may become unstable due to metal NPs, which would decrease their contagiousness. For instance, AgNPs lower the rate of viral replication by disrupting the particles. Viral proliferation and host cell entrance are inhibited by ZnO NPs. Herpes simplex virus (HSV) cannot adhere to the membranes of host cells when AgNPs are present [17, 18, 73].

10.4.13.7 Antiplatelet Activity

Significant antiplatelet efficacy is demonstrated by onion-mediated biosynthesized NPs, specifically AuNPs, AgNPs, and ZnO NPs. Their disruption of the pathways involved in platelet activation, aggregation, and clot formation is the main cause of this action. These effects are amplified when onion extract contains phytochemicals such as sulfur compounds, flavonoids, and quercetin. This characteristic is essential for avoiding thrombotic events, which are brought on by aberrant platelet aggregation and include heart attacks and strokes [41, 80, 81].

10.5 Conclusion

It is currently believed that NPs are very beneficial compounds. For the synthesis of NPs, the green chemistry technique is a valuable alternative to both physical and chemical approaches because it is non-nanotoxicological, safe, economical, and environmentally friendly. As a result, it attracts the attention of additional specialists for future advancements and uses in the biomedical field, which could encourage the large-scale production of goods based on NPs. For a number of nutraceutical components, onions can be utilized as a safe, natural alternative. The overview of bioinspired, onion-based bioinspired NPs and their synthesis process advantages are presented in the current work. Various sources of manufactured NPs and their medical uses, including antibacterial, antifungal, antiviral, anti-inflammatory, antidiabetic, antioxidant, and other uses, are well discussed in the current review. Therefore, the knowledge in this chapter may help readers comprehend how onions are used to synthesize NPs. It also paves the path for further research into other plants for this purpose. The development of dependable, economical, and environmentally friendly methods for the synthesis of metallic NPs is a crucial necessity in the field of nanotechnology.

References

1 Kesharwani R, Sachan A, Singh S, Patel D. Formulation and evaluation of solid lipid nanoparticle (SLN) based topical gel of etoricoxib. J Appl Pharm Sci. 2016;6(10):124–131.

2 Santhosh PB, Genova J, Chamati H. Review green synthesis of gold nanoparticles: an eco-friendly approach. Chemistry (Switzerland). 2022;4.

3 Dhir R, Chauhan S, Subham P, Kumar S, Sharma P, Shidiki A, Kumar G. Plant-mediated synthesis of silver nanoparticles: unlocking their pharmacological potential – a comprehensive review. Front Bioeng Biotechnol. 2024;11.

4 Patel DK, Kesharwani R, Kumar V. Lipid nanoparticle topical and transdermal delivery: a review on production, penetration mechanism to skin. Int J Pharm Investig. 2019;9(4):148–153.

5 Patil S, Chandrasekaran R. Biogenic nanoparticles: a comprehensive perspective in synthesis, characterization, application and its challenges. J Genet Eng Biotechnol. 2020;18:67.

6 Khalilzadeh MA, Borzoo M. Green synthesis of silver nanoparticles using onion extract and their application for the preparation of a modified electrode for determination of ascorbic acid. J Food Drug Anal. 2016;24(4):796–803.

7 Khan F, Shariq M, Asif M, Siddiqui MA, Malan P, Ahmad F. Green nanotechnology: plant-mediated nanoparticle synthesis and application. Nanomaterials. 2022;12.

8 Khandel P, Yadaw RK, Soni DK, Kanwar L, Shahi SK. Biogenesis of metal nanoparticles and their pharmacological applications: present status and application prospects. J Nanostructure Chem. 2018;8:217–254.

9 Kulkarni D, Sherkar R, Shirsathe C, Sonwane R, Varpe N, Shelke S, et al. Biofabrication of nanoparticles: sources, synthesis, and biomedical applications. Front Bioeng Biotechnol. 2023;11.

10 Sportelli MC, Gaudiuso C, Volpe A, Izzi M, Picca RA, Ancona A, Cioffi N. Biogenic synthesis of ZnO nanoparticles and their application as bioactive agents: a critical overview. Reactions. 2022;3:423–441.

11 Kesharwani R, Jaiswal P, Patel DK, Yadav PK. Lipid-based drug delivery system (LBDDS): an emerging paradigm to enhance oral bioavailability of poorly soluble drugs. Biomed Mater Devices. 2023;1(2):648–663.

12 Mani M, Okla MK, Selvaraj S, Ram Kumar A, Kumaresan S, Muthukumaran A, et al. A novel biogenic *Allium cepa* leaf mediated silver nanoparticles for antimicrobial, antioxidant, and anticancer effects on MCF-7 cell line. Environ Res. 2021;198:111199.

13 Patel DK, Kesharwani R, Kumar V. Etodolac loaded solid lipid nanoparticle based topical gel for enhanced skin delivery. Biocatal Agric Biotechnol. 2020;29:101810.

14 Giri R, Sharma D, Ledwani L. Biogenic synthesis of metallic nanomaterials and its antimicrobial and cytotoxic potential: a review. NanoWorld J. 2022;8:S168–S175.

15 Parida UK, Bindhani BK, Nayak P. Green Synthesis and Characterization of Gold Nanoparticles Using Onion (*Allium cepa*) Extract. World J Nano Sci Eng. 2011;1(4):93–98.

16 Aswathi VP, Meera S, Maria CGA, Nidhin M. Green synthesis of nanoparticles from biodegradable waste extracts and their applications: a critical review. Nanotechnol Environ Eng. 2023;8:377–397.

17 Chakraborty AJ, Uddin TM, Matin Zidan BMR, Mitra S, Das R, Nainu F, et al. *Allium cepa*: a treasure of bioactive phytochemicals with prospective health benefits. Evid Based Complement Alternat Med. 2022;2022:1–27.

18 Masud S, Munir H, Irfan M, Tayyab M. Allium cepa -based zinc oxide nanoparticles: characterization and biochemical potentials. Bioinspired Biomimetic Nanobiomater. 2023;12(2):82–93.

19 Santas J, Almajano MP, Carbó R. Antimicrobial and antioxidant activity of crude onion (*Allium cepa*, L.) extracts. Int J Food Sci Technol. 2010;45(2):403–409.

20 Griffiths G, Trueman L, Crowther T, Thomas B, Smith B. Onions – a global benefit to health. Phytother Res. 2002;16:603–615.

21 Singh S, Verma D, Singh H, Singh N, Singh V. Integrated nutrient management for higher yield, quality and profitability of onion (*Allium cepa*). Indian J Agric Sci. 2015;85(9):1214–1218.

22 Jeong CH, Heo HJ, Choi SG, Shim KH. Antioxidant and anticancer properties of methanolic extracts from different parts of white, yellow, and red onion. Food Sci Biotechnol. 2009;18(1).

23 Slimestad R, Fossen T, Vågen IM. Onions: A source of unique dietary flavonoids. J Agric Food Chem. 2007;55:10067–10080.

24 Akram M, Mandal A, Iqbal M, Mukherjee A, Bandyopadhyay R, Rangasamy S, et al. Inhibitory potential of medicinal plants against viral pathogens and their putative modes of action. Hosts Viruses. 2023;10:10.

25 Roldán-Marín E, Jensen RI, Krath BN, Kristensen M, Poulsen M, Cano MP, et al. An onion byproduct affects plasma lipids in healthy rats. J Agric Food Chem. 2010;58(9):5308–5314.

26 Rai S, Rani P, Kumar M, Rai A, Shahi SK. Effect of integrated nutrient management on nutrients uptake and productivity of onion. Nat Environ Pollut Technol. 2016;15(2).

27 Abhang P, Dadasaheb Mane L, Nanaware J. Green synthesis of silver nanoparticles using *Allium CEPA* extract and their application as fertilizers in agriculture. Bull Pure Appl Sci Z;43.

28 Khamis M, Gouda GA, Nagiub AM. Biosynthesis and antibacterial evaluation of zinc oxide nanoparticles from onion extract (*Allium cepa*). Bull Pharm Sci. 2023;46:955–967.

29 Baran MF, Keskin C, Baran A, Hatipoğlu A, Yildiztekin M, Küçükaydin S, et al. Green synthesis of silver nanoparticles from *Allium cepa* L. peel extract, their antioxidant, antipathogenic, and anticholinesterase activity. Molecules. 2023;28(5).

30 Chaudhary S, Shivalkar S, Sahoo AK. Biosynthesis of Zinc Oxide Nanoparticles and Major Applications. In: Mycosynthesis of Nanomaterials: Perspectives and Challenges. 2023. CRC Press, Boca Raton.

31 Ghosh S, Bloch K, Webster TJ. Functionalized biogenic nanoparticles and their applications. In: Nanobiotechnology: Microbes and Plant Assisted Synthesis of Nanoparticles, Mechanisms and Applications. 2021.

32 Jasrotia T, Chaudhary S, Kaushik A, Kumar R, Chaudhary GR. Green chemistry-assisted synthesis of biocompatible Ag, Cu, and Fe2O3 nanoparticles. Mater Today Chem. 2020;15.

33 Kumar LS, Indirajith P, Tetala KKR. Onion based zinc oxide nanoparticles ability as electrochemical sensor. Indian J Chem Sect A. 2020;59A(8).

34 Naseer A, Iqbal M, Ali S, Nazir A, Abbas M, Ahmad N. Green synthesis of silver nanoparticles using *Allium cepa* extract and their antimicrobial activity evaluation. Chem Int. 2022;8(3).

35 Rajoriya P, Misra P, Shukla PK, Ramteke PW. Light-regulatory effect on the phytosynthesis of silver nanoparticles using aqueous extract of garlic (*Allium sativum*) and onion (*Allium cepa*) bulb. Curr Sci. 2016;111(8):1364.

36 Bhakya S, Muthukrishnan S, Sukumaran M, Muthukumar M. Biogenic synthesis of silver nanoparticles and their antioxidant and antibacterial activity. Appl Nanosci. 2016;6(5).

37 Patel D, Kesharwani R, ProfS G. Development & screening approach for lipid nanoparticle: a review. Asian J Pharm Res Dev. 2015;3(6).

38 Akhtar MS, Panwar J, Yun YS. Biogenic synthesis of metallic nanoparticles by plant extracts. ACS Sustain Chem Eng. 2013;1(6):591 602.

39 Al-Gburi N, Al-Hassnawi A, Al-Bayati LA. Biosynthesis of silver nanoparticles and their roles in the biomedical field: a review. Med J Babylon. 2024; 21:493–499.

40 BioRender. Biorender templates. BioRender. 2023.

41 Ojo OA, Olayide II, Akalabu MC, Ajiboye BO, Ojo AB, Oyinloye BE, et al. Nanoparticles and their biomedical applications. Biointerface Res Appl Chem. 2021;11(1):8431–8445.

42 Vijayaram S, Razafindralambo H, Sun YZ, Vasantharaj S, Ghafarifarsani H, Hoseinifar SH, Raeeszadeh M. Applications of green synthesized metal nanoparticles – a review. Biol Trace Elem Res. 2024;202:360–386.

43 Subbanna S, Ts G, Basalingappa KM. Biogenic nanoparticles from *Allium Sativum* and its bioactives applications. Eur J Mol Clin Med. Available from: https://www.researchgate.net/publication/348659725.

44 Fagier MA. Plant-mediated biosynthesis and photocatalysis activities of zinc oxide nanoparticles: a prospect towards dyes mineralization. J Nanotechnol. 2021;2021:1–15.

45 Nabilah Mohd Noor N, Hazirah Kamaruzaman N, Al-Gheethi A, Mohamed MSR, R, Hossain MS. Degradation of antibiotics in aquaculture wastewater by bio-nanoparticles: a critical review. Ain Shams Eng J. 2023;14:101981.

46 Phukan K, Devi R, Chowdhury D. Green synthesis of gold nano-bioconjugates from onion peel extract and evaluation of their antioxidant, anti-inflammatory, and cytotoxic studies. ACS Omega. 2021;6(28):17811–17823.

47 Kesharwani R, Ansari MS, Patel DK. Novel technology used in the preformulation study: a review. J Drug Deliv Ther. 2017;7(4).

48 Patel DK, Kumar V, Kesharwani R. Lipid nanoparticle topical & transdermal delivery: a review on production, penetration mechanism to skin. Curr Nanomed. 2019;9.

49 Saxena EA, Tripathi RM, Singh RP, Saxena A, Tripathi RM, Singh RP. Synthesis of silver nanoparticles using onion (*Allium cepa*) extract and their antibacterial activity biological synthesis of silver nanoparticles by using onion (*Allium cepa*) extract and their antibacterial activity. Digest J Nanomater Biostruct. 2010;5. Available from: https://www.researchgate.net/publication/279555479.

50 Sagar NA, Pareek S. Antimicrobial assessment of polyphenolic extracts from onion (*Allium cepa* L.) skin of fifteen cultivars by sonication-assisted extraction method. Heliyon. 2020;6(11):e05478.

51 Ezealisiji KM, Nwodo CR. Green synthesis of copper oxide nanoparticles using the outer layer of *Allium cepa* L. and evaluation of its antimicrobial properties. GSC Biol Pharmaceut Sci. 2023;22(1):302–311.

52 Sharma VK, Yngard RA, Lin Y, Sharma VK, Yngard RA, Lin Y. Silver nanoparticles: green synthesis and their antimicrobial activities. Adv Colloid Interf Sci. 2009;145:83–96.

53 Susanti D, Haris MS, Taher M, Khotib J. Natural products-based Metallic nanoparticles as antimicrobial agents. Front Pharmacol. 2022;13.

54 Ahmad N, Sharma S. Green Synthesis of Silver Nanoparticles Using Extracts of <i>*Ananas comosus*</i>. Green Sustain Chem. 2012;2(4).

55 Munir H, Bilal M, Mulla SI, Abbas Khan H, Iqbal HMN. Plant-Mediated Green Synthesis of Nanoparticles. In: Advances in Science, Technology and Innovation. 2021.

56 Vanlalveni C, Lallianrawna S, Biswas A, Selvaraj M, Changmai B, Rokhum SL. Green synthesis of silver nanoparticles using plant extracts and their antimicrobial activities: a review of recent literature. RSC Adv. 2021;11:2804–2837.

57 Acharya P, Jayaprakasha GK, Crosby KM, Jifon JL, Patil BS. Green-synthesized nanoparticles enhanced seedling growth, yield, and quality of onion (*Allium cepa* L.). ACS Sustain Chem Eng. 2019;7(17):14580–14590.

58 Katas H, Moden NZ, Lim CS, Celesistinus T, Chan JY, Ganasan P, Suleman Ismail Abdalla S. Biosynthesis and potential applications of silver and gold nanoparticles and their chitosan-based nanocomposites in nanomedicine. J Nanotechnol. 2018;2018:1–13.

59 Pandit S, Bhowal AC, Kundu S. Green synthesis of silver nanoparticles and its potential applications. In: Applications of Silver Nanoparticles. 2023.

60 Puišo J, Paškevičius A, Žvirgždas J, Dimitrova TL, Litvakas A, Adliene D. Application of red onion peel extract for green synthesis of silver nanoparticles in hydrogels exhibiting antimicrobial properties. Gels. 2023;9(6).

61 Degefa A, Bekele B, Jule LT, Fikadu B, Ramaswamy S, Dwarampudi LP, et al. Green Synthesis, Characterization of Zinc Oxide Nanoparticles, and Examination of Properties for Dye-Sensitive Solar Cells Using Various Vegetable Extracts. J Nanomater. 2021;2021:1–9.

62 Islam MF, Islam S, MAS M, AKO H, Saha AK, Mou ZJ, et al. Green synthesis of zinc oxide nano particles using *Allium cepa* L. waste peel extracts and its antioxidant and antibacterial activities. Heliyon. 2024;10(3):e25430.

63 Letchumanan D, Sok SPM, Ibrahim S, Nagoor NH, Arshad NM. Plant-based biosynthesis of copper/copper oxide nanoparticles: an update on their applications in biomedicine, mechanisms, and toxicity. Biomolecules. 2021;11.

64 Bale VK, Katreddi HR. Green synthesis, characterization and antimicrobial activity of nanosized Cuprous Oxide fabricated using aqueous extracts of *Allium cepa* and *Raphanus sativus*. Int J Nano Dimens. 2022;13(2).

65 Mansi B, Sandesh J, Arun G. Copper nanoparticles using onion (*Allium cepa*) extract and their application in plant growth. Int J Life Sci. 2017;5.

66 Rafique MA, Kiran S, Jamal A, Anjum MN, Jalal F, Munir B, et al. Green synthesis of copper nanoparticles using *Allium cepa* (onion) peels for removal of Disperse Yellow 3 dye. Desalin Water Treat. 2022;272:259–265.

67 Shao H, Zhang X, Huang H, Zhang K, Wang M, Zhang C, et al. Magnetron sputtering deposition Cu@Onion-like N—C as high-performance electrocatalysts for oxygen reduction reaction. ACS Appl Mater Interfaces. 2017;9(48):41945–41954.

68 Aboyewa JA, Sibuyi NRS, Meyer M, Oguntibeju OO. Green synthesis of metallic nanoparticles using some selected medicinal plants from southern Africa and their biological applications. Plants. 2021;10.

69 Singh RP. Potential of biogenic plant-mediated copper and copper oxide nanostructured nanoparticles and their utility. In: Nanotechnology in the Life Sciences. 2019.

70 Some S, Mondal R, Dam P, Mandal AK. Synthesis of biogenic silver nanoparticles using medicinal plant extract: a new age in nanomedicine to combat multidrug-resistant pathogens. In: Green Synthesis of Silver Nanomaterials. 2022. Elsevier.

71 Karunakaran G, Sudha KG, Ali S, Cho EB. Biosynthesis of nanoparticles from various biological sources and its biomedical applications. Molecules. 2023;28.

72 Ahmed S, Ahmad M, Swami BL, Ikram S. A review on plants extract mediated synthesis of silver nanoparticles for antimicrobial applications: a green expertise. J Adv Res. 2016;7:17–28.

73 Ren Y, Zhang Y, Li X. Application of AgNPs in biomedicine: an overview and current trends. Nanotechnol Rev. 2024;13.

74 Kumar V, Kesharwani R, Patel DK, Verma A, Mehanna MG, Mohammad A, et al. Epigenetic impact of curcumin and thymoquinone on cancer therapeutics. Curr Med Chem. 2024;31.

75 Patel DK, Kesharwani R, Tripathy S, Kumar V, Kesharwani S, Das MK. Multifunctional silica nanoparticle as a promising cancer theranostics. In: Multifunctional Theranostic Nanomedicines in Cancer. 2021. Elsevier.

76 Kesharwani R, Patel DK, Verma P, Kumar V. Reactive oxygen species induced hepatic cellular carcinoma: a protective role of nano-antioxidant. Curr Cancer Ther Rev. 2024;20(2):126–134.

77 Jini D, Sharmila S. Green synthesis of silver nanoparticles from *Allium cepa* and its *in vitro* antidiabetic activity. Mater Today Proc. 2020;22:432–438.

78 Shanmugam J, Dhayalan M, Savaas Umar MR, Gopal M, Khan MA, Simal-Gandara J, et al. Green synthesis of silver nanoparticles using *Allium cepa var. Aggregatum* natural extract: antibacterial and cytotoxic properties. Nanomaterials. 2022;12(10).

79 Patel DK, Kesharwani R, Kumar V. Nanoparticles: an emerging platform for medical imaging. In: Nanoparticles in Analytical and Medical Devices. 2021. Elsevier.

80 Goldman IL, Kopelberg M, Debaene JEP, Schwartz BS. Antiplatelet activity in onion (*Allium cepa*) is sulfur dependent. Thromb Haemost. 1996;76(3):450–452.

81 Khan N, Ali S, Latif S, Mehmood A. Biological synthesis of nanoparticles and their applications in sustainable agriculture production. Nat Sci (Irvine). 2022;14(6):226–234.

11

Biogenic Synthesis of Nanoparticles Using Orange and Green Nanomedicine

Bhupen Kalita[1], Malay K. Das[2], Fazid Ali[3], Nihalini Kalita[1], and Md. Latifur Rahman[1]

[1] NEF College of Pharmacy, Guwahati, Assam, India
[2] Drug Delivery Research Laboratory, Department of Pharmaceutical Sciences, Dibrugarh University, Dibrugarh, Assam, India
[3] National Institute of Pharmaceutical Education and Research (NIPER) S. A. S. Nagar, Punjab, India

11.1 Introduction

11.1.1 Biogenic Methods in Nanotechnology and Their Significance

Nanotechnology has revolutionized various scientific fields over the last five decades, offering unprecedented solutions to complex problems. The field of nanomedicine explores the unique properties of nanometer-sized particles, bringing significant advancements in diagnostics, drug delivery, and medical imaging. Among the several methods of nanoparticle synthesis, biogenic synthesis has emerged as a promising eco-friendly alternative. Biogenic nanoparticle synthesis utilizes biological organisms or their derivatives to produce nanoparticles, minimizing environmental impact and enhancing sustainability [1, 2]. These methods commonly incorporate biological entities such as microorganisms, plant parts, and enzymes to produce nanoparticles with desirable properties for medical applications [3, 4]. These organisms act as natural nanofactories, capable of reducing metal ions to form nanoparticles with specific sizes and shapes [5]. The biological approach not only eliminates the need for toxic chemicals and high-energy processes but also often results in nanoparticles that are biocompatible and suitable for various applications [4]. Microorganisms like bacteria, fungi, and algae serve as biological factories for nanoparticle production. These organisms possess intrinsic mechanisms that reduce metal ions to nanoparticles. For instance, bacteria such as *Pseudomonas aeruginosa* and *Lactobacillus* species have been used to synthesize silver and gold nanoparticles with significant

Biogenic Synthesis of Nanoparticles and Sustainable Nanotherapeutics,
First Edition. Edited by Malay K. Das.

antimicrobial properties [6]. Plant-based synthesis involves using plant extracts, which contain natural reducing agents, to convert metal ions into nanoparticles. This method is advantageous due to its simplicity, cost-effectiveness, and scalability. For example, *Azadirachta indica* (neem) and *Camellia sinensis* (tea) extracts have been employed to produce nanoparticles with excellent biocompatibility and therapeutic potential [7, 8]. Enzymes derived from biological sources can catalyze the reduction of metal ions, resulting in nanoparticles. This method offers high specificity and can be fine-tuned to produce nanoparticles with desired sizes and shapes. Enzymatic synthesis is particularly useful for producing nanoparticles for targeted drug delivery and bioimaging [9].

11.1.2 Advantages

Biogenic nanoparticle synthesis offers numerous scientific advantages and has promising industrial applications.

- *Eco-friendly*: In the last few years, researchers have shown interest in the preparation of nanoparticles from the point of view of biological and environmental safety. Biogenic nanoparticles are regarded as eco-friendly because they are derived from biological sources. Most chemicals used in nanoparticle synthesis have deleterious effects on the environment. As far as physical synthesis is concerned, it involves the use of radiation (e.g. microwave, UV light, etc.), high temperature, and pressure, which might be hazardous. The use of biological agents eliminates the need for toxic chemicals and high-energy physical methods. Green synthesis of nanoparticles makes use of natural reducing and stabilizing agents thereby reducing the environmental footprint. The waste generated in the green synthesis process can be decomposed in easier ways without much harm to the environment [10]. Biodegradable nanoparticles synthesized through green methods can be safely degraded and absorbed by the environment, minimizing long-term ecological impacts. This contrasts with non-biodegradable materials that can persist in the environment and cause harm [4]. Green nanomedicine eliminates the use of toxic chemicals in nanoparticle synthesis, thereby reducing hazardous waste production and preventing environmental contamination [11].
- *Cost-effective*: This method utilizes readily available natural resources like plant extracts and microorganisms and often requires less energy compared to traditional methods. The nondependence on a large number of synthetic chemicals, reagents, and high-energy physical processes reduces the overall cost involved in nanoparticle synthesis [12]. Also, the time required for some biogenic synthesis processes is much shorter in comparison to other methods. The synthesis of silver nanoparticles using red alga *Laurencia catarinensis* is

completed within two minutes [13]. High yield is also an additional advantage of the biosynthesis of nanoparticles. By utilizing natural and renewable resources, green nanomedicine promotes sustainable resource use and reduces dependency on finite materials. This aligns with the principles of sustainable development and conservation of natural resources [10].

- *Biocompatibility:* Biogenic nanoparticles are generally more compatible with biological systems, making them ideal for pharmaceutical applications [14]. The idea of biogenic nanoparticle synthesis fits well into the principles of green chemistry, which is the basis of sustainable development. It is usually performed by using various microorganisms such as fungi, bacteria, green algae, and also plants. Biosynthesis can be executed at ambient temperature and pressure, with enhanced bioactivity, and less toxicity. It was demonstrated that the proteins and enzymes secreted by the abovementioned biosystems reduce bulk metal salts to their respective nanoparticles and also act as capping agents, making them bio-compatible for various biological applications. The coating of biological molecules on the surface of NPs makes them biocompatible as opposed to the NPs prepared by chemical methods. The biocompatibility of bioinspired NPs offers very interesting applications in biomedicine and related fields [15]. Biogenic AgNP prepared using *S. cymosum* leaf extract showed excellent hemocompatibility with rat and human RBCs. Biogenic AgNPs also demonstrated excellent *in vivo* biocompatibility in terms of liver and kidney function biomarkers (i.e. serum ALT, AST, ALP, γ-GT, albumin, creatinine, and uric acid) [16]. Another benefit of green nanomedicine is that the use of biocompatible and biodegradable materials reduces the risk of toxicity and adverse reactions in patients [8].

11.2 Biological Precursors

Nature acts like a large "biolaboratory" comprising of plants, algae, fungi, yeast, etc., which are composed of biomolecules. These naturally occurring biomolecules have been identified to play an active role in the formation of nanoparticles with distinct shapes and sizes thereby acting as a driving force in the designing of greener, safe, and environmentally benign protocols for the synthesis of nanoparticles. Various microorganisms, such as bacteria, fungi, and yeast, as well as plant extracts and waste materials have acted as eco-friendly precursors for the synthesis of NPs with potential applications. Algae are also used as a bio-precursor as they have a tremendous ability to bioremediate toxic metals, thereby converting them into more pliable forms. Plant leaf, root, latex, seed, and stem extracts have also been used for the synthesis of NPs as they act as stabilizing or reducing agents [17]. Green nanomedicine has resulted from renewable resources such as plant extracts and microbial metabolites, making it a sustainable choice over conventional methods that often rely on nonrenewable resources [18].

11.2.1 The Role of Citrus Fruits in Nanoparticle Synthesis

Biological precursors for nanoparticle synthesis include plants, algae, bacteria, fungi, and yeast. Orange peel is a common plant source due to its rich phytochemical content. The functional groups in the extracts, such as hydroxyl, carboxyl, and phenyl groups, are responsible for the reduction of metal ions to form nanoparticles. Citrus fruits, known for their rich phytochemical content and antioxidant properties, play a significant role in the green synthesis of nanoparticles [19]. Among them, oranges, lemons, limes, and grapefruits have been particularly explored in depth.

1) *Orange Extracts*: Orange peel contains various phytochemicals such as flavonoids, phenols, and terpenoids, which are crucial in reducing metal ions and stabilizing nanoparticles. Studies have shown that silver nanoparticles synthesized using orange peel extracts exhibit remarkable antioxidant and antitumor properties [8].
2) *Lemon Extracts*: Lemon extracts, rich in ascorbic acid and citric acid, act as powerful reducing agents in the synthesis of metal nanoparticles. These extracts facilitate the formation of nanoparticles with controlled size and shape, offering enhanced stability and biocompatibility. Silver and gold nanoparticles synthesized using lemon extracts have shown excellent antibacterial and anticancer activities [4].
3) *Lime Extracts*: Lime extracts, similar to lemon, are rich in vitamin C and organic acids. These components are effective in reducing metal ions to nanoparticles. Lime-derived nanoparticles have demonstrated significant potential in applications such as drug delivery and environmental remediation [20].
4) *Grapefruit Extracts*: Grapefruit extracts contain bioactive compounds like naringin and limonoids, which are potent antioxidants. These compounds aid in the synthesis of nanoparticles, providing them with unique properties for use in biomedicine and environmental applications. Silver nanoparticles synthesized using grapefruit extracts have shown promising results in antimicrobial assays [21].

11.2.2 Unique Properties of Orange Extract

- *Biological Activities:* The unique phytochemical composition of orange peel seems to be the core reason behind its diverse health benefits. These benefits were found to rely on its antioxidant, antimicrobial, and anti-inflammatory properties [22]. This wonder fruit is beneficial for the skin, helps in losing weight, and improves immune response [23]. Orange peel extracts have been shown to mediate the synthesis of silver nanoparticles (AgNPs) with significant antioxidant and antitumor properties [24]. Among various biogenic nanoparticles mediated by orange peel extract, the cobalt and cobalt oxide nanoparticles have a wide range of biomedical applications such as antioxidant, antimicrobial, anticancer, insecticidal, antiparasitic, antidepressant, wound healing, as well as antidiabetic activities [25].

- *Phytochemical Profile:* The peel contains various phytochemicals like flavonoids, phenols, and terpenoids, soluble and insoluble fiber, as well as proteins that play a crucial role in the reduction of silver ions and stabilization of nanoparticles. The orange peel is reported to contain 7.1% protein, 12.79% crude fiber, and bioflavonoids (27–103 mg/g of dry peel); however, these contents vary depending on factors such as the orange variety, cultivation conditions, and extraction methods [26]. Orange juice and peel are also rich in phenolic compounds, pectin, calcium, vitamin B6, sugars, including sucrose, glucose, and fructose, and folic, citric, malic, and ascorbic acids. Additionally, orange juice contains phenolic substances such as flavanones, hydroxybenzoic acids, hydroxycinnamic acids, ferulic acid, hesperidin, and narirutin [23, 27]. Orange peels are primarily composed of polyphenolic and flavonoid compounds. Prominent among these flavonoids are hesperidin, narirutin, naringin, and eriocitrin [28]. The antioxidant activities of orange peel extract is primarily due to the presence of the hesperidin and naringin glycosides [29]. Also, the flavonoid compounds in orange peel (fractions such as *O*-glycosylated flavanones, *O*-glycosylated flavonols, polymethoxylated flavones, phenolic acids, *O*-glycosylated flavones, and *C*-glycosylated flavones) and their ester derivatives possess considerable antioxidant (DPPH) properties [30]. In addition, orange peel molasses contains coniferin and phlorin, which enhance its radical scavenging capacity [31, 32]. Because of the existence of limonene, a naturally present compound, it also possesses fine anti cancerous attributes. Besides that, the essential oil extracted from orange peel has anti-inflammatory characteristics that boost our immune system [33], reported that saponins, chalcones, and anthraquinones were detected only in the extract prepared by boiling. The other phytochemicals (carbohydrates, reducing sugars, proteins, amino acids, phenols, flavonoids, alkaloids, terpenoids, steroids, and cardiac glycosides) were identified in all extracts, with the concentrations of these phytochemicals recorded for the extract prepared by boiling. However, the extract prepared by blending was almost the same as that prepared by boiling, showing the highest titer of phenols. Therefore, orange peel extract prepared by boiling was proposed as the most promising based on the qualitative screening of phytochemicals.
- *Green Synthesis*: Green chemistry is a judicious scientific strategy in which bioresources are used as a template for the synthesis of nanoparticles. Green

nanomedicine is an emerging field that integrates the principles of green chemistry and nanotechnology. It emphasizes the use of biocompatible and biodegradable materials, renewable resources, and environmentally benign processes for the synthesis of nanoparticles that have demonstrated potential application in medical treatments and diagnostics [34, 35]. The synthesis processes in green nanomedicine avoid the use of toxic solvents and harsh chemicals. This reduces the environmental footprint of nanoparticle production [6]. Utilizing orange peel for nanoparticle synthesis is an eco-friendly and cost-effective method compared to conventional chemical and physical methods [19].

11.3 Mechanisms of Biogenic Nanoparticle Formation

The mechanisms underlying biogenic nanoparticle formation are complex and involve multiple biochemical and molecular pathways. Biomolecules such as proteins, enzymes, and secondary metabolites have important roles in the biogenic synthesis of nanoparticles (Figure 11.1). These molecules act as reducing and stabilizing agents, facilitating the conversion of metal ions into nanoparticles, and ensuring their stability [36]. Biogenic nanoparticle synthesis refers to the natural production of nanoparticles by living organisms, including bacteria, fungi, plants, and algae. The synthesis process is typically driven by complex biochemical mechanisms including the reduction, stabilization, and formation of metal nanoparticles [37]. The following is an overview of the key biochemical processes involved in nanoparticle synthesis.

11.3.1 Reduction of Metal Ions

Reducing metal ions (e.g. Ag^{+}, Au^{3+}, Cu^{2+}) to their elemental nanoparticle forms is a key step in biogenic nanoparticle synthesis. This reduction is typically carried

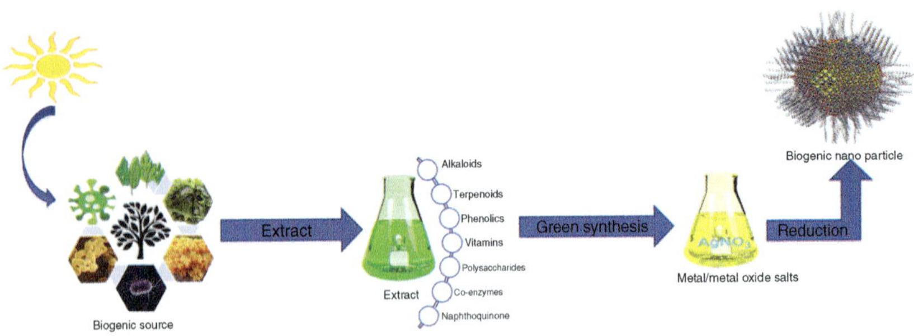

Figure 11.1 Schematic representation of the stages of biogenic nanoparticle synthesis. *Source:* sbtlneet / Pixabay.

out by biological agents such as enzymes, proteins, or reducing agents produced by organisms like bacteria, fungi, or extracts from plant parts [38]. Many organisms produce enzymes that can reduce metal ions to nanoparticles. For example, Pseudomonas species can reduce silver ions (Ag^+) to silver nanoparticles (AgNPs) through the action of reductase enzymes [39].

11.3.2 Biomineralization

In this process, specific proteins, peptides, or polysaccharides serve as templates for nanoparticle formation [40]. Here, specialized proteins like metallothioneins, phytochelatins, or enzymes can bind metal ions, preventing toxic buildup and aiding in the formation of nanoparticles. The metal ions are reduced to their nanoparticulate form, often in highly organized, controlled structures [41]. Biomineralization often leads to the formation of nanoparticles with distinct size distributions, shapes, and crystallinity. For instance, magneto-tactic bacteria biomineralize iron oxide (magnetite) nanoparticles in magnetosomes, using proteins that guide the formation of these nanoparticles [42].

11.3.3 Biosorption and Bioaccumulation

Biosorption and bioaccumulation are the processes by which organisms absorb metal ions from their surroundings, and, depending on the organism, these ions may later be reduced to nanoparticles. While biosorption refers to the adsorption of metal ions onto the surface of cells, bioaccumulation involves the uptake of these ions into cellular compartments, often for storage or detoxification [43]. The metals are typically bound to extracellular or intracellular biomolecules such as proteins, enzymes, peptides, or polysaccharides, which help stabilize the metal ions. In many cases, this binding facilitates the reduction of these ions to nanoparticles [44]. The accumulated metal ions can form nanoparticles inside vacuoles, organelles, or outside the cells, often through the activity of metabolic enzymes. For example, *Brassica juncea* plants bioaccumulate gold ions and form gold nanoparticles in their tissues, which are stabilized by proteins and polysaccharides [45].

11.3.4 Stabilization of Nanoparticles

After the reduction of metal ions to nanoparticles, stabilization is critical to prevent aggregation or uncontrolled growth. Biological organisms often use proteins, lipids, carbohydrates, or secondary metabolites to stabilize and prevent the accumulation of nanoparticles [43]. Some bacteria and fungi excrete extracellular polymeric substances (EPS), which act as capping agents, binding to the nanoparticles and preventing their aggregation. For example, in the synthesis of silver nanoparticles by Fusarium oxysporum, polysaccharides from the cell wall or proteins secreted into the medium act as stabilizing agents. In some cases, peptides (e.g. phytochelatins in plants) or organic acids can serve as capping agents, stabilizing the nanoparticles, and ensuring their dispersion in aqueous solutions [46].

11.3.5 Enzyme-Driven Reduction and Synthesis

Enzymes play a critical role in the enzymatic reduction of metal ions to nanoparticles, often through the transfer of electrons from coenzymes like NADH or NADPH. These enzymes include dehydrogenases, reductases, and oxidases, which facilitate the conversion of metal ions to nanoparticles. In microbial systems, electron transfer often occurs through the cell's redox enzymes. The enzymes involved in nanoparticle synthesis often contain metal-binding sites that provide the necessary interaction with metal ions, allowing for their reduction to nanoparticulate forms [47].

11.3.6 Photochemical Synthesis

Photochemical synthesis involves the use of light energy to drive the reduction of metal ions to nanoparticles. Some microorganisms, such as algae or cyanobacteria, can utilize light energy to activate enzymes that reduce metal ions, resulting in nanoparticle formation. In photosynthetic organisms, light energy is harnessed by chlorophyll or other light-harvesting proteins, which generate high-energy electrons that are used to reduce metal ions. This process may be facilitated by specific enzymes or by the direct interaction of light with metal ions in the medium [48].

11.4 Orange-Derived Nanoparticles and Their Applications

Orange (*Citrus sinensis*; Family Rutaceae) is one of the most common fruits crops whose medicinal and aromatic properties have been known since ancient times. The orange processing beverage industry produces large quantities of waste from orange peels. In recent years, various strategies have been proposed for the extraction and use of the bioactive fractions from orange peel. Antioxidant capacity of the bioactive compounds in orange peel is very well known. The primary metabolites available in orange peel are proteins and carbohydrates and the key secondary metabolites identified in orange peel include phenols, carotenoids, and essential oils. Flavonoids are the highest fraction of phenolic compounds in orange peel. In addition, ascorbic acid is a main phyto-constituent in orange peel with

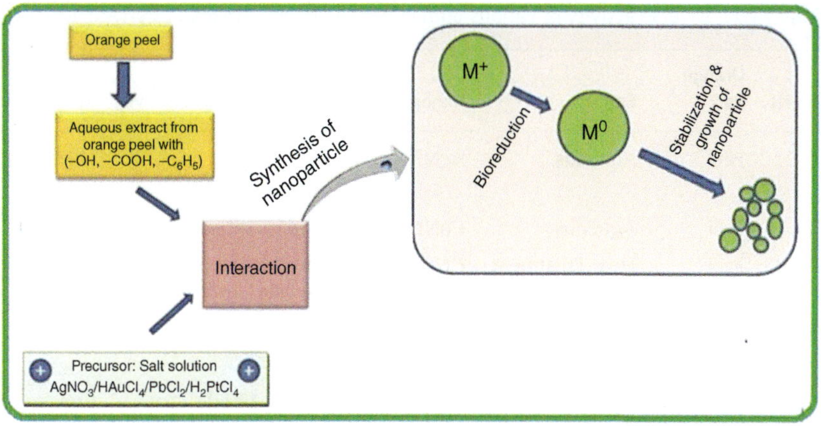

Figure 11.2 Mechanism of synthesis of orange-derived green nanoparticles.

well-documented free radical-scavenging ability, reducing power, and antioxidant activity [19, 49]. Due to the excellent reducing power of the bioactive components of orange peel, its extract has become a choice in the design and development of biogenic nanoparticles (Figure 11.2). The rich phytochemical profile of orange peel allows for its use in the synthesis of various nanoparticles (Table 11.1).

11.4.1 Antimicrobial Applications

1) *Antiseptics and Disinfectants*: The ability of AgNPs to impede bacterial growth make them suitable for application in the creation of topical antiseptics, sanitizers, and disinfectants for wound healing. The antibacterial activity of AgNPs may potentially be strengthened by orange peel extract, which is abundant in bioactive substances like flavonoids and alkaloids [60]. These AgNPs exhibit notable antibacterial properties, making them suitable for incorporation into packaging materials to extend the shelf life of food products [61]; advocated the antimicrobial applications of orange peel-derived CoNPs against common bacterial (*Staphylococcus aureus, Bacillus subtilis, Escherichia coli, Klebsiella pneumonia*) and fungal (*Aspergillus niger*) pathogens. The effectiveness of such AgNPs in inhibiting growth and biofilm formation of *E. coli, S. aureus*, and *Candida albicans* has been investigated. The minimum inhibitory concentration (MIC) for *E. coli* and *S. aureus* was 32 μg/ml for both, and for *C. albicans*, it was 128 μg/ml. At 250 μg/ml of AgNPs, 94% and 92% biofilm inhibition were observed against *E. coli* and *S. aureus*, respectively [62].
2) *Catheter Development*: Silver nanoparticles with antibacterial qualities derived from orange peel may be utilized to stop infections in catheters, which are frequently used in surgical procedures [63]. Such nanoparticles have been

Table 11.1 Diverse activity of various orange-derived nanoparticles.

Sr. No.	Orange Part	Extract	Nanoparticles	Pharmacological Activity	References
1	Peel	Aqueous	AgNP	Antioxidant, Anti-tumor	[19]
2	Peel	Aqueous	CoNP	Antimicrobial Activity	[25]
3	Peel	Heat Treatment to Form Char	CNP	Anti-hemolytic Activity, Antioxidant	[50]
4	Peel	Aqueous	AgNP	Antidiabetic, Anticancer	[51]
5	Peel	Aqueous	ZnNP	Antibacterial,	[52]
6	Peel	Aqueous	AgNP	Anti-platelet, and Wound-healing	[53]
7	Trunk	Soda Pulping	LigninNP	Anti-hepatocarcinoma	[54]
8	Peel	Aqueous	AuNP	Biomedical Application	[55]
9	Peel	Aqueous	SeNP	Antimicrobial Activity Against Methicillin-resistant *Staphylococcus aureus*	[56]
10	Peel	Aqueous	SilicaNP	Antioxidant	[57]
11	Orange Juice	Aqueous	Fe_3O_4NP	Antibacterial	[58]
12	Peel	Aqueous	Ag-Fe-Ni Trimetallic NP	Anticancer, Antioxidant	[59]

proposed as coating agents for medical devices like catheters. Biofilms are a major cause of catheter-associated infections. Coated devices were found to be effective against *S. aureus*, *E. coli*, and *P. aeruginosa*. Furthermore, the slow release of silver ions from the catheter surface offers prolonged protection [62].

11.4.2 Cosmetics and Personal Care Products

1) *Antibacterial Cosmetics*: To improve skin care, AgNPs can be used in the composition of cosmetics such as face creams, sunscreens, and shampoos, as AgNPs prepared using orange peel extract have demonstrated significant antioxidant and antimicrobial efficacy. Studies have demonstrated that silver nanoparticles synthesized from orange peel extracts exhibit significant antibacterial activity

against common pathogens such as *E. coli*, *P. aeruginosa*, and *S. aureus*. These findings highlight the potential of orange-derived biogenic nanoparticles as natural antimicrobial agents in cosmetic formulations, offering a sustainable and effective alternative to synthetic preservatives [64, 65].

2) *Anti-aging Goods*: Silver nanoparticles and bioactive substances from orange peel may be used to develop skin-rejuvenating or anti-aging goods for preserving the youthful appearance of the skin and shielding it from oxidative stress [60]. Orange peels are rich in bioactive compounds such as flavonoids, carotenoids, and ascorbic acid, which contribute to their antioxidant capacity. When these compounds are present as a coating on silver nanoparticles (AgNPs) derived through eco-friendly green synthesis methods, the resultant nanoparticles exhibit notable antioxidant activity. Such properties are beneficial in combating oxidative stress, a key factor in skin aging [19].

11.4.3 Food Industry

1) *Antioxidant Additives*: Because orange peel extract is rich in antioxidants (like flavonoids, limonene), it may be utilized in the food sector as a natural additive to prolong foods' shelf life by preventing oxidative spoiling. ZnONPs synthesized with orange peel extract are used as an edible coating on fresh fruits to delay browning and spoilage [66].
2) *Bioactive Food Packaging*: By combining the bioactive chemicals from orange peel with AgNPs, active food packaging materials that not only inhibit microbial proliferation but also release beneficial substances to extend freshness of the contents can be developed [60]. Orange-derived AgNPs and ZnONPs exhibit potent antimicrobial activity against foodborne pathogens like *E. coli*, *S. aureus*, and *Listeria monocytogenes*. Biodegradable films made of starch or chitosan incorporated with orange peel-mediated AgNPs have been used to extend shelf life and improve food safety in meat and dairy packaging [67]. Orange-derived biogenic nanoparticles have been proposed to encapsulate and protect sensitive nutrients, flavors, or probiotics from degradation. Curcumin or vitamin C encapsulated in NPs synthesized using orange peel extract have demonstrated enhanced bioavailability and controlled release in functional beverages or yogurts [68].

11.4.4 Drug Delivery

1) Biogenic nanoparticles have been found to be highly effective in targeted drug delivery systems. Their biocompatibility and ability to encapsulate therapeutic agents ensure precise delivery to specific sites within the body, minimizing side effects and enhancing therapeutic efficacy [18]. Researchers succeeded in

developing Nanobiochar (NBC), endowed with a graphene structure from the hydrothermal carbonization of orange peels. This NBC was functionalized with targeting ligands like biotin and riboflavin and loaded with a hydrophobic anti-cancer agent DHF (5,5-dimethyl-6*a*-phenyl-3-(trimethylsilyl)-6,6*a*-dihydrofuro [3,2-*b*]furan-2(5*H*)-one). The functionalized NBC demonstrated enhanced uptake by A549 lung cancer cells and induced significant cell death. Notably, the biotin-functionalized NBC exhibited over twice the cytotoxicity compared to the free drug, highlighting its potential in targeted chemotherapy [69]. ZnO nanoparticles obtained by green synthesis using orange peel demonstrated pH-dependent drug release which is an ideal factor for cancer therapy [70]. Curcumin-loaded biogenic nanoparticles synthesized with orange peel extract as a stabilizer enhanced anti-inflammatory effects *in vitro*, showing promise for treating conditions like rheumatoid arthritis or inflammatory bowel disease [71].

2) *Antimicrobial Agents*: Biogenic nanoparticles exhibit potent antimicrobial properties, making them suitable for treating infections caused by drug-resistant bacteria. Silver nanoparticles synthesized using biological methods have shown remarkable effectiveness against a wide range of pathogens [18, 72].
3) *Cancer Management*: Biogenic nanoparticles hold promise in cancer therapy, particularly in hyperthermia treatment and photodynamic therapy. Their ability to accumulate in tumor tissues and induce localized heating or generate reactive oxygen species enhances the efficacy of cancer treatments [73, 74]. In a novel approach, orange juice was used to biosynthesize nanoparticles of 5-fluorouracil (5-FU), a chemotherapeutic agent. These biosynthesized 5-FU nanoparticles retained anticancer efficacy and showed enhanced therapeutic effects against chemically induced skin fibrosarcoma in mice compared to conventional 5-FU formulations [75]. Silver nanoparticles synthesized using orange peel extract have been used to deliver anticancer drugs such as doxorubicin. These green-synthesized nanoparticles exhibited controlled release and higher cytotoxicity against cancer cells due to their synergistic effect with bioactive phytochemicals from the orange peel [72].

11.4.5 Imaging and Diagnostics

The unique optical and magnetic properties of biogenic nanoparticles make them ideal for medical imaging and diagnostics. Gold nanoparticles, for example, are used in imaging techniques such as photoacoustic imaging and surface-enhanced Raman spectroscopy [76, 77]. Silver nanoparticles synthesized using *C. sinensis* peels have been investigated for enhancing contrast in OCT imaging. Studies indicate that these nanoparticles can penetrate tissue layers, resulting in improved imaging depth and contrast, which is beneficial for soft tissue imaging applications [78]. Orange peel-mediated selenium nanoparticles (SeNPs) have demonstrated intrinsic fluorescence properties, enabling their use as diagnostic agents. These nanoparticles can serve as fluorescent probes for cellular imaging, aiding in the visualization of cellular structures and processes.

11.4.6 Catalytic Activity

AgNPs synthesized from orange peel extract have demonstrated efficacy in the A3 coupling reaction, facilitating the formation of propargylamines, which are valuable intermediates in pharmaceutical synthesis [79]. Silver nanoparticles synthesized using orange peel extract demonstrated strong catalytic activity in the reduction of 4-nitrophenol (4-NP) to 4-aminophenol in the presence of $NaBH_4$ – a model reaction in environmental catalysis [80]. Additionally, these nanoparticles have shown promise in environmental applications, such as the catalytic degradation of methylene blue dye, highlighting their potential in wastewater treatment [81]. Biogenic nanoparticles can also catalyze organic transformations such as the oxidation of alcohols and the reduction of nitro compounds. *Gold nanoparticles* synthesized from orange juice extract have been used to catalyze alcohol oxidation reactions efficiently under mild conditions [82].

11.5 Patents and Commercialization of Orange-Derived Biogenic Nanoparticles

Remarkable research outcomes in the field of orange derived biogenic nanoparticles have been achieved in recent times. The development of orange-derived biogenic nanoparticles presents a promising avenue for sustainable nanotechnology. Patents in this domain indicate a growing interest in harnessing agricultural waste for high-value applications. A few of these are discussed below:

1) A Chinese patent (CN114226746B) describes a method for synthesizing gold nanoparticles using ultrasonic-assisted orange peel extract. The process emphasizes eco-friendly techniques for nanoparticle production by combining orange peel extract with chloroauric acid solution, followed by ultrasonic treatment. The process enhances the yield, stability, and anti-inflammatory properties of the nanoparticles [83].
2) Tropicana Products, Inc. obtained patent (US4497838A) for a process for the production of useful products from orange peel, which details the process for producing valuable products from orange peel, including the extraction of essential oils and other compounds [84].

3) A US patent (US9491947B1) outlines a method for synthesizing metal and metal oxide nanoparticles using plant extracts, including orange peel extract. The process involves combining a metal salt solution with an orange peel extract to produce metal nanoparticles or metal oxide nanoparticles, which can be used to create nanoparticle powders or metal-polymer nanocomposites. The metal salt used in the method is silver nitrate ($AgNO_3$) and the metal nanoparticles are silver nanoparticles, whereas the metal oxide is titanium (IV) isopropoxide and the metal oxide nanoparticles are titanium dioxide (TiO_2) nanoparticles [85].
4) Another patent filed in China (CN105057692A) describes a method for synthesizing gold nanoparticles using orange juice. The process involves diluting orange juice with deionized water, mixing it with a chloroauric acid solution, and stirring to initiate the reduction of Au^{3+} ions to form gold nanoparticles. The resulting nanoparticles have an average size of approximately 11.8 nm. The inventor states that the synthesis method of this invention is simple and fast, has low cost, and causes little environmental pollution [86].

Though it may take some time for the commercialization of specific products developed from orange peel-based biogenic nanoparticles, the potential applications are vast. Industries such as pharmaceuticals and cosmetics, as well as environmental sectors are exploring these nanoparticles for drug delivery systems, antioxidant therapies, and waste management solutions. Projection of these orange-derived biogenic nanoparticles into commercial products is an area of ongoing research and development.

Many investigations in the preclinical stage on orange-derived biogenic nanoparticles can be considered as significant developments in nanotechnology that show growing efforts toward commercialization. In a study, silver nanoparticles (AgNPs) synthesized from bitter orange peel extract have demonstrated significant antibacterial activity against pathogens like *E. coli* and *K. pneumoniae*, with minimum inhibitory concentrations (MICs) of 125 μg/ml and 250 μg/ml, respectively. Additionally, these AgNPs exhibited anticancer activity against the MCF-7 human breast cancer cell line, with an IC50 value of 4.28 μg/ml, demonstrating their potential as anticancer agents [87]. AgNPs prepared using orange peel extract have shown 94% and 92% biofilm inhibition against *E. coli* and *S. aureus* at a concentration of 250 μg/ml, respectively. The minimum inhibitory concentration (MIC) for both *E. coli* and *S. aureus* was 32 μg/ml, and for *C. albicans*, it was 128 μg/ml. These nanoparticles also exhibited significant toxic effects against the DU145 human prostate cancer cell line, as investigated by anti-apoptotic, 4′,6-diamidino-2-phenylindole (DAPI), reactive oxygen species (ROS), and acridine orange/ethidium bromide (AO/EtBr) assays, indicating their potential as anticancer therapeutic agents [62]. Gelatin nanoparticles encapsulating bitter orange peel extract have been evaluated for their antioxidant and antibacterial properties. These nanoparticles were found to be effective against various bacterial strains, suggesting their potential application in food preservation and pharmaceutical formulations [88]. A randomized clinical trial assessed the effectiveness of a solid dispersion of orange peel polymethoxy-flavonoids (OPMF) as a palatal dressing material following free gingival grafts. The study found that OPMF facilitated hemostasis, reduced pain, and promoted wound healing compared to placebo treatments, indicating its potential for clinical applications in dental surgeries [89].

11.6 Challenges and Limitations in Orange-Mediated Biogenic Nanoparticle Synthesis

Orange-derived green nanoparticles, synthesized using orange peel extracts, present an eco-friendly alternative to conventional nanoparticle production methods. However, several challenges and limitations hinder their widespread application as discussed further.

1) Variability in Nanoparticle Characteristics
   The biogenic synthesis process often results in nanoparticles with inconsistent sizes, shapes, and surface properties. For instance, Niluxsshun et al. [90] reported that silver nanoparticles synthesized from orange peel extract exhibited a range of morphologies, including spherical, rod-like, and triangular shapes, with sizes varying from 5 nm to 80 nm. This variability can affect the reproducibility and reliability of their performance in applications such as drug delivery and antimicrobial treatments.
2) Limited Scalability and Yield
   The green synthesis of nanoparticles typically involves small-scale laboratory procedures, which may not be easily scalable for industrial production. Additionally, the yield of nanoparticles can be low, and the purification processes can be labor-intensive, making large-scale manufacturing challenging [91].
3) Stability and Shelf-Life Concerns
   Nanoparticles synthesized from orange peel extracts may exhibit limited stability over time. Aggregation, degradation, and loss of bioactivity can occur, particularly under varying environmental conditions. This instability can compromise their efficacy and safety in long-term applications [62].
4) Complex Reaction Conditions
   The synthesis of nanoparticles using orange peel extracts often requires precise control over reaction parameters, including pH, temperature, and concentration. Small deviations in these conditions can lead to significant changes in the

properties of the nanoparticles, complicating the standardization of the synthesis process [92].

5) Presence of Contaminants
   Orange peels may contain residual pesticides, heavy metals, or other contaminants that can affect the purity and safety of the synthesized nanoparticles. These impurities can pose risks in biomedical and environmental applications, necessitating thorough purification and quality control measures [92].
6) Use of Waste Materials in Nanoparticle Synthesis
   Though waste materials can be cost-effective, the overall process, including waste pre-treatment and nanoparticle purification, needs to be optimized to ensure economic viability [93].
7) Nanoparticles in Agriculture
   In agriculture, biosynthesized nanoparticles have many applications for improving plant hormone regulation, nutrient delivery, and insect control effectiveness. However, there are drawbacks and restrictions, such as the requirement for specialized infrastructure for their application, possible environmental effects, and scalability problems when generating nanoparticles in large quantities [94].
8) Environmental Impact
   Th removal of heavy metals such as mercury, lead, cadmium, arsenic, and thallium is a challenging task since they are harmful to the environment and human health. This is being accomplished by the use of nanomaterials such as metal oxides, carbon nanotubes, and nano-zeolites. Nanoscale Zero Valent Iron (nZVI) is especially promising because of its nontoxicity, low cost, and simplicity of production. Bimetallic nanoparticles, which mix iron with metals like nickel, silver, or gold, can also improve metal removal efficiency by accelerating the reduction of these dangerous metals [95].
9) Waste Management
   Although employing waste materials to synthesize nanoparticles is thought to be environmentally friendly, there may still be issues with how to dispose of or recycle any leftover materials following the manufacturing of nanoparticles. To prevent environmental contamination or toxicity, any unreacted metal ions or leftover biomolecules must be disposed away carefully [17].

Addressing all these challenges is vital for advancing the practical applications of orange-derived green nanoparticles in various fields, including medicine, agriculture, and environmental management. The safety rules and toxicological impacts of biogenic metal nanoparticles are still not very well documented. Advancements in safety precautions, and reasonably priced scalability are required to make green nanotechnology based on orange peel and other similar natural sources a viable and competitive substitute for conventional methods [96].

11.7 Research Gaps and Future Perspectives in Green Nanotechnology

Biogenic nanoparticles (BNPs) have gained significant attention for their potential applications in fields such as medicine, environmental remediation, agriculture, and electronics. However, while much progress has been made, several research gaps persist in the field of biogenic nanoparticles. These gaps primarily relate to the mechanisms of nanoparticle synthesis, scalability, characterization, stability, and their specific applications. Addressing these gaps is essential for harnessing the full potential of biogenic nanoparticles.

1) Mechanisms of Biogenic Nanoparticle Synthesis
   Although numerous natural sources have been employed in the synthesis of nanoparticles, the precise biochemical and biophysical mechanisms behind their formation remain largely unexplored. For example, although *Bacillus* species, *Pseudomonas*, and *Fusarium* fungi can reduce metal ions to form nanoparticles, the exact pathways and enzymes involved are not fully elucidated [97]. Similarly, in plant-mediated synthesis, the roles of specific metabolites, proteins, and enzymes in nanoparticle formation are poorly explained [17]. The research gap in understanding the mechanisms hinders the optimization of nanoparticle synthesis for industrial-scale applications and limits the ability to control particle size, shape, and stability.
2) Scalability and Cost-Effectiveness of Production
   Despite the promising potential of biogenic nanoparticles, the scalability of their production remains a significant challenge. While laboratory-scale studies have successfully demonstrated the synthesis of nanoparticles with desired properties, translating these methods to large-scale production has been limited. Many biogenic synthesis methods rely on the growth of microorganisms or plants, which require controlled environmental conditions and significant time investments [95]. Additionally, the cost of raw materials, such as plant extracts or specific microbial cultures, can make these processes economically unfeasible for large-scale applications.
3) Characterization and Standardization
   The characterization of biogenic nanoparticles presents a major challenge. While various techniques, such as transmission electron microscopy (TEM), scanning electron microscopy (SEM), X-ray diffraction (XRD), and dynamic light scattering (DLS), are commonly used to characterize the size, shape, and surface charge of nanoparticles, comprehensive standardization protocols are still lacking. Diversity in the composition of biological templates (e.g. plant species, microbial strains) complicates the ability to produce uniform nanoparticles, making it difficult to correlate particle properties with specific

biological activities. Furthermore, the heterogeneity of nanoparticles produced by biological methods compared to those produced chemically poses challenges in replicating results and ensuring quality control for industrial applications [98].

4) Stability and Environmental Impact
   One of the major advantages of biogenic nanoparticles is their potential for biodegradability and low toxicity compared to their synthetic counterparts. However, the stability of these nanoparticles, especially under different environmental conditions, remains a research gap. Many biogenic nanoparticles exhibit limited stability when exposed to factors such as temperature fluctuations, UV radiation, or changes in pH [34]. Additionally, the long-term environmental impact of biogenic nanoparticles, particularly in terms of their bioaccumulation and potential toxicity to aquatic and terrestrial ecosystems, has not been extensively studied.
5) Application-Specific Challenges
   Biogenic nanoparticles hold promise for a wide range of applications, from drug delivery and imaging to environmental remediation. However, their application in these fields is still in its infancy. In drug delivery, for example, while some biogenic nanoparticles have shown promise in terms of biocompatibility and targeting ability, their low loading capacity and limited controlled release properties are significant barriers. Similarly, the use of biogenic nanoparticles for environmental remediation requires a better understanding of their interaction with pollutants, as well as their potential to catalyze degradation reactions [99].

As a future perspective, green chemistry is well defined, and in a span of about 25 years, it has been recognized as a formal sub-discipline of chemistry. Green nanotechnology is a new area of designing new products that are harmless to human health and the environment. Green synthesis through green nanotechnology protocols has the potential to transform nanoparticle synthesis and benefit the future of nanotechnology. This new concept – along with the basics of green chemistry – is setting new standards for producing safe and sustainable nanomaterials [96]. With the use of nanotechnology in renewable energy, green technology appears to have a bright future for innovations [100]. Green nanotechnology has several potentialities, such as energy storage, nanosensing, biomedical, and nanomedicine [101]. Some of the points are as follows:

- Nanomaterials can improve solar cells and wind turbines, increasing their efficiency and reducing their cost [102].
- Turbine blades can be strengthened and made lighter with carbon nanotubes. A sustainable hydrogen economy is developed in part through nanotechnology [103].

- In order to contribute to a cleaner and more sustainable energy future, nanotechnology can increase the fuel economy of hybrid vehicles [102]. The constraints in traditional energy resources and environmental problems are addressed by these developments.
- To overcome issues like drug resistance, green-synthesized nanoparticles are being employed more and more in medication delivery, cancer treatment, and diagnostics. These nanoparticles offer safer and more effective substitutes for traditional therapies.

The use of green nanotechnology for the stability and controlled release of pharmaceutical substances is gaining popularity [96].

11.8 Future Prospects of Orange-Mediated Green Nanotechnology

A more detailed investigation into the molecular and cellular mechanisms of nanoparticle biosynthesis will be very supportive for future innovations in the field of green nanotechnology based on orange and other such plant extracts. Advanced techniques such as high-throughput sequencing, proteomics, and transcriptomics could provide insights into the specific genes and pathways involved in nanoparticle formation. Additionally, understanding the role of plant metabolites and secondary metabolites in nanoparticle synthesis could lead to more sustainable and efficient methods of production [104]. There need to be continuous efforts to establish standardized protocols for characterizing biogenic nanoparticles. This may include standardizing the size distribution, shape, and surface properties of nanoparticles synthesized using various biological sources. Advancements in analytical tools that can more accurately quantify the involvement of biomolecules in the nanoparticle synthesis process can be accelerated to achieve better outcomes [35]. Research efforts are of great worth to understand the stability of biogenic nanoparticles under various environmental conditions. This includes investigating the effects of aging, degradation, and interaction with environmental contaminants. Furthermore, comprehensive ecotoxicological studies can be conducted to evaluate the potential risks associated with the release of biogenic nanoparticles into the environment [105]. Future research should focus on optimizing the functionalization of biogenic nanoparticles to enhance their application in drug delivery, biosensing, and environmental remediation. Such attempts may include surface modification strategies to improve nanoparticle stability, loading capacity, and specificity [34]. However, it is understandable that large-scale field studies and clinical trials will be essential to validate the practical utility of biogenic nanoparticles. There is a rising need for cost-cutting strategies for scaling up the synthesis of biogenic nanoparticles. This could include the use of genetically engineered microorganisms that can grow faster or produce nanoparticles in larger quantities. Optimization of reaction parameters, such as temperature, pH, and metal-ion concentration, will be crucial to enhancing the efficiency and cost-effectiveness of the process [95].

11.9 Conclusion

The eco-friendly approach of nanoparticle synthesis based on plant and microbial sources has gained wider attention among researchers in various fields. Plant-derived green nanomedicines, especially those derived from orange peel, have been found to be a cost-effective and easy approach in comparison to those produced from microbial sources. Metal nanoparticles derived from green sources have been successfully developed in recent years with demonstrated effectiveness in drug delivery. Among various biogenic nanoparticles, AgNPs have emerged as a promising pharmaceutical technology. The field of biogenic nanoparticles holds significant promise for sustainable and environmentally friendly technologies, yet several research gaps must be addressed before their widespread application can be realized. These include better understanding of the mechanisms behind nanoparticle synthesis, development of scalable production methods, standardization of characterization protocols, enhancement of nanoparticle stability, and improvement of nanoparticles' application potential. By addressing these gaps, biogenic nanoparticles can be optimized for industrial-scale production and a wide array of applications, contributing to advancements in materials science, medicine, and environmental sustainability. The volume of literatures and diverse strategies adopted by researchers in the field of green nanotechnology based on orange and other similar bio resources in the last two decades signifies that, in the near future, green nanotechnology may serve as a substitute for chemical and physical approaches to nanoparticle synthesis. Also, it can be realized that the large number of biogenically synthesized nanoparticle prototypes, which are at the preclinical or scale-up stage shall pave the way for sustainable development and have potential applications in antimicrobial, anticancer, and antioxidant therapeutics; functional cosmetics, novel drug delivery, biomedicine, and biomedical device development; and also add value to products such as packaging materials and food and beverages.

Abbreviations

AgNP	Silver nanoparticle
AuNP	Gold nanoparticle

BNP	Biogenic nanoparticle
CoNP	Cobalt nanoparticle
CNP	Carbon nanoparticle
NPs	Nanoparticles
ZnNP	Zinc nanoparticle
SeNP	Selenium nanoparticle

References

1 Parashar UK, Saxena PS, Srivastava A. Bioinspired synthesis of silver nanoparticles. Digest J Nanomater Biostruct. 2009;4(1):159–166.

2 Sastry M, Ahmad A, Khan MI, Kumar R. Microbial nanoparticle production. J Chem Educ. 2003;80(1):30–35.

3 Jain KK. The role of nanobiotechnology in drug discovery. Drug Discov Today. 2012;10(21):1435–1442.

4 Singh P, Kim YJ, Zhang D, Yang DC. Biological synthesis of nanoparticles from plants and microorganisms. Trends Biotechnol. 2018;34(7):504–517.

5 Ahmed S, Ahmad SM, Swami BL, Ikram S. Green synthesis of silver nanoparticles using Azadirachta indica aqueous leaf extract. J Radiat Res Appl Sci. 2016;9(1):1–7.

6 Narayanan KB, Sakthivel N. Biological synthesis of metal nanoparticles by microbes. Adv Colloid Interf Sci. 2010;156(1–2):1–13.

7 Iravani S. Green synthesis of metal nanoparticles using plants. Green Chem. 2011;13(10):2638–2650.

8 Sharma VK, Yngard RA, Lin Y. Silver nanoparticles: green synthesis and their antimicrobial activities. Adv Colloid Interf Sci. 2015;145(1–2):83–96.

9 Prabhu S, Poulose EK. Silver nanoparticles: mechanism of antimicrobial action, synthesis, medical applications, and toxicity effects. Int Nano Lett. 2012;2(1):32.

10 Kowshik M, Ashtaputre S, Kharrazi S, Vogel W, Urban J, Kulkarni SK, Paknikar KM. Extracellular synthesis of silver nanoparticles by a silver-tolerant yeast strain MKY3. Nanotechnology. 2003;14(1):95.

11 Gade A, Bonde P, Ingle A, Marcato P, Durán N, Rai M. Exploitation of Aspergillus niger for synthesis of silver nanoparticles. J Biobased Mater Bioenergy. 2008;2(3):243–247.

12 Mukherjee P, Ahmad A, Mandal D, Senapati S, Sainkar SR, et al. Fungus-mediated synthesis of silver nanoparticles and their immobilization in the mycelial matrix: a novel biological approach to nanoparticle synthesis. Nano Lett. 2001;1(10):515–519.

13 Abdel-Raouf N, Alharbi RM, Al-Enazi NM, Alkhulaifi MM, Ibraheem IBM. Rapid biosynthesis of silver nanoparticles using the marine red alga Laurencia catarinensis and their characterization. Beni Suef Univ J Basic Appl Sci. 2018;7:150–157.

14 Shah M, Fawcett D, Sharma S, Tripathy SK, Poinern GEJ. Green synthesis of metallic nanoparticles via biological entities. Materials. 2014;8(11): 7278–7308.

15 Huang J, Lin L, Sun D, Chen H, Yang D, Li Q. Bio-inspired synthesis of metal nanomaterials and applications. Chem Soc Rev. 2015;44(17):6330–6374.

16 Mahmud KM, Hossain MM, Polash SA, Takikawa M, Shakil MS, et al. Investigation of antimicrobial activity and biocompatibility of biogenic silver nanoparticles synthesized using *Syzigyum cymosum* extract. ACS Omega. 2022;7(31):27216–27229.

17 Sharma D, Kanchi S, Bisetty K. Biogenic synthesis of nanoparticles: a review. Arab J Chem. 2019;12(8):3576–3600.

18 Rai M, Yadav A, Gade A. Silver nanoparticles as a new generation of antimicrobials. Biotechnol Adv. 2009;27(1):76–83.

19 Mickky B, Elsaka H, Abbas M, Gebreil A, Shams ER. Orange peel-mediated synthesis of silver nanoparticles with antioxidant and antitumor activities. BMC Biotechnol. 2024;24(1):66.

20 Malik AR, Aziz MH, Atif M, Irshad MS, Ullah H, Gia TN, et al. Lime peel extract induced $NiFe_2O_4$ NPs: synthesis to applications and oxidative stress mechanism for anticancer, antibiotic activity. J Saudi Chem Soc. 2022; 26(2):101422.

21 Arsene MMJ, Podoprigora IV, Davares AKL, Razan M, Das MS, Senyagin AN. Antibacterial activity of grapefruit peel extracts and green-synthesized silver nanoparticles. Vet World. 2021;14(5):1330–1341.

22 Barreca D, Mandalari G, Calderaro A, Smeriglio A, Trombetta D, Felice MR, Gattuso G. *Citrus* flavones: an update on sources, biological functions, and health promoting properties. Plants. 2020;9:288.

23 Sharmila VG, Kavitha S, Obulisamy PK, Banu JR. Production of fine chemicals from food wastes. In: Banu JR, Kumar G, Gunasekaran M, Kavitha S (eds). Food Waste to Valuable Resources. 2020. Academic Press, p. 163–188.

24 Kaur N, Singh A, Ahmad W. Microwave assisted green synthesis of silver nanoparticles and its application: a review. J Inorg Organomet Polym Mater. 2023;33:663–672.

25 Anupong W, On-Uma R, Jutamas K, Joshi D, Salmen SH, Alahmadi TA, Jhanani GK. Cobalt nanoparticles synthesizing potential of orange peel aqueous extract and their antimicrobial and antioxidant activity. Environ Res. 2023;216:114594.

26 Abuzeid HM, Julien CM, Zhu L, Hashem AM. Green synthesis of nanoparticles and their energy storage, environmental, and biomedical applications. Crystals. 2023;13(11):1576.

27 Roussos PA. Phytochemicals and antioxidant capacity of orange (Citrus sinensis (l.) Osbeck cv. Salustiana) juice produced under organic and integrated farming system in Greece. Sci Hortic. 2011;129(2):253–258.

28 Bampidis VA, Robinson PH. Citrus byproducts as ruminant feeds: a review. Anim Feed Sci Technol. 2006;128:175–217.

29 Williams CA. Specialized dietary supplements. In: Geor RJ, Harris PA, Coenen M (eds). Equine Applied and Clinical Nutrition. 2013. W.B. Saunders, p. 351–366.

30 Kanaze FI, Termentzi A, Gabrieli C, Niopas I, Georgarakis M, Kokkalou E. The phytochemical analysis and antioxidant activity assessment of orange peel (Citrus sinensis) cultivated in Greece-Crete indicates a new commercial source of hesperidin. Biomed Chromatogr. 2009;23(3):239–249.

31 Negro V, Ruggeri B, Fino D, Tonini D. Life cycle assessment of orange peel waste management. Resour Conserv Recycl. 2017;127:148–158.

32 Olabinjo OO, Ogunlowo AS, Ajayi OO, Olalusi AP. Analysis of physical and chemical composition of sweet orange (Citrus sinensis) peels. Int J Environ Agric Biotechnol. 2017;2:2201–2206.

33 de la Torre I, Martin-Dominguez V, Acedos MG, Esteban J, Santos VE, Ladero M. Utilisation/upgrading of orange peel waste from a biological biorefinery perspective. Appl Microbiol Biotechnol. 2019;103:5975–5991.

34 Osman AI, Zhang Y, Farghali M, Rashwan AK, Eltaweil AS, Abd El-Monaem EM, et al. Synthesis of green nanoparticles for energy, biomedical, environmental, agricultural, and food applications: a review. Environ Chem Lett. 2024;22:841–887.

35 Patil S, Chandrasekaran R. Biogenic nanoparticles: a comprehensive perspective in synthesis, characterization, application and its challenges. J Genet Eng Biotechnol. 2020;18(1):67.

36 Ghosh S, Bhagwat T, Chopade BA, Webster TJ. Patents, technology transfer, and commercialization aspects of biogenic nanoparticles. NanoBiotechnology. 2021:323–339.

37 Thompson IP, Van Der Gast CJ, Ciric L, Singer AC. Bioaugmentation for bioremediation: the challenge of strain selection. Environ Microbiol. 2005;7(7):909–915.

38 Bhati M. Biogenic synthesis of metallic nanoparticles: principles and applications. Mater Today Proc. 2023;81(2):882–887.

39 Arakaki A, Nakazawa H, Nemoto M, Mori T, Matsunaga T. Formation of magnetite by bacteria and its application. J R Soc Interface. 2008;5(26):977–999.

40 Adekunle KF, Okolie JA. A review of biochemical process of anaerobic digestion. Adv Biosci Biotechnol. 2015;6(3):205–212.

41 Mei XL, Zheng QF. Role of cellular biomolecules in screening, diagnosis and treatment of colorectal cancer. Curr Drug Metab. 2019;20(11):880–888.

42 Boskey AL. Biomineralization: an overview. Connect Tissue Res. 2003;44(1):5–9.

43 El-Ansary A, Faddah LM. Nanoparticles as biochemical sensors. Nanotechnol Sci Appl. 2010;3(1):65–76.

44 Zheng M, Huang X. Nanoparticles comprising a mixed monolayer for specific bindings with biomolecules. J Am Chem Soc. 2004;126(38):12047–12054.

45 Becker N, Oroudjev E, Mutz S, Cleveland JP, Hansma PK, Hayashi CY, et al. Molecular nanosprings in spider capture-silk threads. Nat Mater. 2003;2(4):278–283.

46 Liu R, Ma GH, Wan YH, Su ZG. Influence of process parameters on the size distribution of PLA microcapsules prepared by combining membrane emulsification technique and double emulsion-solvent evaporation method. Colloids Surf B: Biointerfaces. 2005;45(3–4):144–153.

47 Gao Y, Gong C, Chen M, Huan S, Zhang XB, Ke G. Endogenous enzyme-driven amplified DNA Nanocage probe for selective and sensitive imaging of mature MicroRNAs in living cancer cells. Anal Chem. 2024;96(23): 9453–9459.

48 Pietrobon B, Kitaev V. Photochemical synthesis of monodisperse size-controlled silver decahedral nanoparticles and their remarkable optical properties. Chem Mater. 2008;20(16):5186–5190.

49 Suri S, Singh A, Nema PK. Current applications of citrus fruit processing waste: a scientific outlook. Appl Food Res. 2022;2(1):100050.

50 Ben Amor A, Hemmami H, Gherbi MT, Seghir BB, Zeghoud S, Gharbi AH, et al. Synthesis of spherical carbon nanoparticles from orange peel and their surface modification with chitosan: evaluation of optical properties, biocompatibility, antioxidant and anti-hemolytic activity. Biomass Convers Biorefinery. 2024;11:1–4.

51 Gupta S, Choudhary DK, Sundaram S. Green synthesis and characterization of silver nanoparticles using Citrus sinensis (Orange peel) extract and their antidiabetic, antioxidant, antimicrobial and anticancer activity. Waste Biomass Valorization. 2024;16:1101–1114.

52 Thi TU, Nguyen TT, Thi YD, Thi KH, Phan BT, Pham KN. Green synthesis of ZnO nanoparticles using orange fruit peel extract for antibacterial activities. RSC Adv. 2020;10(40):23899–23907.

53 Abdelmoneim D, Porter G, Duncan W, Lim K, Easingwood R, Woodfield T, Coates D. Three-dimensional evaluation of the cytotoxicity and antibacterial properties of alpha lipoic acid-capped silver nanoparticle constructs for oral applications. J Nanomater. 2023;13:705.

54 Porto DS, Estevão BM, Lins PMP, Rissi NC, Zucolotto V, Maria FGF d S. Orange trunk waste-based lignin nanoparticles encapsulating curcumin as a photodynamic therapy agent against liver cancer. ACS Appl Polym Mater. 2021;3(10):5061–5072.

55 Yang B, Qi F, Tan J, Yu T, Qu C. Study of green synthesis of ultrasmall gold nanoparticles using *citrus Sinensis* Peel. Appl Sci. 2019;9(12):2423.

56 Dang-Bao T, Gia-Thien TH, Long BD, Nguyen PA, Phan TDT, et al. Green orange peel-mediated bioinspired synthesis of nanoselenium and its antibacterial activity against methicillin-resistant Staphylococcus aureus. ACS Omega. 2022;7(40):36037–36046.

57 Abomughaid MM. Bio-fabrication of bio-inspired silica Nanomaterials from Orange peels in combating oxidative stress. Nanomaterials. 2022;12(18):3236.

58 Susithra V, Sindhu Kavi S, Abd El-Rehim AF, Ranjith Kumar ER. Citrus sinensis assisted biogenic synthesis and physicochemical properties of Fe3O4 nanoparticles for antibacterial activity. Ceram Int. 2024;50(7):Part A:10225-31.

59 Gupta A, Haque B, Roy A, Malik A, Khan AA, Kaur K, Roy A. Antioxidant, dye degradation, and molecular docking studies of orange peel extract derived Ag-Fe-Ni nanoparticles. Inorg Chem Commun. 2024;166:112599.

60 Nnemeka I, Godwin E, Olakunle F, Olushola O, Moses O, Chidozie OP, Rufus S. Microwave enhanced synthesis of silver nanoparticles using orange peel extracts from Nigeria. Chem Biomol Eng. 2016;1(1):5–11.

61 Quiroz-Hernández JE, Kharissova OV, Aguirre-Arzola VE, Martinez-Avila GCG, Castillo-Velazquez U. Evaluation of the conditions for the synthesis of silver nanoparticles from orange peels and its antibacterial effect. Recent Pat Nanotechnol. 2020;14(3):250–258.

62 Hani U, Kidwan FN, Albarqi LA, Al-Qahtani SA, AlHadi RM, AlZaid HA, et al. Biogenic silver nanoparticle synthesis using orange peel extract and its multifaceted biomedical application. Bioprocess Biosyst Eng. 2024;47(8):1363–1375.

63 Konwarh R, Gogoi B, Philip R, Laskar MA, Karak N. Biomimetic preparation of polymer-supported free radical scavenging, cytocompatible and antimicrobial "green" silver nanoparticles using aqueous extract of Citrus sinensis peel. Colloids Surf B: Biointerfaces. 2011;84(2):338–345.

64 Doan VD, Nguyen VC, Pham HAL, Tran TLA, Pham TV, Nguyen TLH. Orange peel essential oil nanoemulsions supported by nanosilver for antibacterial application. Indonesian Journal of Chemistry. 2020;20(2):430–439.

65 MubarakAli D, Thajuddin N, Jeganathan K, Gunasekaran M. Plant extract mediated synthesis of silver and gold nanoparticles and its antibacterial activity against clinically isolated pathogens. Colloids Surf B: Biointerfaces. 2011;85(2):360–365.

66 Kumar N, Upadhyay A, Shukla S, Bajpai VK, Keiliszek M, Yadav A, Kumaravel V. Next generation edible nanoformulations for improving post-harvest shelf-life of citrus fruits. Food Measure. 2014;18:1825–1856.

67 Terzioglu P, Guney F, Parın FN, Sen I, Tuna S. Biowaste orange peel incorporated chitosan/polyvinyl alcohol composite films for food packaging applications. Food Packag Shelf Life. 2021;30:100742.

68 Chudasama M, Goyary J. Nanostructured materials in food science: current progress and future prospects. Next Mater. 2024;5:100206.

69 Iannazzo D, Celesti C, Espro C, Ferlazzo A, Giofrè SV, Scuderi M, et al. Orange-Peel-derived nanobiochar for targeted cancer therapy. Pharmaceutics. 2022;14(10):2249.

70 Anjum S, Hashim M, Malik SA, Khan M, Lorenzo JM, Abbasi BH, Hano C. Recent advances in zinc oxide nanoparticles (ZnO NPs) for cancer diagnosis, target drug delivery, and treatment. Cancers (Basel). 2021;13(18):4570.

71 Quispe C, Cruz-Martins N, Manca ML, Manconi M, Sytar O, et al. Nano-derived therapeutic formulations with curcumin in inflammation-related diseases. Oxidative Med Cell Longev. 2021;1:3149223.

72 Fahim M, Shahzaib A, Nishat N, Jahan A, Bhat TA, Inam A. Green synthesis of silver nanoparticles: a comprehensive review of methods, influencing factors, and applications. JCIS Open. 2024;16:100125.

73 Huang X, Jain PK, El-Sayed IH, El-Sayed MA. Gold nanoparticles: interesting optical properties and recent applications in cancer diagnostics and therapy. Nanomedicine (London). 2007;2(5):681–693.

74 Thakkar KN, Mhatre SS, Parikh RY. Biological synthesis of metallic nanoparticles. Nanomedicine (London). 2010;6(2):257–262.

75 Mohammad O, Faisal SM, Ahmad N, Rauf MA, Umar MS, Mujeeb AA, et al. Bio-mediated synthesis of 5-FU based nanoparticles employing orange fruit juice: a novel drug delivery system to treat skin fibrosarcoma in model animals. Sci Rep. 2019;9(1):12288.

76 Chanda N, Shukla R, Zambre A, Mekapothula S, Kulkarni RR, Katti K, Katti KV. An aqueous route to biofunctionalized luminescent gold quantum clusters at room temperature. Chem Commun. 2011;47(37):10078–10080.

77 Dreaden EC, Alkilany AM, Huang X, Murphy CJ, El-Sayed MA. The golden age: gold nanoparticles for biomedicine. Chem Soc Rev. 2012;41(7):2740–2779.

78 Fatikasari T, Nurhasanah I, Khumaeni A. Green synthesis of silver nanoparticles using *Citrus sinensis* peels assisted by microwave irradiation as a contrast agents for computed tomography imaging. Indonesian Journal of Chemistry. 2024;24(5):1573–1582.

79 Sharma NK, Vishwakarma J, Rai S, Alomar TS, AlMasoud N, Bhattarai A. Green route synthesis and characterization techniques of silver nanoparticles and their biological adeptness. ACS Omega. 2022;7(31):27004–27020.

80 Mulu M, Tefera M, Guadie A, Basavaiah K. Biosynthesis, characterization and study of the application of silver nanoparticle for 4-nitrophenol reduction, and antimicrobial activities. Biotechnol Rep (Amst). 2024;42:e00838.

81 Simatupang C, Jindal VK, Jindal R. Biosynthesis of silver nanoparticles using orange peel extract for application in catalytic degradation of methylene blue dye. Environ Nat Resour J. 2021;19:468–480.

82 Dikshit PK, Kumar J, Das AK, Sadhu S, Sharma S, Singh S, et al. Green synthesis of metallic nanoparticles: applications and limitations. Catalysts. 2021;11(8):902.

83 Chen X, Gao Y. Method for green synthesis of gold nanoparticles by using ultrasonic-assisted orange peel extract. Patent No. CN114226746B, 2017, China. Available at: https://patents.google.com/patent/CN114226746B/en#patentCitations.

84 Bonnell JM. Process for the production of useful products from orange peel. Patent No. US4497838A, 1983, United States. Available at: https://patents.google.com/patent/US4497838A/en.

85 Awad MA, Hendi AA, Eisa NE, Ortashi KM, Merghani NM. Method of synthesizing nanoparticles and a nanoparticle-polymer composite using a plant extract. Patent No. US9491947B1, 2015, United States. Available at: https://patents.google.com/patent/US9491947B1/en#patentCitations.

86 Yu J, Guan H, Wu S, Chi D. Green synthesis method of gold nanoparticles. Patent No. CN105057692A, 2015, China. Available at: https://patents.google.com/patent/CN105057692A/en.

87 Ghoreishi SM, Mortazavi-Derazkola S. Biological evaluation of silver nanoparticles synthesized from bitter orange peel extract: antibacterial and anticancer properties on human breast cancer cell line. J Nanostruct. 2024;14(2):505–512.

88 Garcia-Juarez A, Garzon-Garcia AM, Ramos-Enriquez JR, Tapia-Hernandez JA, Ruiz-Cruz S, et al. Evaluation of antioxidant and antibacterial activity of gelatin nanoparticles with bitter orange peel extract for food applications. Foods. 2024;13(23):3838.

89 Alghriany AA, Ali AU, Khallaf ISA, Hassan AS, Sayed MA, Fikry AM. Clinical effectiveness of orange peel polymethoxy-flavonoids rich fraction as a palatal dressing material compared to Alveogyl: randomized clinical trial. Sci Rep. 2024;14(1):3067.

90 Niluxsshun MCD, Masilamani K, Mathiventhan U. Green synthesis of silver nanoparticles from the extracts of fruit peel of *Citrus tangerina*, *Citrus sinensis*, and *Citrus limon* for antibacterial activities. Bioinorg Chem Appl. 2021;2: 6695734.

91 Castro L, Blazquez ML, Gonzalez F, Munoz JA, Ballester A. Biosynthesis of silver and platinum nanoparticles using orange peel extract: characterisation and applications. IET Nanobiotechnol. 2015;9(5):252–258.

92 Salem SS, Badawy MSEM, Al-Askar AA, Arishi AA, Elkady FM, Hashem AH. Green biosynthesis of selenium nanoparticles using orange peel waste: characterization, antibacterial and antibiofilm activities against multidrug-resistant bacteria. Life (Basel). 2022;12(6):893.

93 Mazari SA, Ali E, Abro R, Khan FSA, Ahmed I, Ahmed M, et al. Nanomaterials: applications, waste-handling, environmental toxicities, and future challenges – a review. J Environ Chem Eng. 2021;9(2):105028.

94 Ingale AG, Chaudhari AN. Biogenic synthesis of nanoparticles and potential applications: an eco-friendly approach. J Nanomed Nanotechol. 2013; 4(165):1–7.

95 Mughal B, Zaidi SZ, Zhang X, Hassan SU. Biogenic nanoparticles: synthesis, characterisation and applications. Appl Sci. 2021;11(6):2598.

96 Mahdavi Dehkharghani F, Ghahremanlou M, Zandi Z, Jalili M, Mozafari MR, Mardani P. Future energy and therapeutic perspectives of green nano-technology: recent advances and challenges. Nano Micro Biosyst. 2023;2(1):11–21.

97 Sood R, Chopra DS. Metal–plant frameworks in nanotechnology: an overview. Phytomedicine. 2018;50:148–156.

98 Khan I, Saeed K, Khan I. Nanoparticles: properties, applications and toxicities. Arab J Chem. 2019;12(7):908–931.

99 Abdel-Megeed RM. Biogenic nanoparticles as a promising drug delivery system. Toxicol Rep. 2024;14:101887.

100 Rita E, Chizoo E, Cyril US. Sustaining COVID-19 pandemic lockdown era air pollution impact through utilization of more renewable energy resources. Heliyon. 2021;7(7):e07455.

101 Aithal PS, Aithal S. Opportunities and challenges for green and eco-friendly nanotechnology in twenty-first century. In: Sustainable Nanotechnology: Strategies, Products, and Applications, Vol. 16. 2022, p. 31–50.

102 Wang J, He N, Fei J, Ma Z, Ji Z, Chen Z, et al. Flexible and wearable fuel cells: a review of configurations and applications. J Power Sources. 2022;551:232190.

103 Lin C, Du Y, Li XL, Zhou WL, Zhang CY, Xie H, et al. Efficient fabrication of low-density polyethylene/polyethylene oxide/carbon nanotubes films with robust shape stability and photothermal property for thermal management and afterheat utilization. Polymer. 2022;260:125381.

104 Mortimer M, Wang Y, Holden PA. Molecular mechanisms of nanomaterial-bacterial interactions revealed by omics-the role of nanomaterial effect level. Front Bioeng Biotechnol. 2021;9:683520.

105 Shah ST, Sari IP, Yanto DHY, Chowdhury ZZ, Bashir MN, et al. Nature's nanofactories: biogenic synthesis of metal nanoparticles for sustainable technologies. Green Chem Lett Rev. 2025;18(1).

12

Sustainable Appproaches for the Synthesis of Biogenic Nanoparticles Using Egg and Biogenic Nanomedicine

Priyanka Das, Moumita Gope, Jyoti Kumari Mishra, Jack Patgiri, and Malay K. Das

Drug Delivery Research Laboratory, Department of Pharmaceutical Sciences, Dibrugarh University, Dibrugarh, Assam, India

12.1 Introduction

Eggs have long been a staple in the human diet, offering a rich source of essential nutrients, especially high-quality proteins [1]. Alongside milk and meat, eggs are considered a valuable protein source, containing amino acids, carbohydrates, easily digestible fats, minerals, and vital vitamins [2]. Common edible egg types include those from chickens, quails, and guinea fowls, all of which share similar nutritional values and culinary applications [3]. Among these, chicken eggs are particularly popular because of their affordability and nutritional density, which has resulted in widespread consumption and a significant amount of eggshell waste [4]. Structurally, a chicken egg comprises three main parts: shell, albumen (white), and yolk. While extensive research has been conducted on albumen and yolk, comparatively less attention has been paid to eggshells. The shell itself features multiple layers, including the cuticle, crystalline layer, spongy calcareous layer, pores, cores, and mammillary layer [5]. Inside the shell, two fibrous membranes, often appearing as a single layer, can be distinguished. One encloses the albumen, while the other attaches to the calcified layer of the shell, known as the inner and outer shell membranes, respectively [6]. These membranes are known for their adsorptive properties and are effective in removing reactive dyes and heavy metal ions from wastewater. Additionally, they contain bioactive compounds that have potential applications across various sectors, such as medicine, pharmaceuticals, food, cosmetics, and biochemistry [7] (Figure 12.1).

Biogenic synthesis has gained considerable attention for its environmentally friendly and non-toxic approach. This method employs natural biological

Biogenic Synthesis of Nanoparticles and Sustainable Nanotherapeutics,
First Edition. Edited by Malay K. Das.

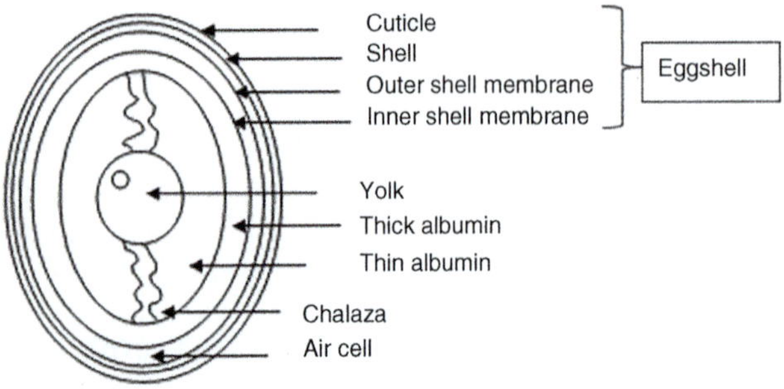

Figure 12.1 Components of egg.

resources such as plants, biomolecules, microorganisms, and enzymes as both reducing and stabilizing agents [8]. Often it is referred to as green synthesis, which also involves organisms like algae, fungi, animal cell cultures, and even eggs in nanoparticle production [9]. With the rapid development of nanotechnology globally, nanoparticles (NPs) are now widely researched and applied in numerous areas, including medicine, drug delivery, cell labeling, plant tissue culture, diagnostics, the automobile industry, and energy systems. Although physical and chemical techniques are commonly used for synthesizing nanoparticles with varying shapes and sizes, biological synthesis is emerging as a promising alternative [10]. Utilizing biological entities such as plant parts, algae, fungi, eggs, and microbes offers an edge over chemical methods, which can leave behind toxic residues on nanoparticle surfaces, limiting their suitability for biomedical use [11]. Research has also shown that biological systems release significant amounts of proteins, which help reduce metal ions and influence the size and shape of nanoparticles formed [12]. Overall, biogenic nanoparticle synthesis stands out as a sustainable and eco-conscious approach in nanotechnology, relying on renewable and biodegradable inputs [13].

12.2 Egg and Biogenic Nanoparticles

While making nanoparticles using egg parts like shells, egg white, or yolk, everything must be tracked from gathering materials to when they are used and thrown away. Using eggshells turns kitchen waste into something valuable but mining metals like silver costs a lot more than just money; it drains energy, water, and harms the environment. Green synthesis helps to reduce impacts by cutting out toxic chemicals and lowering energy, but it's not a total fix. For truly sustainable nanoparticles, several methods have been developed that operate at room temperature, powered by renewables and using waste materials like egg yolk, aiming for cleaner, greener production. If the particles work well and last long, then we can make the products from nanoparticles environmentally friendly. However, once they are thrown out, they might build up in nature or harm living things, so it is important to know how they break down or if we can reuse them.

Eggshell waste contains 8–11% of an egg's mass that often accumulates in landfills, causing odor, pests, microbial contamination, and extremely slow natural breakdown. Yet eggshells are ~94–95% calcium carbonate ($CaCO_3$) with small amounts of calcium phosphate, magnesium carbonate, and proteins. This makes them an eco-friendly, low-cost biomaterial with multiple high-value uses. Eggshell powder (ESP) serves as filler in biodegradable packaging films (e.g. polylactic acid/polypropylene carbonate) supporting sustainable materials. Its porous, multilayered structure with functional C=O/C—O groups enables efficient heavy-metal immobilization, like cadmium and lead, in soil and water. Additionally, ESP can be transformed via calcination into CaO and $Ca(OH)_2$ for use in catalysts, soil remediation, and photocatalysis. Composites such as CuSe/$CaCO_3$-loaded calcium silicate boards have been created that exhibit antibacterial and mechanical strength. High-purity eggshell-derived $CaCO_3$ is converted into calcium phosphates and hydroxyapatite for bone implants and bioceramics, while CaO nanoparticles are used in dye degradation, CO_2 capture, and antibacterial water treatment. This underscores the pivotal role of eggshell in sustainable materials and nanotechnology [14, 15].

Egg white proteins (like ovalbumin, ovotransferrin, and ovomucoid) are remarkable natural agents, serving as reducers and stabilizers during the green, water-based synthesis of nanoparticles. In the sol–gel production of titanium dioxide (TiO_2), egg white acts as a biotemplate and gelling agent, enabling the formation of small, uniform anatase TiO_2 nanoparticles (~9 nm) under mild, room-temperature conditions ideal for applications in photocatalysis, solar cells, sensors, self-cleaning surfaces, and antibacterial materials. Similarly, egg white proteins reduce silver ions (Ag^+) and cap newly formed particles to prevent aggregation, yielding stable, biocompatible Ag–protein nanoparticles (~20–26 nm) under sunlight or at ambient temperature. These particles show excellent colloidal stability and low toxicity, enhancing cancer radiation therapy. As the egg white proteins act as both reducer and capping shell, they offer an eco-friendly, non-toxic route to uniformly sized functional TiO_2 and Ag nanoparticles that provide significant strides in sustainable nanomaterial design with diverse biomedical and environmental applications [16, 17].

Egg yolk, a rich source of phospholipids, especially lecithin, contains ~73% phosphatidylcholine, making it superior to plant-derived lecithin. It supports cell membranes, brain function, and vitamin absorption, offering benefits like anti-aging and liver protection. Lecithin is traditionally extracted using ethanol or ethanol-acetone mixtures that may leave solvent residues. Advanced methods include supercritical CO_2 extraction for high purity, enzymatic hydrolysis with proteases for improved yield, and column chromatography for precise separation. Emerging techniques like membrane separation and cryo-precipitation preserve lecithin's bioactivity. These methods enhance lecithin extraction efficiency, purity, and applicability in pharmaceuticals and functional foods [18].

12.3 General Egg-Based Nanoparticle Synthesis

12.3.1 Extraction and Preparation of Egg-Based Precursors

12.3.1.1 Eggshell Powder

Shells are meticulously cleaned, dried, and milled into a fine particulate. This eggshell powder is abundant in calcium carbonate ($CaCO_3$), forming a porous, biocompatible scaffold. It also retains minor organic constituents like collagen and glycoproteins, which contribute to its biological compatibility and mechanical robustness [19].

12.3.1.2 Egg White (Albumen)

Egg white consists mostly of water and proteins such as ovalbumin, lysozyme, ovotransferrin, and ovomucin. These proteins are abundant in functional groups like hydroxyl (—OH), amino (—NH_2), and carboxyl (—COOH) that readily bind and reduce metal ions. As a result, egg white acts efficiently as both a reducing agent and a nanoparticle stabilizer under gentle reaction conditions [17].

12.3.1.3 Egg Yolk

Yolk is rich in lipids (phospholipids and triglycerides), phosphoproteins, cholesterol, and vitamins (A, D, E, and K). This composition enhances its metal-reducing capacity and enables it to act as an effective capping agent. The amphiphilic nature of yolk lipids enables them to adhere to nanoparticle surfaces, reducing aggregation and controlling morphology [20].

12.3.2 Gentle Reduction of Metal Ions

In green synthesis, aqueous extracts from egg components are mixed with metal salts such as Fe^{3+}, Ag^{+}, Pt^{4+}, or Zn^{2+} under ambient temperature and neutral pH. Egg-derived biomolecules donate electrons to these ions, reducing them to elemental or oxide nanoparticles. This eco-friendly process avoids harsh conditions and noxious byproducts, making it ideal for biomedical and environmental applications [21–23] (Figure 12.2).

12.3.3 Nucleation and Inherent Capping

Following reduction, metal atoms cluster to form nuclei, resulting in nanoparticles typically ranging from 40 to 200 nm. Proteins and lipids naturally present in

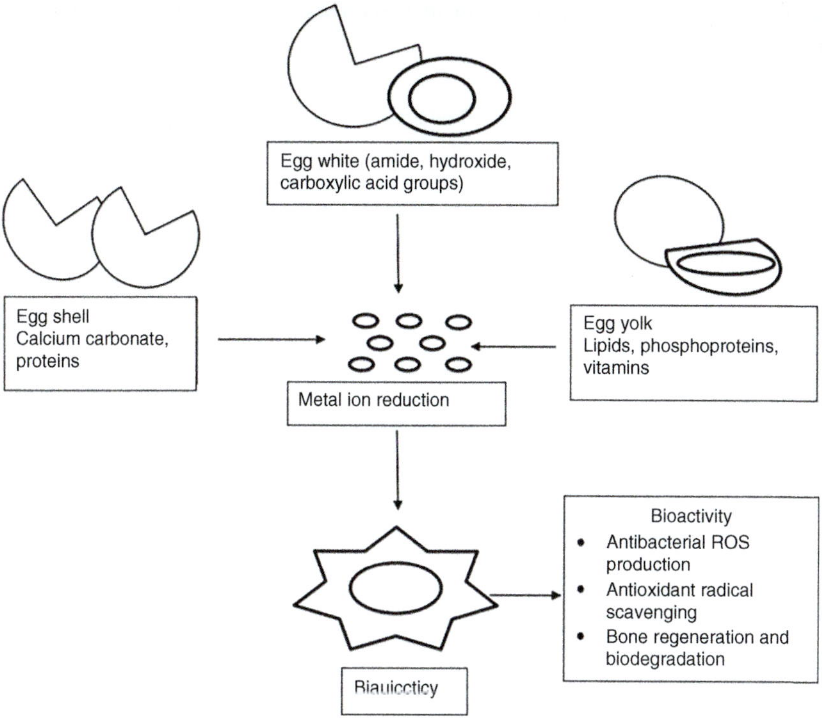

Figure 12.2 Mechanism of Egg-based nanoparticle synthesis.

the egg mixture immediately coat these particles, providing a protective capping layer. This natural encapsulation prevents aggregation, controls particle size and shape, and enhances biocompatibility by providing functional surfaces [24, 25] (Figure 12.2).

12.3.4 Diverse Bio Functional Properties

12.3.4.1 Antimicrobial Activity

Nanoparticles especially those of silver and zinc oxide produce reactive oxygen species (ROS) like superoxide anions and hydroxyl radicals. These ROS compromise bacterial cell integrity by damaging cell membranes, proteins, and deoxyribonucleic acid (DNA), effectively killing microbes, including resistant strains.

12.3.4.2 Antioxidant Function

Egg-based nanoparticles often retain proteins and polyphenolic substances that neutralize free radicals. This antioxidant activity helps protect cells from oxidative damage, making them suitable for use in food preservation, skincare formulations, and anti-aging products [26].

12.3.4.3 Bone Tissue Support and Biodegradability

Eggshell-derived $CaCO_3$ scaffolds encourage the deposition of hydroxyapatite, a mineral essential for bone formation. Enzymes like lysozyme from egg white facilitate slow scaffold degradation, enabling safe, gradual integration into bone tissue. These properties are valuable in orthopedic and dental tissue engineering [5, 27].

Table 12.1 represents the various egg-derived nanoparticles with their applications.

12.4 Factors Affecting Egg-Based Nanoparticle Formation

12.4.1 Biomolecules as Natural Reducing and Stabilizing Agents

Egg proteins such as ovalbumin, ovotransferrin, and lysozyme act as green bio-reductants and stabilizers. Because of their rich amino acid content and surface chemical functionality, they effectively coordinate and reduce metal ions. For instance, Lu et al. [17] demonstrated that egg white proteins formed biocompatible silver nanoparticles via sunlight irradiation, showing excellent stability owing to capping by ovotransferrin and ovomucoid.

12.4.2 Effect of pH

The reaction medium's pH dramatically influences protein reactivity and nanoparticle formation. Alkaline conditions tend to boost the reduction capability of egg proteins, promoting faster nucleation and yielding smaller, more homogeneous nanoparticles [33]. For example, the biogenic synthesis of ZnO and MgO nanoparticles using ovalbumin showed that adjusting pH alongside metal ion concentration tuned nanoparticle size from ~8 to 14 nm [34]

12.4.3 Temperature and Light Exposure

Ambient warmth and light especially natural sunlight can drive the electron transfer from egg biomolecules to metal precursors. These mild energy inputs eliminate the need for chemical catalysts, marking the process sustainable and energy efficient. For example, sunlight-mediated reduction was used in egg-white based silver nanoparticle fabrication, yielding stable AgNPs without additional heat [35].

Table 12.1 Various egg-mediated synthesized nanoparticles.

Egg-based material	Reaction condition	Nanoparticle (size nm, shape)	Application	References
Eggshell membrane	Dried ESM + $HAuCl_4$ (0.1 mM – 0.1 M) at room temperature	AuNPs (less than 20 nm, spherical/ hexagonal/ triangular)	Bio-imaging and biolabeling	[28]
Egg white	Egg white combined with 1–2 M zinc nitrate, which was then sonicated for 10 minutes and dried at 120 °C, and finally calcined at 550 °C for 2 hours	ZnO nanowhiskers (≈10–100 nm, avg. ~40 nm)	Antibacterial, antibiofilm, dermal infection treatment	[29]
Calcined eggshell (biowaste)	Calcination at ≤120 °C; catalyst under mild solvent-free/ aqueous conditions	CaO–MOFs, Ag/CaO, Fe_3O_4–CaO (32–40 nm, porous/irregular)	Organic carbonate synthesis; pharmaceuticals; agricultural catalysis	[30]
Quail egg yolk	0.5 mM Pt ions, pH 6.0, 20 °C for four hours	PtNPs (2–50 nm, spherical)	Catalysis; fuel cells; antibacterial; sensors; drug delivery; antitumor	[31]
Egg white	Egg white mixed with nickel nitrate hexahydrate ($Ni(NO_3)_2 \cdot 6H_2O$) and heated at 80 °C for 12 hours, and then dried at 120 °C for 6 hours, and finally calcined between 300 and 600 °C for 3 hours.	NiO NPs (10–30 nm, cubic)	Photocatalysis; antimicrobial; cancer therapy; biosensing diagnostics	[25]
Egg white or yolk	Mild room temperature aqueous; proteins as reducing/ stabilizing agents	AuNPs (~2.4 nm, spherical)	Bio-imaging; photothermal therapy; drug delivery; antibacterial; antioxidant; biosensors	[32]

12.4.4 Concentration of Reactants

The ratio between egg extract and metal precursor determines the speed of nucleation and growth. Higher extract-to-metal ratios can enhance particle stability by increasing capping coverage, but may slow down the reduction process [36]. In the ovalbumin-mediated ZnO/MgO synthesis, variation of metal-ion molarity (1 M versus 2 M) resulted in predictable size control and morphology, confirming this factor's impact [34].

12.4.5 Reaction Time

Longer incubation periods support gradual reduction and full nanoparticle development. Extended times often lead to greater particle maturation and improved colloidal stability due to enhanced coverage by biomolecules. The formation of nanotubes from ovalbumin and lysozyme required hours of incubation, ultimately producing ordered stable structures.

12.4.6 Functional Groups in Egg Proteins

The carboxyl, amino, and thiol groups in egg-derived proteins bind metal ions, aiding chelation and reduction. These chemical functionalities also serve as capping moieties, impacting nanoparticle size distribution, dispersion, and biological activity. In-the-field Fourier transform infrared spectroscopy (FTIR) analysis of egg-albumin-stabilized Fe_3O_4 NPs confirmed involvement of phenol, alcohol, and amine groups in reduction and capping.

12.4.7 Eggshell Membrane as a Bio-Scaffold

Using eggshell membrane (ESM) in the synthesis adds structural integrity and biocompatibility. Due to sustained ion release from $CaCO_3$ matrices, ESM-based particles are well-suited for wound-healing and drug-delivery systems.

12.4.8 Combined Optimization

Together pH, temperature, reaction time, reactant concentrations, and egg material type must be fine-tuned. Properly balancing these factors ensures nanoparticles that are uniform in size and morphology, biologically active, eco-friendly, cost-effective, and tailored for biomedical use [37]. For example, Fe_3O_4 nanoparticles synthesized with egg albumin displayed high colloidal stability, controlled size (10–100 nm), and enhanced antifungal and photocatalytic activity when all these factors were optimized [38].

12.5 Sustainability and Green Chemistry Aspects

Egg-derived biomaterials have emerged as versatile, eco-friendly platforms for synthesizing high-value nanoparticles under mild and sustainable conditions, seamlessly aligning with green chemistry principles. **ESM**, when exposed to aqueous $HAuCl_4$ (0.1 mM–0.1 M) at room temperature, yields diverse AuNPs that are suitable for bioimaging and biolabeling without the need for toxic reducing agents or energy-intensive equipment [39]. **Egg white templated ZnO nanowhiskers** produced via ultrasonication, low-temperature drying, and calcination at 550 °C, exhibit potent antibacterial and antibiofilm properties in dermal infection models, avoiding harsh chemicals and favoring aqueous media, confirming a biocompatible, energy-efficient synthesis route [29]. **Calcined eggshell biowaste** processed at modest temperatures (approx. 120 °C) and applied as catalyst supports in aqueous or solvent-free environments that enable porous CaO MOFs, Ag/CaO, and Fe_3O_4—CaO nanocomposite formation for applications spanning organic-carbonate synthesis, sustainable pharmaceuticals, and agricultural catalysis, turning biowaste into high-utility materials. **PtNPs** synthesized using **quail egg yolk** without additional chemicals highlight a low-energy, protein-driven process suitable for catalysis, fuel cells, biosensing, antibacterial treatment, and drug delivery applications [31]. Moreover, **egg white-derived NiO nanoparticles**, produced via low-temperature heating and calcination, show promise in photocatalysis, antimicrobial action, cancer therapy, and biosensing, which are all synthesized through benign, aqueous-based, renewable routes [40]. Room-temperature protein-mediated reactions using egg white or yolk synthesize ultra-small AuNPs that are suitable for bioimaging, photothermal therapy, drug delivery, antibacterial, antioxidant, and biosensor applications, once again without toxic agents or high-energy input [41].

These methods showcase core sustainability features, such as the use of **renewable feedstocks, aqueous and low-energy processes, waste valorization** (especially eggshell conversion to catalysts), and the **elimination of harmful chemicals.** They embody the 12 principles of green chemistry like prevention, use of renewable feedstocks, energy efficiency, and designing safer chemicals, while delivering multifunctional nanomaterials tailored for biomedical, environmental, and catalytic applications. The integration of waste-derived egg components into nanoparticle synthesis not only reduces environmental footprints but also offers scalable, decentralized routes that support circular economies and accessible green manufacturing [42].

12.6 Future Perspectives

Egg-based nanoparticle synthesis should emphasize enhancing **process efficiency and reproducibility**. Adopting **continuous-flow reactors** or integrating **microwave/ultrasound-assisted synthesis** can stabilize key parameters like protein concentration and reaction duration to consistently produce nanoparticles of well-defined sizes and morphologies. Embedding **life-cycle impact assessment (LCIA)** tools (e.g. ReCiPe, USEtox) early in development phases will help to identify and mitigate any environmental trade-offs, ensuring that egg-mediated routes remain greener than their conventional counterparts. Researchers should also pivot toward **multi-functional, stimuli-responsive nanoparticles** like pH or light-triggered drug-delivery systems using egg proteins as cost-effective and bio-safe building blocks. Establishing **eco-labels and certification standards** specific to biological nanotechnologies could enhance trust among consumers and regulators in sectors like pharmaceuticals, cosmetics, and food packaging. Finally, overcoming challenges such as colloidal instability, aggregation, and batch variability will require deeper physicochemical characterization like hydrodynamic measurements, zeta potential analysis, and long-term shelf-life studies, while fostering cross-disciplinary collaboration among chemists, engineers, toxicologists, and eco-design specialists to translate benchtop ingenuity into scalable, commercial-grade manufacturing.

12.7 Conclusion

Egg-mediated synthesis exemplifies a holistic approach to green chemistry, leveraging **renewable byproducts**, **aqueous, ambient conditions,** and **non-toxic reagents** to produce functional nanomaterials like AuNPs, ZnO nanowhiskers, CaO/MOFs composites, PtNPs, NiO, and ultra-small AuNPs. These strategies align tightly with green chemistry's 12 core tenets particularly regarding prevention, renewable feedstocks, energy efficiency, and safer chemicals. By eliminating hazardous substances like $NaBH_4$ and avoiding high-energy processing, these bioprocesses significantly reduce environmental toxicity and carbon emissions. To realize full commercial and societal impact, future efforts should broaden the standardization efforts, secure green certifications, and embed LCIA frameworks into product development. When this occurs, egg-derived nanomaterials rooted in circular bio economy stand to springboard into a new generation of **environmentally responsible, biocompatible, and cost-effective** technologies across sectors such as biomedicine, environmental remediation, and catalysis.

Conflict of interest: Authors declare that there is no conflict of interest. All the Tables and Figures are self-made and original.

Abbreviations

Abbreviation / Unit	Full Form
ESM	Eggshell Membrane
ESP	Eggshell Protein
NP	Nanoparticle/Nanoparticles
ROS	Reactive Oxygen Species
$CaCO_3$	Calcium Carbonate
ZnO	Zinc Oxide
Ag	Silver (ion)
Fe	Iron (ion)
Pt	Platinum (ion)
Ni	Nickel (ion)
Au	Gold
$HAuCl_4$	Hydrogen Tetrachloroaurate (gold precursor)
$Zn(NO_3)_2$	Zinc Nitrate
$Ni(NO_3)_2{\cdot}6H_2O$	Nickel Nitrate Hexahydrate
$NaBH_4$	Sodium Borohydride
LCIA	Life Cycle Impact Assessment
–OH	Hydroxyl Group
$-NH_2$	Amino Group
–COOH	Carboxyl Group
CaO/MOFs	Calcium Oxide/Metal–Organic Frameworks
mM	Millimolar (10^{-3} mol per liter)
M	Molar (1 mol per liter)
°C	Degrees Celsius (temperature unit)
hr	Hour (time unit)
nm	Nanometer (10^{-9} m; unit for nanoparticle size)

References

1 Gunnars K. 6 reasons why eggs are the healthiest food on the planet. Available from: https://www.healthline.com/nutrition/6-reasons-why-eggs-are-the-healthiest-food-on-the-planet (http://healthline.com); 2023.

2 Kuang H, Yang F, Zhang Y, Wang T, Chen G. The impact of egg nutrient composition and its consumption on cholesterol homeostasis. Cholesterol. 2018; 2018: 6303810.

3 Ajala EO, Eletta OAA, Ajala MA, Oyeniyi SK. Characterization and evaluation of chicken eggshell for use as a bio-resource. Arid Zone J Eng Technol Environ. 2018;14(1):26–40.

4 Réhault-Godbert S, Guyot N, Nys Y. The golden egg: nutritional value, bioactivities, and emerging benefits for human health. Nutrients. 2019;11(3):684.

5 Torres-Mansilla A, Hincke M, Voltes A, López-Ruiz E, Baldión PA, Marchal JA, et al. Eggshell membrane as a biomaterial for bone regeneration. Polymers (Basel). 2023;15(6):1342.

6 Saratale RG, Sun Q, Munagapati VS, Saratale GD, Park J, Kim DS. The use of eggshell membrane for the treatment of dye-containing wastewater: batch, kinetics and reusability studies. Chemosphere. 2021;281:130777.

7 **Eggnovo.** *Eggshell membrane collagen: why choose it over others*. Available from: https://eggnovo.com/en/eggshell-membrane-collagen; 2021.

8 Menon S, Rajeshkumar S, Venkat KS. A review on biogenic synthesis of gold nanoparticles, characterization, and its applications. Resour-Effic Technol. 2017;3(4):516–527.

9 Rozhin A, Batasheva S, Kruychkova M, Cherednichenko Y, Rozhina E, Fakhrullin R. Biogenic silver nanoparticles: synthesis and application as antibacterial and antifungal agents. Micromachines. 2021;12(12):1480.

10 Kalpana VN, Devi RV. A review on green synthesis, biomedical applications, and toxicity studies of ZnO NPs. Bioinorg Chem Appl. 2018;2018:3569758.

11 Owoseni-Fagbenro KA, Saifullah S, Imran M, Perveen S, Rao K, Fasina TM, et al. Egg proteins stabilized green silver nanoparticles as delivery system for hesperidin enhanced bactericidal potential against resistant *Staphylococcus aureus*. J Drug Deliv Sci Technol. 2019;50:347–354.

12 Murthy HCA, Desalegn Zeleke T, Ravikumar CR, Anil Kumar MR, Nagaswarupa HP. Electrochemical properties of biogenic silver nanoparticles synthesized using *Hagenia abyssinica* (Brace) JF Gmel. medicinal plant leaf extract. Mater Res Express. 2020;7(5):055016.

13 Gunasena MDKM, Galpaya GDCP, Abeygunawardena CJ, Induranga DKA, Priyadarshana HVV, Millavithanachchi SS, et al. Advancements in bio-nanotechnology: green synthesis and emerging applications of bio-nanoparticles. Nanomaterials. 2025;15(7):528.

14 Ahmed NS, Kamil FH, Hasso AA, Abduljawaad AN, Saleh TF, Mahmood SK. Calcium carbonate nanoparticles of quail's egg shells: synthesis and characterizations. J Mech Behav Mater. 2022;31(1):1–7.

15 Sharma P, Ganguly M, Doi A. Application of the synergism between eggshells and copper in nanotechnology. Nanoscale Adv. 2025;7(13):3914.

16 Bagheri S, Shameli K, Abd Hamid SB. Synthesis and characterization of anatase titanium dioxide nanoparticles using egg white solution via sol–gel method. J Chem. 2013;2013:848205.

17 Lu R, Yang D, Cui D, Wang Z, Guo L. Egg white-mediated green synthesis of silver nanoparticles with excellent biocompatibility and enhanced radiation effects on cancer cells. Int J Nanomedicine. 2012;7:2101–2107.

18 Zhao F, Li R, Liu Y, Chen H. Perspectives on lecithin from egg yolk: extraction, physicochemical properties, modification, and applications. Front Nutr. 2023:9.

19 Ahmed TAE, Wu L, Younes M, Hincke M. Biotechnological applications of eggshell: recent advances. Front bioeng biotechnol. 2021;9:675364.

20 Abeyrathne EDN, Nam KC, Huang X, Ahn DU. Egg yolk lipids: separation, characterization, and utilization. Food Sci Biotechnol. 2022;31(10):1243–1256.

21 Hmamouchi S, El Yacoubi A, El Idrissi BC. Using egg ovalbumin to synthesize pure α-Fe_2O_3 and cobalt doped α-Fe_2O_3: structural, morphological, optical and photocatalytic properties. Heliyon. 2022;8(2):e08953.

22 Kaushik A, Gola D, Raghav J, Gupta D, Chauhan R, Chhokar V, et al. Synthesis of silver nanoparticles using egg white: dye degradation and antimicrobial potential. Biointerface Res Appl Chem. 2021;12(2):2361–2372.

23 Vijan EA, Modan EM, Moga SG, Negrea DA, Schiopu AG, Oproescu M, et al. Assisted egg white biogenic synthesis for elaboration of ZnO nanoparticles. Crystals. 2025;15(1):71.

24 Gu L, Liu Y, Zhang W, Li J, Chang C, Su Y, et al. Novel extraction technologies and potential applications of egg yolk proteins. Food Sci Biotechnol. 2023;32(2):121–133.

25 Huang P, Yang DP, Zhang C, Lin J, He M, Bao L, et al. Protein-directed one-pot synthesis of Ag microspheres with good biocompatibility and enhancement of radiation effects on gastric cancer cells. Nanoscale. 2011;3(9):3623–3626.

26 Arya A, Tyagi PK, Bhatnagar S, Bachheti RK, Bachheti A, Ghorbanpour M. Biosynthesis and assessment of antibacterial and antioxidant activities of silver nanoparticles utilizing *Cassia occidentalis* L. seed. Sci Rep. 2024;14(1):7243.

27 Torres Gouveia ME, Milhans C, Gezek M, Camci UG. Eggshell based unconventional biomaterials for medical applications. Adv Nanobiomed Res. 2025;5(2):2400120.

28 Zheng B, Qian L, Yuan H, Xiao D, Yang X, Paau MC, et al. Preparation of gold nanoparticles on eggshell membrane and their biosensing application. Talanta. 2010;82(1):177–183.

29 Abdikakharovich SA, Rauf MA, Khattak S, Shah JA, Al Keridis LA, Alshammari N, et al. Exploring the antibacterial and dermatitis mitigating properties of chicken egg white synthesized zinc oxide nano whiskers. Front Cell Infect Microbiol. 2023;13:1295593.

30 Naik NS, Divakar S, S JM, Budagumpi S, Balakrishna RG, Padaki M. Biogenic derived egg shell and its derivatives as solid base heterogeneous catalysts for organic transformations: a comprehensive review. RSC Sustain. 2024;2(5):1246–1268.

31 Siddiqi KS, Husen A. Green synthesis, characterization and uses of palladium/platinum nanoparticles. Nanoscale Res Lett. 2016;11(1):482.

32 Iwata T, Doi H, Okumura K, Horio T, Hattori T, Takahashi K, et al. Comparative study on the deposition of enzyme-entrapped membranes with spatial homogeneity for bioimaging. Sensors Actuators B Chem. 2017;239:800–806.

33 Jain D, Kumar Daima H, Kachhwaha S, Kothari SL. Synthesis of plant-mediated silver nanoparticles using papaya fruit extract and evaluation of their antimicrobial activities. Dig J Nanomater Biostruct. 2009;4(3):557–563.

34 Schiopu AG, Vijan EA, Modan EM, Moga SG, Negrea DA, Istrate D, et al. Ovalbumin-mediated biogenic synthesis of ZnO and MgO nanostructures: a path toward green nanotechnology. Molecules. 2025;30(5):1164.

35 Kaviya S, Santhanalakshmi J, Viswanathan B, Muthumary J, Srinivasan K. Biosynthesis of silver nanoparticles using *Citrus sinensis* peel extract and its antibacterial activity. Spectrochim Acta A Mol Biomol Spectrosc. 2011;79(3):594–598.

36 Song JY, Kim BS. Rapid biological synthesis of silver nanoparticles using plant leaf extracts. Bioprocess Biosyst Eng. 2009;32(1):79–84.

37 A. M. King ori. A review of the uses of poultry eggshells and shell membranes. Int J Poult Sci. 2011;10:908–912.

38 Raza A, Khadrawy SM, Ahmad I, Imran M, Khuwaja G, Parveen H, et al. Biosynthesis of Fe^3O^4 nanoparticles using egg albumin: antifungal, dielectric analysis and photocatalytic activity. Catalysts. 2025;15(6):505.

39 Malode U, Patil YS, Selokar YN, Yadav PR, Bhagat RP, Nikose VM, et al. Sustainable approaches for the synthesis of biogenic platinum nanoparticles. Bull Natl Res Cent. 2023;47(1):1–14.

40 Sabouri Z, Akbari A, Hosseini HA, Khatami M, Darroudi M. Egg white-mediated green synthesis of NiO nanoparticles and study of their cytotoxicity and photocatalytic activity. Polyhedron. 2020;178:114351.

41 Pandit C, Roy A, Ghotekar S, Khusro A, Islam MN, Emran TB, et al. Biological agents for synthesis of nanoparticles and their applications. J King Saud Univ Sci. 2022;34(3):101869.

42 Federico T. Composition and method. Great Britain patent *GB2598715A*. Available from: https://patents.google.com/patent/GB2598715A/en?oq=GB2598715A+egg; 2022.

13

An Overview of Recent Patents on Biogenic Nanopharmaceutical Products

Arindam Chatterjee[1], Pratap Kalita[2], Parveen Kumar[3], Raja Chakraborty[1], and Saikat Sen[4]

[1] *Institute of Pharmacy, Assam Don Bosco University, Tepesia, Assam, India*
[2] *Pratiksha Institute of Pharmaceutical Sciences, Guwahati, Assam, India*
[3] *NIMS Institute of Pharmacy, NIMS University, Jaipur, Rajasthan, India*
[4] *Faculty of Pharmaceutical Science, Assam Down Town University, Guwahati, Assam, India*

13.1 Introduction

13.1.1 Definition of Biogenic Nanopharmaceuticals

Currently, biogenic nanopharmaceuticals (the word "nano" refers to a particle size that has a size range between 1 and 100 nm in at least one of the three dimensions) are of great interest in novel drug delivery systems, where active and functional materials can be delivered in a mannered approach to improving the bioavailability, bioaccessibility and bioactivity, and also improve the release kinetics of the active moieties to assert the pharmacological activities [1]. Considering the biomedical field to the energy field, nanotechnology covers a vast area in modern daily lives due to developing nanopharmaceuticals with unique physicochemical properties over bulk materials such as size, shape, increased surface area to volume, stability, low polydispersity, colloidal stability with Brownian movement, tunable surface charge and chemistry, easy modification functionality, and other properties like electronic and magnetic [2].

Metal nanoparticles such as silver, gold, iron, platinum, and selenium nanoparticles have potential applications in diverse fields such as electronics, agriculture, medicine, water treatment, and many more fields, and these metal nanoparticles are used to generate a series of physical and chemical methods which has some limitations [3]. Therefore, to overcome these drawbacks, biologically active agents have been utilized to fabricate biogenic nanoparticles, which are abundantly used in medicine, science, and technology. Also, these biogenic nanoparticles are

Biogenic Synthesis of Nanoparticles and Sustainable Nanotherapeutics,
First Edition. Edited by Malay K. Das.

safe, nontoxic, and eco-friendly, and are moving toward a greener approach. The biogenic nanoparticles have been fabricated by using several metal ions that stabilized, capped and polished by using biologically active materials such as plants, bacteria, algae, and fungi, which assert extended pharmacological activities due to their stability, unique size, biodistribution, improve bioavailability, improve morphological behavior (zero-dimensional spherical shape, one-dimensional rod shape, and two-dimensional thin films) [4].

The nanopharmaceuticals used to synthesize by using several physical (in which high energy radiations, mechanical pressure, and thermal energy induce condensation, material abrasion, melting, evaporation, etc., applied to fabricate nanopharmaceuticals) and chemical (in which organic/inorganic reducing agents such as sodium citrate, ascorbate, hydrazine hydrate, potassium auro chlorate, etc., have been used to fabricate the nanopharmaceuticals) methods utilized. Numerous physical and chemical methods were high-energy ball milling, laser ablation, electro-spraying, laser pyrolysis, flash spray pyrolysis, inert gas condensation and sol–gel method, micro-emulsion technique, hydrothermal synthesis, polyol synthesis, chemical vapor synthesis, and chemical vapor deposition technique, respectively [1, 4]. Although these methods are rapidly used to synthesize the nanoparticles, parallelly, these methods also have many limitations due to the use of hazardous chemicals that restrict the uses of these technologies and also not eco-friendly to the environment, and therefore biological methods such as biogenic synthesis using microorganisms, biogenic synthesis using biomolecules, biogenic synthesis using plants have been developed to fabricate the biogenic nanopharmaceuticals that are safe, minimal risk of toxicity, economical, easy and successful [1, 5].

13.1.2 Importance of Patents in Pharmaceutical Innovation

Patents are essential in fostering pharmaceutical innovation. They confer exclusive rights to innovators for their innovations, promoting a secure atmosphere for investment in research and development (R&D). The development of new pharmaceuticals is a protracted and expensive endeavor, frequently requiring years and billions of dollars to introduce a single medicine to the market. Patents confer a temporary monopoly, allowing corporations to recover their investments by prohibiting competitors from replicating their innovations.

This exclusivity promotes innovation by guaranteeing that pharmaceutical companies can achieve financial rewards, hence stimulating additional R&D initiatives. Patents facilitate the diffusion of knowledge, as inventors must publicly reveal the specifics of their ideas. This establishes a basis for subsequent progress.

Furthermore, patents stimulate the advancement of therapies for rare diseases and disorders, when market demand alone may not warrant expenditure. They reconcile public health requirements with economic sustainability, facilitating advancements in medical science.

13.2 Biogenic Nanopharmaceuticals

13.2.1 Sources of Biogenic Nanoparticles

Currently, nanomedicine fabrication, along with plant sources, has been considered as a potential biomaterial for treating various diseases with significant results and releasing behavior. Several metals or ions have been stabilized and capped with diverse plant metabolites or phytochemicals in the plant extract using different simple, easy, and cost-effective methods [6, 7]. Green synthesis of nanoparticles always remains a novel area in research to generate metallic nanoparticles, which is an easy and economical procedure that gives biocompatible nanoparticles due to its cost-effectiveness and favorable environmental impact. Several plants were used to fabricate the biogenic nanoparticles with the help of different polymers that improved the properties such as stability, delivery patterns, bioavailability, and release kinetics [8, 9]. The *Polygonum salicifolium* leaves aqueous extract was used in the green synthesis of platinum nanoparticles by a one-step method, and the prepared nanoparticles were spherical with a size range from 1 to 3 nm. In addition, the prepared nanoparticle also showed high stability due to its zeta potential of −25.9 mV, confirmed by its stability for more than 3 months [10]. In another study, the magnesium oxide nanoparticles were prepared by using the aqueous extract of *Manihot esculenta* leaf as a source of bioactive phytochemicals, and fabrication of nanoparticles was confirmed by several sophisticated studies like ultraviolet–visible spectrophotometry, scanning electron microscopy (SEM), energy-dispersive X-ray analysis (EDX), transmission electron microscopy (TEM), X-ray diffractometry (XRD), Fourier transform infrared spectroscopy (FTIR), and the prepared nanoparticles produced monodisperse, hexagonal in shape with particle size of 36.7 nm [11].

The iron oxide nanoparticles were prepared using *Rhamnella gilgitica* leaves extract, which was used as a reducing and stabilizing agent in the preparation, and the prepared nanoparticles were analyzed through a series of studies like SEM, TEM, XRD, DLS, FTIR, EDX, and Raman spectroscopy. The iron oxide nanoparticles capped with *R. gilgitica* leaf extract produced 269 nm with moderate polydispersity, spherical in shape, various functional groups of leaf extract involved in the formation of the iron oxide nanoparticles, and a highly crystalline nature of the iron oxide nanoparticles [12]. The aqueous extract of *Cinnamomum tamala* leaf was used for the biosynthesis of silver nanoparticles (AgNPs) without adding any extra stabilizing and capping agents, and a strong interaction was observed between AgNPs and leaf extract. Spherical-shaped nanoparticles with 10–12 nm were reported, and the stabilized nanoparticles produced significant therapeutic responses [7].

Biogenic silver chloride nanoparticles successfully capped with the ethanolic extract of *Pulicaria vulgaris* Gaertn. aerial part revealed nanoparticles' stabilized structure due to plant metabolites such as phenolic and flavonoid compounds. Also, it displayed average crystalline size and spherical size particles of 14.3–50 nm in size [13]. The potential AgNPs were prepared using *Lysiloma acapulcensis* tree extract containing many antimicrobial compounds. The compounds with chemical groups available in the *L. acapulcensis* tree extract stabilized the AgNP by acting as stabilizing and antimicrobial capping agents. They showed a crystalline nature, spherical and quasi-spherical in shape, with 1.2–62 nm in size [6]. The gold nanoparticles (AuNPs) were prepared with *Jasminum auriculatum* leaf extract, which works as a stabilizing and reducing agent in the preparation, and the prepared nanoparticles were well-stable, spherical, and 8–37 nm in size, confirmed by SEM and TEM. Also, the incorporation of *J. auriculatum* leaf extract stabilized the AuNPs due to the presence of flavonoids, polyalcohol, and terpenoids in *J. auriculatum* leaf, which showed a more crystalline nature [14].

Earlier studies reported that biogenic AgNPs were prepared using *Linum usitatissimum* seed plant extract and confirmed by TEM, XRD, and FTIR that the *L. usitatissimum* seed plant contains numerous functional agents that work as a reducing agent to stabilize the AgNPs [9]. The AuNPs were prepared through a simple photosynthesis process where the aqueous peel extract of *Spondias dulcis* was applied. The prepared AuNPs were crystalline and spherical, with an average particle size of 36.75 nm, confirmed by XRD, TEM, and SEM. They produced the most stable nanoparticles due to the reducing and stabilizing properties of the compounds present in the peel extract [8]. The fresh fruit extract of *Vaccinium arctostaphylos* L. was used to prepare the functional selenium nanoparticles, which show a better crystalline nature, regular spherical in shape, narrow size distribution with desired zeta-potential values, and appropriate biocompatibility [15]. The AgNP was prepared by a biogenic method where *Cotyledon orbiculata* extract was used to cap the AgNP, and the extract contained higher levels of polyphenols, flavanols, flavonols, and tannins, which are directly involved in the stabilization of the nanoparticles [16].

The photosynthesized AgNPs with methanolic extract of *Phagnalon niveum* plant, and the capped AgNPs produced spherical and monodispersed in shape, 12–28 nm in size, sharp crystalline nature, as reducing and capping agents due to the presence of several cyclic aromatic compounds such as phenols, tannins, flavonoids, terpenoids, coumarins and quinones in the plant extract [17]. The *Hibiscus sabdariffa* calyces extract was applied to synthesize the biogenic iron oxide nanoparticles that exhibited crystalline structure, spherical in shape, 61 nm

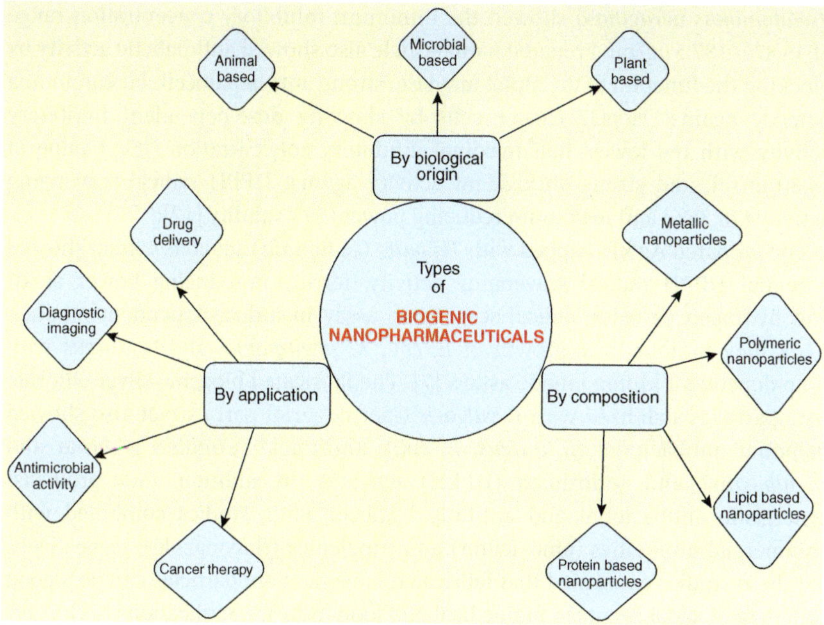

Figure 13.1 Types of biogenic nanopharmaceuticals.

in average particle size, absorption peak at 290–300 nm, and produced stable iron oxide nanoparticles confirmed by FTIR studies that the extract contains several functional groups that successfully capped the iron oxide nanoparticles [18]. In addition, the *Heliotropium eichwaldi* plant extract fully capped the AgNP by the functional groups present in the plant extract and also produced irregular in shape, 19.79 nm in average particle size confirmed by SEM, the crystalline region at several diffraction angles and the silver content was 30.68% while tested the capped AgNPs confirmed by EDX studies (Figure 13.1) [19].

13.2.2 Applications in Medicine

Polygonum salicifolium leaves aqueous extract platinum nanoparticles were tested against gram-negative bacteria, and antioxidant activity was investigated. The results showed that the highly stable nanoparticle produced high antibacterial efficacy against gram-negative bacteria (*Escherichia coli*), scavenged ~90% of 2,2-diphenyl-1-picrylhydrazyl (DPPH) radical, and also produced powerful antidiabetic activity [10]. Iron oxide nanoparticles capped with *R. gilgitica* leaf extract produced antimicrobial activity against pathogenic bacterial strains such as *Bacillus subtilis*, *Staphylococcus aureus*, *E. coli*, *Klebsiella pneumonia*, and

Pseudomonas aeruginosa showed the minimum inhibitory concentration range of 46.875–187.5 μg/ml. Prepared nanoparticle also showed antidiabetic activity by blocking the function of the alpha-amylase, strong anti-hepatocellular carcinoma potency against HepG2 cancer cells by showing dose-dependent inhibitory activity with the lowest half maximal inhibitory concentration (IC_{50}) value of 14.30 μg/ml, and strong antioxidant activity against DPPH radical scavenging activity (78.36%) and maximum reducing power (59%) ability [12].

The prepared AgNPs capped with *Tej patta* (*C. tamala*) aqueous extract showed potential DPPH radical scavenging activity, ferric ion-reducing power assay, and hydrogen peroxide radical scavenging assay in a dose-dependent manner. Also, they inhibited the growth of *E. coli*, *K. pneumonia*, and *S. aureus* with time-dependent killing kinetic assays [7]. The fabricated biogenic silver chloride nanoparticles stabilized with *P. vulgaris* Gaertn. aerial part extract also showed potential antibacterial (*S. aureus*, *E. coli*), antifungal (*Candida glabrata* and *C. albicans*), and antioxidant (DPPH) activates. In addition, they produced synergistic antibacterial and antifungal effects while studies combined with commercial antibiotics (amoxicillin) and fungicides (fluconazole), respectively. All these studies confirmed that fabricated biogenic nanoparticles can be a good source of biocidal agents in biomedical and food industry applications [12].

AgNPs were prepared using aqueous extract of *L. acapulcensis* tree, and the antimicrobial activity was increased after capping the AgNPs with the extract. The disk diffusion method was used to study the antimicrobial activity of the prepared nanoparticle and showed 19, 18, 16, and 15 mm as zones of inhibition against *C. albicans*, *E. coli*, *S. aureus*, *P. aeruginosa*, respectively, and prepared nanoparticles induce a decrease in cell viability in human peripheral blood lymphocytes that confirmed the synthesized biogenic nanoparticles were nontoxic and biocompatible [6]. Fabricated AuNPs with *J. auriculatum* leaf extract produced significant antifungal and antimicrobial activity against several human bacterial pathogens, such as *Streptococcus pyogenes*, *S. aureus*, *E. coli*, and *K. pneumonia*, and fungi such as *C. albicans*, *Aspergillus fumigatus*, *Lecanicillium lecanii*, and *Trichoderma viride*. The zone of inhibition range was reported from 7 to 12 mm and 4–5 mm against pathogenic bacteria and fungi at 30 μl. Nanoparticles suppressed the proliferation of the HeLa cancer cell and displayed a significant IC_{50} value of 104 μg/ml against the proliferation of the human cervical cancer cell line [14].

Linum usitatissimum seed plant extract was introduced in the fabrication of AgNPs to polish the bioactivity of the AgNPs by encapsulating them with the functional moieties available in *L. usitatissimum* seed. The nanoparticles produced antibacterial activity with high lethal effect against *B. subtilis*, *S. aureus*, *E. coli*, and Methicillin-resistant *S. aureus* (MRSA). In addition, the nanoparticle asserted anticancer activity at several concentration ranges from 1 to 160 μg/ml while tested against the HT-29 colon cancer cell line, MDA-MB-231 breast cancer cell line, healthy cell line L929-murine fibroblast cell lines, and MIA PaCa-2 human pancreatic cancer cell line, and confirmed the antioxidant activity against DPPH radicals and DNA cleavage activity [9]. Significant cytotoxic activity was reported for AuNPs stabilized by the aqueous peel extract of *S. dulcis*, and the stabilized AuNPs showed time and dose-dependent potential anticancer activity in Michigan cancer foundation – 7 (MCF-7) human breast cancer cells, and also confirmed its nontoxicity [8].

In addition, the synthesized selenium nanoparticles with *V. arctostaphylos* L. fruit extract showed IC_{50} of 84.19 μg/ml after the treatment of 72 hours while tested against 4T1 breast cancer cells. Also, they confirmed that fabricated selenium nanoparticles were attractive, nontoxic, and eco-friendly candidates for drug delivery [15]. *Cotyledon orbiculata* extract was used to stabilize the AgNP that showed potent antioxidant by significant ferric reducing power (127.28 μmol AAE/g) and 2,2′-Azino-bis (3-ethylbenzthiazoline-6-sulfonic acid) (ABTS) radical scavenging (134.54 μmol TE/g) ability that promotes the wound healing activity. While conducting the scratch assay, the extract and AgNPs showed growth of HaCaT, KMST-6 and CHO cells at low concentrations and promoted cell migration. Phytochemicals such as polyphenols, flavanols, flavanol, and tannins present in nanoparticles enhanced the bioavailability and bioactivity and promoted wound healing activity through activation of different pathways like phosphatidylinositol 3-kinase (PI3K)-protein kinase B (Akt), ras-proximate-1 (Rap1), interleukin-17 (IL-17), and tumor necrosis factor (TNF) signaling pathways, which confirmed that capped AgNPs with *C. orbiculata* extract are potential novel wound healing agents [16].

The methanolic extract of *P. niveum* plant was applied for stabilizing and capping the AgNPs that were studied for their antidiabetic activity in the alloxan-induced diabetic models, where the capped AgNPs produced significant antidiabetic activity by reducing the body weight, blood glucose level, triglycerides, cholesterol, low-density lipoprotein, and improving the level of high-density lipoprotein, kidney function, and liver function profiles. The phytoconstituents such as phenols, tannins, flavonoids, terpenoids, coumarins, alkaloids, sterols, and quinones present in the extract were only responsible for generating and displaying all the possible bioactivities, which confirmed that capped AgNPs with *P. niveum* extract are a promising antidiabetic agent [17]. The fabricated iron oxide nanoparticles with *H. sabdariffa* calyces extract revealed significant efficacy in inhibiting the growth of pathogenic bacteria such as *E. coli*, *K. pneumonia*, and *P. aeruginosa*. They produced minimum effective concentrations of 6.25, 12.5, and 25 μg/ml for *K. pneumonia*, *E. coli*, and *P. aeruginosa*, respectively [18].

Encapsulated AgNPs with the *H. eichwaldi* plant extract asserted potential activity against free radical scavenging, such as alpha-amylase and acetylcholinesterase inhibition. The capped AgNPs showed excellent % scavenging activity of 73%, 72%, 64%, 63.6%, 67.36%, and 67% against ferric chloride, ammonium molybdenum, DPPH radical, H_2O_2, ABTS radicals, and hydroxyl radical, respectively. In addition, the nanoparticle also showed maximum inhibition of 71.4% (IC_{50} value of 34.43 μg/ml) and 64% (IC_{50} value of 16.81 μg/ml) against alpha-amylase and acetylcholinesterase, respectively. All these findings confirmed that AgNPs capped with *H. eichwaldi* plant extract are potent antioxidant, antidiabetic, and anti-Alzheimer agents [19]. Biogenic synthesis of AgNPs using *Persicaria odorata* leaves extract produced dose-dependent antimicrobial potential against *S. epidermidis* and MRSA through time-kill kinetic assay. The nanoparticles also displayed potential wound-healing activity and cytocompatibility against normal human fibroblast cells [20].

13.3 Recent Advances in Biogenic Nanopharmaceutical Patents

Recent advancements in biogenic nanopharmaceutical patents underscore the increasing interest in leveraging natural resources and nanotechnology for therapeutic development. Biogenic nanoparticles, originating from flora, microorganisms, and other biological sources, demonstrate biocompatibility, improved bioavailability, and precise drug delivery potential. Recent patents emphasize novel techniques for producing these nanoparticles through environmentally sustainable technologies, utilizing their antibacterial, anticancer, and anti-inflammatory attributes. Innovative delivery technologies, such as nanoemulsions and nanocarriers, enhance treatment effectiveness and minimize adverse effects. Emerging trends encompass the integration of artificial intelligence (AI) in nanoparticle design and the investigation of synergistic effects with current medicines, highlighting customized medicine. These innovations are revolutionizing pharmaceutical innovation and sustainable healthcare [21–23].

13.3.1 Trends in Patent Filing and Approval

Current developments in patent applications and approvals for biogenic nanopharmaceuticals indicate a transition towards environmentally sustainable and highly efficient nanotechnology-based therapies. Biogenic nanoparticles, produced through biological entities such as plants, fungi, and bacteria, have garnered interest due to their biocompatibility, lower toxicity, and cost-effectiveness relative to chemically synthesized counterparts. The increase in patents in this field is propelled by progress in green synthesis techniques, functionalization approaches, and the investigation of distinctive biological sources [24].

Recent patents have concentrated on employing plant-derived nanoparticles for cancer therapy, owing to their inherent targeting ability and reduced side effects. A significant patent features a nanoparticle system developed from *Moringa oleifera* for the delivery of anticancer medicines with improved efficacy [25]. Likewise, the bacterial-mediated production of AuNPs for antimicrobial purposes and antioxidant properties has experienced a significant increase in intellectual property (IP) registrations [26].

There is a rise in hybrid biogenic nanopharmaceuticals that integrate bio-based nanoparticles with polymers for regulated medication release. AI-driven technologies facilitating accurate design and prediction of nanoparticle behavior represent an expanding domain of patent activity [27].

Patent filings are predominantly concentrated in the United States, Europe, and emerging economies such as India, indicative of worldwide rivalry in this sector. Regulatory frameworks are advancing to optimize approval processes, enabling expedited commercialization of these innovative treatments [28].

13.3.2 Key Players in Biogenic Nanopharmaceutical Patenting

Biogenic nanopharmaceuticals signify an advanced domain in drug delivery and treatments, utilizing nanoparticles sourced from biological processes. Prominent entities in this industry prioritize the creation of innovative patents to protect their advancements, frequently partnering with academic institutions, biotechnology companies, and major pharmaceutical corporations.

13.3.2.1 Academic and Research Institutions

Universities like Massachusetts Institute of Technology (MIT) and Stanford have led advancements in biogenic nanotechnology, securing patents for nanoparticles produced using microbes or plant extracts. MIT has patented nanocarriers for precise drug delivery utilizing biologically-derived polymers [29–31].

13.3.2.2 Pharmaceutical Corporations

Prominent global pharmaceutical entities such as Novartis and Pfizer are making substantial investments in biogenic nanotechnology. Pfizer has patented lipid-based nanoparticles, essential for the delivery of mRNA vaccines, exemplified by their COVID-19 vaccine [32–34].

13.3.2.3 Biotech Startups

Companies such as Nanospectra Biosciences concentrate on biogenic nanoparticles for therapeutic purposes. Nanospectra's proprietary nanoparticle technology utilizes laser-activated gold nanoshells to eradicate cancer cells [35].

13.3.2.4 Collaborations

Initiatives and collaborations backed by the government are essential. India's CSIR-Institute of Himalayan Bioresource Technology partners with entrepreneurs to patent eco-friendly synthesis techniques utilizing plant extracts for nanoparticles exhibiting antibacterial capabilities [36, 37].

13.3.3 Geographical Distribution of Patents

The geographical distribution of patents on biogenic nanoparticles reflects the global race for innovation in nanotechnology and biopharmaceuticals, with certain regions dominating due to their strong research ecosystems, funding, and industrial collaborations.

13.3.3.1 North America

The United States leads in patents related to biogenic nanoparticles, driven by significant investment in R&D, renowned academic institutions, and biotech startups. The National Institutes of Health (NIH) and large pharmaceutical companies like Pfizer and Moderna have bolstered patent activity, particularly in drug delivery and vaccine technologies. Canada also contributes, focusing on green synthesis and medical applications.

13.3.3.2 Europe

Countries like Germany, the United Kingdom, and Switzerland are prominent in this field. European patents often emphasize sustainability, reflecting the region's focus on green chemistry. Institutions like ETH Zurich and companies like Novartis and AstraZeneca have filed patents on biogenic nanocarriers for therapeutics and diagnostics.

13.3.3.3 Asia

India and China are emerging leaders, leveraging biodiversity for green synthesis. India's CSIR labs and universities file patents on plant-based nanoparticles with antimicrobial properties. China, with robust government funding, focuses on biomedical and industrial applications.

13.3.3.4 Others

Japan and South Korea contribute to high-tech applications, while Australia explores agricultural uses.

Regulatory frameworks are advancing to optimize approval processes, enabling expedited commercialization of these innovative treatments [28].

13.4 Patent Categories and Innovations

Patents on biogenic nanoparticles span various categories reflecting their versatile applications. **Drug Delivery Systems** dominate, with innovations in lipid or polymer-based nanoparticles for targeted therapies, as patented by companies like Pfizer. **Green Synthesis Methods**, employing plant extracts or microorganisms for eco-friendly nanoparticle production, are increasingly patented, especially in India and Europe. **Theranostics**, combining diagnostics and therapeutics, represents another major category, exemplified by gold or AgNPs for cancer treatment. The diversity of patent categories highlights biogenic nanoparticles' potential to revolutionize medicine, industry, and environmental management, driving sustainable and cutting-edge innovations globally [38].

13.4.1 Patents on Nanoparticle Synthesis

Patents on the biogenic synthesis of nanoparticles highlight eco-friendly, sustainable methods that utilize biological agents like plants, microorganisms, and enzymes. These methods reduce the need for toxic chemicals and energy-intensive processes, aligning with green chemistry principles.

13.4.1.1 Plant-Based Synthesis

Patents on plant-based biogenic synthesis of nanoparticles focus on harnessing plant extracts as eco-friendly reducing and stabilizing agents to produce nanoparticles with diverse applications. This method is gaining popularity due to its sustainability, cost-effectiveness, and the elimination of hazardous chemicals.

Many patents focus on using plant extracts as reducing and stabilizing agents. For example, a patent filed in India describes the synthesis of Cobalt Dioxide using neem (*Azadirachta indica*) leaf extract. These nanoparticles exhibit potent antimicrobial properties, suitable for medical and environmental applications [39, 40].

Synthesis Using Leaf Extracts Patents often involve leaves of medicinal plants rich in phytochemicals such as flavonoids, alkaloids, and phenolics. These nanoparticles demonstrate significant antimicrobial properties, ideal for medical and water purification purposes [40, 41].

Fruit and Flower Extracts Fruits and flowers are also used in nanoparticle synthesis. A patent describes the use of banana peel extract for AuNP synthesis, which finds applications in cancer therapy and biosensing. Similarly, hibiscus flower extract has been patented for producing iron oxide nanoparticles used in drug delivery [42–44].

Seed and Bark Extracts Extracts from seeds and barks, like *M. oleifera* seeds or cinnamon bark, are patented for their role in synthesizing metallic nanoparticles with antioxidant and antibacterial properties [45, 46].

These patents emphasize green methods that utilize plants as a rich source of bioactive compounds. Applications range from targeted drug delivery and diagnostics to environmental cleanup, supporting sustainable nanotechnology innovation.

13.4.1.2 Microbial Synthesis

Patents on microbe-based biogenic synthesis of nanoparticles explore the use of bacteria, fungi, algae, and yeast as biological factories to produce eco-friendly nanoparticles. This approach leverages the natural metabolic processes of microorganisms to reduce metal ions into nanoparticles, offering a sustainable alternative to chemical methods. A notable patent involves the use of *B. subtilis* to produce AuNPs with controlled morphology, ideal for drug delivery and diagnostics.

Bacteria-Mediated Synthesis Bacteria such as *B. subtilis* and *E. coli* are widely patented for synthesizing gold and AgNPs. A notable example is a patent utilizing *B. subtilis* to produce AuNPs with controlled size and shape, suitable for drug delivery and catalysis [47, 48].

Fungi-Based Synthesis Fungi like *A. niger* and *Fusarium oxysporum* are efficient at producing nanoparticles due to their high metal uptake capacity. A patent describes the synthesis of AgNPs using *F. oxysporum*, which show significant antibacterial activity [49, 50].

Algae and Cyanobacteria Microalgae like *Chlorella vulgaris* are patented for synthesizing nanoparticles such as zinc oxide and titanium dioxide. These nanoparticles are used in sunscreens and water purification [51–54].

Yeast-Based Synthesis Yeasts like *Saccharomyces cerevisiae* are patented for producing selenium nanoparticles with antioxidant properties for nutraceuticals and medicine [55, 56].

13.4.1.3 Other Biogenic Methods

Beyond plant and microbe-based methods, biogenic synthesis of nanoparticles extends to approaches using **enzymes**, **viruses**, and **biomolecules** like proteins or polysaccharides. Patents in this domain explore innovative ways to leverage biological components for sustainable and controlled nanoparticle production.

Enzyme-Assisted Synthesis Enzymes derived from biological systems offer precision in nanoparticle synthesis. Enzymes such as nitrate reductase and laccase are patented for their ability to reduce metal ions into nanoparticles. For example, a patent describes using nitrate reductase to synthesize AgNPs with uniform size and biocompatibility, suitable for drug delivery and biosensing applications [57–60].

Virus-Mediated Synthesis Certain viruses act as biological templates for nanoparticle formation. Patents include methods employing *Tobacco mosaic virus* to create nanostructures for electronics or catalytic applications. The virus's surface proteins help in the precise deposition of metal ions, resulting in nanoparticles with tailored shapes and sizes [61, 62].

Protein- and Polysaccharide-Based Synthesis Proteins such as bovine serum albumin and polysaccharides like chitosan are patented for nanoparticle synthesis due to their stabilizing and reducing properties. A notable example is a patent for chitosan-mediated synthesis of AgNPs with antimicrobial and wound-healing applications [63, 64].

13.4.2 Patents on Drug Delivery Systems

Patents on biogenic nanoparticles in drug delivery focus on their biocompatibility, targeted delivery capabilities, and sustained release properties. Biogenic nanoparticles, synthesized using plants, microbes, or biomolecules, are widely patented for encapsulating therapeutic agents. For example, patents detail AuNPs synthesized via microbial methods for delivering anticancer drugs directly to tumor cells, minimizing side effects. Lipid nanoparticles, patented by Pfizer for mRNA vaccines, demonstrate precision delivery to target cells. Other patents involve silver or polymer-based biogenic nanoparticles for antibacterial drug delivery. These innovations enhance efficacy, reduce toxicity, and enable personalized medicine, transforming the landscape of therapeutic interventions [24, 38].

13.4.2.1 Targeted Drug Delivery

Biogenic nanoparticles, synthesized using biological entities like plants and microorganisms, have garnered significant attention for their applications in targeted drug delivery. Several patents exemplify innovative methods and systems utilizing these nanoparticles to enhance therapeutic efficacy and specificity.

Composite Magnetic Nanoparticle Drug Delivery System This patented system integrates biocompatible and biodegradable polymers with magnetic nanoparticles and human serum albumin as a biological targeting agent. The design facilitates targeted, controlled release of chemotherapeutic agents directly to cancerous tumors and sites of inflammatory diseases, thereby minimizing systemic side effects [65].

Polymeric Nanoparticles for Drug Delivery This invention involves polymeric nanoparticles, ranging from 1 to 400 nm in size, with target molecules covalently bonded to their surfaces. These nanoparticles are designed to disperse homogeneously in aqueous solutions, enabling precise delivery of therapeutic agents to specific cells or tissues, which enhances treatment efficacy and reduces adverse effects [66].

Nanoparticles for Drug Delivery to the Brain Addressing the challenge of delivering therapeutics across the blood–brain barrier, this patent describes coated nanoparticles capable of penetrating brain tissue. The technology holds promise for treating neurological disorders by facilitating the transport of drugs directly to the brain [67]. These patents underscore the potential of biogenic nanoparticles in revolutionizing targeted drug delivery systems, offering more effective and safer therapeutic options across various medical fields.

Sustained Release Mechanisms Biogenic nanoparticles have drawn attention for their potential as a sustained-release drug delivery system. Poly(lactide-co-glycolide) has been used to design the nanoparticles that have the ability to circulate the vascular system for a long duration [68]. Dendrimers obtained otherwise than by reactions only involving carbon-to-carbon unsaturated bonds, e.g. polyethylene glycol, polyamines, and polyanhydrides, have also been utilized for the designing of nanoparticles that are capable of delivering drugs into nervous system [69].

13.4.3 Patents on Therapeutic Applications

Biogenic nanoparticles have shown significant promise in various therapeutic applications due to their unique properties and the environmentally friendly methods used for their synthesis. Several patents have been filed for biogenic nanoparticles in the fields of drug delivery, cancer therapy, and antimicrobial treatment.

13.4.3.1 Cancer Treatment

Biogenic nanoparticles derived from natural sources, such as plants or microorganisms, are specifically designed for targeted drug delivery. The nanoparticles can encapsulate various therapeutic agents, including anticancer drugs, and are engineered to improve bioavailability and reduce side effects [70]. Therapeutic nanoparticles can encapsulate nucleic acids and other therapeutic agents. The nanoparticles are designed for targeted delivery, particularly in cancer treatment, enhancing the efficacy of the therapeutic agents while minimizing side effects [71].

13.4.3.2 Antimicrobial Therapies

Several patents focus on the therapeutic applications of biogenic nanoparticles, particularly in antimicrobial therapy, leveraging the unique properties of nanoparticles produced through biological methods. Some describe antimicrobial coating compositions and antimicrobial properties of AgNPs. These compositions are suitable for coating medical devices and laboratory surfaces, providing a means to prevent microbial infections associated with these surfaces. This technology is particularly relevant for applications in wound healing and infection control [72, 73].

13.4.3.3 Other Diseases

Numerous studies have explored the potential of biogenic nanoparticles in the treatment of various noncancerous and noninfectious diseases.

For instance, AgNPs synthesized using plant extracts have been investigated for their potential in wound healing and tissue regeneration [74]. Additionally, biogenic nanoparticles have shown promising results in the management of neurodegenerative disorders, such as Alzheimer's disease and Parkinson's disease, through their ability to cross the blood–brain barrier and deliver therapeutic agents to targeted areas of the brain [75]. Furthermore, biogenic nanoparticles have been explored for their applications in cardiovascular diseases, where they can be used as drug delivery systems to improve the bioavailability and targeted delivery of therapeutic agents [76].

13.5 Case Studies of Prominent Biogenic Nanopharmaceutical Patents

The combination of biotechnology and nanotechnology has changed the pharmaceutical sector, enabling innovative avenues for medication discovery, development, and delivery. Biogenic nanopharmaceuticals, originating from biological materials or processes, have emerged as a possible solution for complex medical challenges [77]. These nanomedicines utilize the distinctive characteristics of nanoscale materials to enhance therapeutic efficacy, target specificity, and safety profiles, resulting in revolutionary advancements in the treatment of various diseases. The significance of IP in this field is paramount, as patents are essential for safeguarding and monetising these creative solutions. This chapter thoroughly introduces the area by analyzing case studies of notable biogenic nano-pharmaceutical patents [78].

13.5.1 The Rise of Biogenic Nano-Pharmaceuticals

The introduction of nanotechnology has caused a sea change in the pharmaceutical industry, opening the door to creating drug delivery systems on a nanoscale. Compared to conventional synthetic methods, biogenic nanopharmaceuticals have clear benefits from using natural or bioengineered ingredients. Among these are biocompatibility, decreased toxicity, and the capacity to utilize the inherent biological characteristics of materials like microbes, proteins, polysaccharides, and lipids [79].

For example, one way to circumvent the constraints of traditional treatments is by employing nanoparticles made from plant or microbial cultures. These have shown tremendous promise in delivering medications to precise sites. Nanomaterials with specific surface features and functionalization potential enable targeted drug delivery, prolonged release, and improved therapeutic results. Therefore, research into biogenic nanopharmaceuticals is ongoing across several fields, including cancer, infectious illnesses, regenerative medicine, vaccine delivery, and more [80].

13.5.2 The Role of Patents in Biogenic Nano-Pharmaceuticals

Nanomedicine is a very competitive industry. Hence, inventions are essential for encouraging research and securing market exclusivity. They shield innovators and organizations from infringement and give them the legal protection they need to monetize their ideas. In addition to encouraging spending on R&D, patents pave the way for licensing agreements, partnerships between universities and businesses, and the transfer of information [81].

Patents for biogenic nanopharmaceuticals often encompass a wide range of activities along the invention pipeline, such as:

- **Nanoparticle synthesis methods:** Techniques for producing nanostructures using biological systems, such as green synthesis using plant extracts or microbial fermentation [82].
- **Drug-nanoparticle formulations:** Compositions that combine therapeutic agents with biogenic nanoparticles for improved efficacy and delivery.
- **Targeting strategies:** Innovations in surface modification, functionalization, and ligand attachment for site-specific drug delivery.
- **Applications:** Patents highlighting specific therapeutic uses, such as cancer treatment, antimicrobial therapies, or vaccine platforms.

13.5.3 Importance of Case Studies in Understanding Patent Landscapes

Case studies offer significant insights into the dynamics of innovation and IP management within the biogenic nanopharmaceutical industry. Analyzing significant patents reveals insights regarding the following:

- **Research trends and technological advancements:** Identifying areas of active exploration and emerging technologies.
- **Market potential and commercial viability:** Understanding the value proposition of patented technologies and their potential to address unmet medical needs.
- **Collaborations and partnerships:** Examining the role of academic institutions, research organizations, and pharmaceutical companies in driving innovation.
- **Legal and regulatory considerations:** Analyzing how patents navigate the complexities of IP law, regulatory approvals, and ethical considerations [83].

13.5.4 Challenges in Biogenic Nano-Pharmaceutical Development

Despite their immense promise, biogenic nanopharmaceuticals face several challenges that must be addressed to realize their full potential [84]. These include

1) **Scalability and reproducibility:** Ensuring consistent quality and performance of biogenic nanoparticles across large-scale production.
2) **Regulatory hurdles:** Navigating stringent approval processes, particularly for biologically derived materials with inherent variability.
3) **Safety and toxicity concerns:** Evaluating nanomaterials' long-term effects and biocompatibility in complex biological systems.
4) **Economic and infrastructural constraints:** Overcoming high R&D, manufacturing, and commercialization costs.

Understanding how patents address these challenges can provide valuable lessons for researchers, policymakers, and industry stakeholders.

13.5.5 Notable Innovations in Cancer Therapy

Cancer remains a leading cause of global mortality, characterized by uncontrolled cellular growth, tissue invasion, and resistance to conventional therapies. The complexity of cancer treatment has driven the search for innovative methods that enhance efficacy while minimizing side effects. Biogenic nanopharmaceuticals have emerged as a revolutionary option, employing nanotechnology and biogenesis to develop targeted and efficacious cancer therapies. This section analyses the significant breakthroughs in biogenic nanopharmaceuticals that have substantially progressed in the battle against cancer [85].

13.5.5.1 The Possibilities of Biogenic Nano-Pharmaceuticals in Cancer Therapy

Biogenic nanopharmaceuticals are nanoscale drug delivery systems or therapeutic substances derived from natural or biological sources. They are often produced by biological processes or materials, such as plant extracts, microorganisms, and biopolymers, which offer specific advantages. Biogenic nanopharmaceuticals are nanoscale drug delivery systems or therapeutic substances derived from natural or biological sources. They are often produced by biological processes or materials, such as plant extracts, microorganisms, and biopolymers, which offer specific advantages. Biogenic nanopharmaceuticals are nanoscale drug delivery systems or therapeutic substances derived from natural or biological sources [86]. They are often produced by biological processes or materials, such as plant extracts, microorganisms, and biopolymers, which offer specific advantages:

(a) **Biocompatibility:** Reduced toxicity due to the use of naturally derived materials.
(b) **Targeted delivery:** Ability to functionalize nanoparticles for site-specific delivery to cancer cells.
(c) **Multifunctionality:** Potential to combine diagnostic and therapeutic capabilities in a single platform.
(d) **Sustainability:** Green synthesis approaches to minimize environmental impact.

The characteristics of these features have propelled the investigation of biogenic nanopharmaceuticals in oncological therapy, resulting in novel approaches to drug delivery, diagnostics, and treatment strategies.

13.5.5.2 Innovations in Biogenic Nano-Pharmaceuticals for Cancer Therapy

Plant-Based Nanoparticles Plants have been heavily used as biogenic feedstock for nanoparticle production, offering an environmentally acceptable and sustainable alternative to conventional chemical methods. Plant extracts contain bioactive compounds such as polyphenols, flavonoids, and alkaloids, which serve as reducing and stabilizing agents in the synthesis of nanoparticles.

- **AuNPs:** AuNPs derived from plant extracts hold significant promise for cancer treatment. Targeting ligands can be included in these nanoparticles to facilitate the localized delivery of anticancer drugs to cancer cells. Moreover, AuNPs play a significant role in photothermal therapy (PTT), which entails converting light into heat to target and eradicate cancer cells selectively.
- **AgNPs:** During evaluation against various cancer cell lines, AgNPs produced using biogenic methods showed cytotoxic effects. The antibacterial properties are beneficial for infection prevention during cancer treatment.

Microbial-Derived Nanoparticles Microorganisms, including yeast, fungi, and bacteria, substantially enhance the production of biogenic nanoparticles. They enable the synthesis of nanoparticles with controlled morphologies and dimensions, are economical, and can quickly scale in both directions [87].

- **Magnetic nanoparticles:** Magnetic iron oxide nanoparticles generated by microbial systems have been examined for potential application in hyperthermia for cancer treatment. These nanoparticles generate heat capable of selectively targeting and annihilating cancer cells when exposed to an alternating magnetic field.
- **Quantum dots (QDs):** Recently, microbially generated quantum dots have gained prominence in cancer detection and imaging. Their luminescence enables real-time monitoring of cancer cells during therapy.

Polymeric Nanoparticles Biopolymers like chitosan, alginate, and dextran are extensively utilized to formulate nanoparticles for cancer treatment. These polymers, sourced from natural origins, exhibit biodegradability and biocompatibility and are suitable for functionalization.

- **Chitosan-based drug carriers:** Chitosan nanoparticles have been used to deliver chemotherapeutic agents, including doxorubicin and paclitaxel. Their mucoadhesive characteristics augment drug absorption, while their capacity to target specific receptors on cancer cells enhances therapeutic efficacy.
- **Alginate nanoparticles:** Alginate-based systems have encapsulated hydrophobic pharmaceuticals, improving their solubility and bioavailability. These nanoparticles facilitate regulated drug release, thereby minimizing systemic toxicity.

Extracellular Vesicles (EVs) Cells spontaneously generate nanoparticles known as exosomes and microvesicles, categorized as extracellular vesicles. Designed for cancer treatment, they participate in intercellular communication.

- **Drug-loaded exosomes:** Chemotherapeutic agents or small interfering RNA (siRNA) can be encapsulated within exosomes derived from immune cells or mesenchymal stem cells to target neoplastic cells specifically. Their immunosuppressive qualities enhance their efficacy as therapy choices.
- **Tumor-derived vesicles:** Extracellular vesicles produced from tumors investigated as individualized cancer vaccinations. They possess tumor-specific antigens capable of eliciting an immune response that systematically targets cancer cells.

Nanohybrids and Combination Therapies Combining biogenic nanoparticles with other therapeutic agents or modalities has offered new paths in cancer treatment.

- **Nanohybrids for synergistic therapy:** Biogenic nanohybrids that integrate PTT with chemotherapy have improved antitumor efficacy. AuNPs functionalized with phytochemicals can concurrently deliver medications and enable PTT [88].

- **Combination with immunotherapy:** Biogenic nanoparticles are being investigated as vehicles for immune checkpoint inhibitors and cytokines, facilitating a dual approach to target cancer cells via direct cytotoxicity and immune regulation.

13.5.5.3 Translational Impact and Future Directions

Numerous biogenic nanopharmaceutical advancements have transitioned from preclinical studies to clinical trials, indicating their promise for practical applications. Liposomal versions of doxorubicin, such as Doxil, along with other nanomedicines, have facilitated the introduction of biogenic systems into the pharmaceutical market. Improvements in manufacturing technologies and regulatory frameworks are anticipated to expedite the conversion of these biogenic breakthroughs into sanctioned cancer treatments [89].

However, challenges remain, including:

1) **Scalability:** Developing cost-effective methods for large-scale production of biogenic nanoparticles.
2) **Regulatory hurdles:** Ensuring compliance with stringent safety and efficacy standards.
3) **Target specificity:** Improving targeting mechanisms to reduce off-target effects.

Biogenic nanopharmaceuticals possess significant potential to revolutionize cancer treatment by providing safer, more efficient, and eco-friendly alternatives. As the discipline progresses, interdisciplinary relationships and strong IP strategies will be essential in advancing these discoveries in oncology.

13.5.6 Patented Breakthroughs in Antimicrobial Agents

Antimicrobial resistance (AMR) has become a critical global health challenge, rendering many conventional antibiotics ineffective against emerging multidrug-resistant pathogens. Biogenic nanopharmaceuticals have emerged as a transformative solution to combat this crisis. These nanoscale therapeutic agents, synthesized using biological processes or materials, offer unique advantages such as enhanced bioavailability, targeted delivery, and reduced toxicity. Recent patented innovations in this field have opened new avenues for developing potent antimicrobial agents with high efficacy and specificity [90].

13.5.6.1 The Promise of Biogenic Nano-Pharmaceuticals in Antimicrobial Therapy

Biogenic nanopharmaceuticals utilize the inherent characteristics of biological systems to produce nanoparticles exhibiting antibacterial capabilities. The fabrication of these nanoparticles generally uses plant extracts, microbial cultures, or biologically derived compounds as reducing agents. This method renders the process ecologically sustainable and endows the nanoparticles with improved biocompatibility. The nanoscale dimensions of these compounds facilitate direct interaction with bacterial membranes, break biofilms, and penetrate otherwise impervious barriers. Furthermore, the surfaces of these nanoparticles can be modified with antimicrobial peptides (AMPs), medicines, or ligands, facilitating tailored activity against particular bacteria [82].

13.5.6.2 Key Patented Breakthroughs

Silver Nanoparticles (AgNPs) with Biogenic Synthesis A significant innovation is employing plant-based synthesis techniques for producing AgNPs, which have extensive antibacterial efficacy. Patents describe the utilization of medicinal plant extracts to synthesize AgNPs. These nanoparticles have shown effectiveness against drug-resistant bacteria and fungi while reducing toxicity to human cells.

Gold Nanoparticles Functionalized with Antimicrobial Peptides A different advancement pertains to AuNPs produced via microbial templates and modified with AMPs. These nanoparticles have significant efficacy against biofilm-forming bacteria, including *S. aureus* and *P. aeruginosa*.

Chitosan-Based Nanocarriers for Antimicrobial Delivery Applications have been granted for biogenic nanopharmaceuticals employing chitosan nanoparticles due to their ability to deliver antibiotics with exceptional accuracy. These systems augment drug stability, diminish the necessary dosage, and enhance antibacterial effectiveness. An example includes a chitosan nanoparticle system to combat resistant *E. coli* bacteria.

Green-Synthesized Nanocomposites Patents also emphasize developments in nanocomposites that integrate metal nanoparticles with additional antibacterial agents, including graphene oxide or phytochemicals. The composites demonstrate synergistic effects, rendering them effective against resistant infections.

13.5.7 Unique Formulations and Delivery Systems

Biogenic nanopharmaceuticals represent a category of novel therapies obtained from biologically based materials, including plant extracts, microbial cultures, and biopolymers. These nanomedicines attract interest due to their biocompatibility, environmentally sustainable synthesis, and improved therapeutic characteristics. A significant advancement in this subject is the creation of novel formulations and delivery technologies that tackle persistent difficulties in drug administration.

Table 13.1 Some of the recent patents filed in the field of biogenic nanopharmaceuticals.

Year of filing of patent	Patent application number	Type of biogenic nanoparticle	Innovator; current assignee	Current status	References
2016	WO2016102613A1	Gold nanoparticles for cancer therapy	Jon Golding, Phillip Williams, Meike Roskamp; Midatech Ltd, UK	Withdrawn/ Abandoned	[72]
2017	WO2017122224A1	Nanobiocomposite using *Syzygium cumini*	Sudesh Kumar Yadav, Rubbel Singla, Avnesh Kumari; CSIR, India	Active	[36, 37]
2018	US10507249B2	Lipid nanoparticles for mRNA delivery	Braydon Charles Guild, Frank DeRosa, Michael Heartlein; Translate Bio Inc. (Sanofi), USA	Active	[93]
2020	CN111515384A	Silver nanoparticles via *Moringa oleifera*	Zhu Qinchang, He Zhendan, Mohammad Riaz, Hafiza Mahreen Mehwish, Xiong Yongai; Shenzhen University, China	Active	[45]
2022	WO2022070198A1	Cinnamon extract-based nanoparticles	Eliezer J Pinthus, Gilad Agmon, Abraham Jacob Domb; Novid 20 Ltd, Israel	Published	[46]
2023	WO2023058037A1	Biomolecule mixture for biogenic synthesis of metal nanoparticles	Kiran D. Pawar, Megha Prakash Desai; Shivaji University, India	Published	[94]
2023	US20230301930A1	Nanoparticle cancer therapy	Ivan Mark Kempson, Tyron James Turnbull; University of South Australia, Australia	Withdrawn/ Abandoned	[95]
2024	US10828328B1	Method of preparing biogenic silver nanoparticles	R. Jothi Ramalingam, Hamad Al Lohedan; King Saud University, Saudi Arabia	Expired-Fee related	[96]

The nanoscale size of biogenic formulations improves their pharmacokinetic and pharmacodynamic characteristics. These nanoparticles enhance the solubility of hydrophobic medicines, hence assuring superior bioavailability. Functionalized biogenic nanoparticles (including those linked with targeting ligands or antibodies) facilitate site-specific distribution, limiting off-target effects and decreasing systemic toxicity. It is especially beneficial in oncology, where accurate medicine delivery to tumor cells is essential [91].

Controlled-release strategies in biogenic nanopharmaceuticals represent a significant advancement. Researchers have created drug carriers utilizing biologically produced polymers, such as chitosan, alginate, or proteins, that release their payload in response to specific physiological stimuli, such as pH, temperature, or enzyme activity. Stimuli-responsive systems enhance therapeutic efficacy by delivering the medicine at the appropriate time and location.

Moreover, biogenic nanopharmaceuticals enable the traversal of biological barriers, such as the blood–brain barrier, creating new opportunities for treating central nervous system illnesses. Innovations such as liposomal, dendrimer-based, or exosome-inspired delivery systems demonstrate the adaptability of biogenic nanoparticles in addressing delivery difficulties [92].

In summary, distinctive formulations and delivery mechanisms in biogenic nanopharmaceuticals signify a substantial advancement, facilitating safer, more effective, and patient-oriented treatment options (Table 13.1).

13.6 Challenges in Biogenic Nanopharmaceutical Patenting

Biogenic nano-pharmaceuticals, which integrate the principles of biotechnology and nanotechnology, have revolutionized modern medicine. These nanoscale innovations, derived from biological materials or processes, offer unique advantages in biocompatibility, targeted delivery, and therapeutic efficacy. However, the patenting of biogenic nanopharmaceuticals is fraught with complex challenges that stem from their intricate nature, interdisciplinary origins, and regulatory nuances [92, 97].

13.6.1 Definitional and Classification Ambiguities

A primary challenge to patenting biogenic nanopharmaceuticals is the definition and classification of these advances. Biogenic nanoparticles frequently display dual features, merging biological and nanoscale aspects. This duality engenders uncertainties in patent descriptions, complicating the precise definition of the invention. Vague assertions may result in conflicts, denials, or restrictions throughout the patent evaluation process.

Moreover, the novelty and inventive step necessary for patent approval may be challenged when biogenic nanopharmaceuticals use natural or commonly utilized biological components. For example, when a nanoparticle manufacturer utilizes a readily accessible plant extract, demonstrating the technique's or final product's distinctiveness may prove difficult.

13.6.2 Overlap with Existing Patents and Prior Art

The interdisciplinary characteristics of biogenic nanopharmaceuticals frequently lead to overlapping areas of IP, including biotechnology, nanotechnology, and pharmacology. It heightens the probability of patent disputes or infringement allegations. Moreover, locating prior art might be especially difficult owing to the enormous volume of current literature and patents within these intersecting domains.

Patent examiners and applicants may find distinguishing unique ideas from previous art challenging, particularly when minor alterations in synthesis methods or formulations are presented as novel inventions. This overlap generates ambiguities and may result in extended review durations or legal disputes.

13.6.3 Complexities in Biological Variability and Reproducibility

Biogenic nanopharmaceuticals are obtained from biological materials, including plant extracts, microorganisms, or proteins. The intrinsic unpredictability of biological materials might hinder the reproducibility of results, a vital consideration in patenting. Regulatory authorities and patent examiners may require proof of consistent and reproducible results to issue patents.

The manufacture of nanoparticles by plant extracts may lead to batch-to-batch discrepancies owing to variations in the chemical content of the extract. Proving that this diversity does not impact the functioning or efficacy of the final product presents a considerable challenge in the patenting process [98].

13.6.4 Ethical and Legal Constraints

Ethical considerations are important in biogenic nanopharmaceuticals, especially when the materials or processes involve living beings, genetically modified organisms, or rare biological resources. Numerous jurisdictions enforce limitations on patenting naturally occurring chemicals or biological materials, regardless of nanoscale modifications.

The Nagoya Protocol on Access and Benefit-sharing requires equitable distribution of benefits derived from genetic resources. Adherence to international agreements can complicate patent applications, particularly for advances related to traditional knowledge or biodiversity-rich areas.

13.6.5 Regulatory and Patent Office Disparities

The patenting process for biogenic nano-pharmaceuticals varies significantly across jurisdictions, reflecting differences in regulatory frameworks and patent office expertise. Some countries may need more technical infrastructure or expertise to assess the novelty and inventiveness of complex biogenic nanopharmaceuticals, leading to inconsistent patent-granting practices.

Moreover, the lack of harmonization in patent laws can challenge global commercialization. An invention granted a patent in one country may not be eligible in another due to differing standards for patentability, such as exclusions for biological materials or restrictive interpretations of novelty.

13.6.6 Cost and Resource Intensiveness

Securing patents for biogenic nanopharmaceuticals is a resource-intensive process, requiring significant financial and technical investment. The costs can be prohibitive for small-scale innovators or academic institutions, from drafting comprehensive patent applications to addressing legal disputes and conducting infringement analyses.

Defending patents against challenges or infringements requires substantial legal and financial resources. It creates barriers for startups or research-focused entities, potentially stifling innovation.

13.6.7 Balancing Broad and Narrow Claims

Applicants face a strategic dilemma in drafting claims for biogenic nanopharmaceutical patents. While broad claims provide wider protection, they are more susceptible to rejection or invalidation during examination. On the other hand, narrow claims may secure patent approval but limit the scope of protection, reducing the patent's commercial value.

Balancing these competing priorities requires careful drafting, supported by robust experimental data and a thorough understanding of the patent landscape.

13.6.8 Patenting Regulatory Hurdles

Biogenic nanopharmaceuticals, synthesized using biological systems like plants, microorganisms, or bioengineered processes, hold immense potential in revolutionizing drug delivery, diagnostics, and therapeutics. However, navigating the dual pathways of patenting and regulatory approval presents significant hurdles that innovators must overcome to bring these technologies to market [99].

13.6.8.1 Challenges in Patenting Biogenic Nano-Pharmaceuticals

Patents are essential for securing IP rights and fostering innovation, but the patenting process for biogenic nanopharmaceuticals is fraught with complexities. Key challenges include:

1) **Defining novelty and nonobviousness:** The overlap between natural processes and engineered innovations often leads to disputes over what qualifies as patentable. Demonstrating novelty and inventive steps becomes challenging, particularly when leveraging traditional knowledge or natural materials.
2) **Patent scope and claims:** Drafting claims that adequately protect the technology without being overly broad or vague is a delicate balance. In biogenic nanopharmaceuticals, this is complicated by the field's multidisciplinary nature, requiring precise descriptions of synthesis methods, compositions, and applications.
3) **Global patent variability:** Patentability standards vary across jurisdictions, leading to inconsistencies in IP protection. For instance, the European Patent Office (EPO) has stricter guidelines for biotechnological inventions than the United States Patent and Trademark Office (USPTO).

13.6.8.2 Regulatory Hurdles

Regulatory approval for biogenic nanopharmaceuticals involves stringent evaluation due to the novel nature of these materials. Key hurdles include:

1) **Safety and efficacy:** Ensuring that nanomaterials derived from biological sources are nontoxic, biocompatible, and effective across diverse populations.
2) **Characterization standards:** Regulatory agencies demand detailed physicochemical characterization of nanoparticles, including size, shape, surface properties, and batch-to-batch consistency.
3) **Ethical considerations:** Biogenic nanopharmaceuticals may involve materials derived from living organisms, raising concerns about biodiversity preservation and ethical sourcing.

Addressing these patenting and regulatory challenges requires a multidisciplinary approach, fostering collaboration among scientists, legal experts, and policymakers to unlock the full potential of biogenic nano-pharmaceuticals [100].

13.6.9 Ethical and Environmental Considerations

The emergence of biogenic nanopharmaceuticals has opened new frontiers in healthcare, leveraging biological systems to create nanoscale drug delivery systems with superior therapeutic potential; however, alongside the scientific and commercial advancements in this field, ethical and environmental considerations must be carefully addressed, particularly in the context of patenting these innovations.

13.6.9.1 Ethical Considerations in Patenting

Ethical challenges in patenting biogenic nanopharmaceuticals arise from the balance between incentivizing innovation and ensuring equitable access to life-saving treatments. The exclusivity of patents can lead to monopolistic pricing, potentially restricting access for underserved populations. It raises concerns about fairness in the global healthcare landscape, particularly in developing countries where affordability is critical.

Additionally, the source of biogenic materials often presents ethical dilemmas. Innovations based on traditional knowledge or biodiversity must adhere to principles of benefit-sharing, as outlined by frameworks like the Nagoya Protocol. Misappropriation of Indigenous resources or knowledge without adequate acknowledgement or compensation is unethical and legally contentious [79, 81].

13.6.9.2 Environmental Considerations in Development

The biogenic synthesis of nanopharmaceuticals is often touted as eco-friendly compared to chemical methods, as it utilizes biological systems such as plants, microbes, or biomolecules. However, large-scale production can pose unforeseen environmental risks, including biodiversity loss from the overharvesting of natural resources or unintended ecological impacts from the release of nanoparticles.

Patents must incorporate sustainability in their claims, focusing on green synthesis and waste reduction strategies. Encouraging renewable resources and promoting life-cycle assessments in production processes are critical to minimizing environmental footprints [83].

13.6.10 Intellectual Property Conflicts

The rapid advancement of biogenic nanopharmaceuticals, which harness biologically derived nanoparticles for therapeutic purposes, has opened new medical frontiers. However, this innovation is accompanied by significant IP challenges and conflicts. The dynamic interplay between cutting-edge research, competitive market forces, and complex patent landscapes has given rise to disputes that impact the development and commercialization of these technologies [12].

13.6.10.1 Origins of Intellectual Property Conflicts

1) Overlapping Patents
2) In the highly competitive field of nanomedicine, overlapping patent claims are a common source of disputes. Multiple entities may independently develop similar biogenic nanoparticles, leading to conflicts over ownership and exclusivity. For example, patents claiming similar synthesis methods using plant extracts or microbial cultures may clash, creating legal and operational hurdles.
3) Ambiguities in Patent Scope

4) Biogenic nanopharmaceuticals' complex and interdisciplinary nature often leads to ambiguities in patent claims. Questions regarding biogenic nanomaterials' novelty, nonobviousness, and utility can result in disputes during patent examinations or litigations. Additionally, determining the boundaries of a biogenic nanoparticle—such as its biological source or functional properties—can be contentious.
5) Global Regulatory Disparities
6) Variations in patent laws and standards across jurisdictions contribute to IP conflicts. What is patentable in one country may not qualify for protection in another, leading to fragmented market strategies and potential legal disputes for companies operating globally.

13.6.10.2 Resolving Intellectual Property Conflicts

1) Collaborative Frameworks
2) Promoting collaborations and cross-licensing agreements can help resolve IP disputes and ensure that critical technologies are shared. Consortia and public-private partnerships may play a pivotal role in mitigating conflicts.
3) Improved Patent Examination Processes
4) Strengthening the patent examination process, particularly for emerging fields like biogenic nanopharmaceuticals, can reduce ambiguities. Guidelines tailored to this niche domain help clarify patentability criteria.
5) Alternative Dispute Resolution (ADR)
6) Mechanisms such as mediation and arbitration offer cost-effective and time-efficient alternatives to litigation. These approaches can be particularly valuable for resolving disputes collaboratively and constructively [101].

13.7 Future Perspectives of Biogenic Nanoparticles

Biogenic nanoparticles, with their unique physicochemical properties and diverse applications, have garnered significant attention in the scientific community. Emerging areas of research, their potential impact on global healthcare, and strategies for overcoming patent challenges are crucial aspects that warrant exploration.

13.7.1 Emerging Areas of Research

The development of nanotechnology has provided an exciting and novel frontier for various fields, including biogenic nanoparticle synthesis [102].

Recent advancements in this domain have unveiled the potential of plant-based green synthesis as a sustainable and environmentally friendly approach [103].

Researchers have successfully utilized plant extracts to synthesize nanoparticles with diverse applications, such as in biomedical research, diagnostics, and drug discovery [104]. Additionally, the exploration of microbial synthesis of nanoparticles has gained momentum, presenting opportunities for optimization and improved efficiency [105].

13.7.2 Potential Impact on Global Healthcare

The unique properties of biogenic nanoparticles, including their enhanced bioavailability and pharmacological activity, have significant implications for the healthcare sector [104].

These nanoparticles have demonstrated promising applications in areas such as targeted drug delivery, tissue engineering, and disease diagnosis [103, 106]. Furthermore, the environmental and cost-effectiveness of biogenic nanoparticle synthesis offer potential solutions to the challenges faced in traditional nanomaterial production, making them attractive for large-scale healthcare applications.

However, the translation of these laboratory-based findings to real-world clinical settings remains a significant hurdle.

13.7.3 Strategies for Overcoming Patent Challenges

The rapid advancement of biogenic nanoparticle research has led to the emergence of numerous patents and IP claims. Navigating the complex patent landscape and ensuring the protection of innovative technologies are crucial for the successful commercialization and widespread adoption of these nanoparticles. Strategies such as collaborative research, strategic patent filing, and the development of novel green synthesis methods can help overcome these patent-related challenges. Additionally, fostering interdisciplinary partnerships and engaging with regulatory bodies can facilitate the seamless integration of biogenic nanoparticles into the global healthcare system [103–106].

13.8 Conclusion

Biogenic nanopharmaceuticals represent a transformative frontier in modern medicine, combining the benefits of green synthesis methods with innovative therapeutic applications. Using biological sources to produce nanoparticles offers distinct advantages, such as better compatibility with the human body, lower toxicity, and environmentally friendly processes. Advances in areas like targeted drug delivery, antimicrobial treatments, and cancer therapies highlight their potential to address some of the most pressing challenges in healthcare today.

However, realizing this potential comes with challenges. Issues like scaling up production, navigating complex regulations, addressing ethical concerns, and resolving IP disputes need thoughtful solutions. Collaborative efforts across disciplines, improvements in sustainable synthesis techniques, and strategic approaches to managing patents will be key to overcoming these barriers.

As the field continues to evolve, biogenic nanopharmaceuticals have the potential to reshape medical treatment approaches. By offering safer, more efficient, and environmentally sustainable options, this technology can significantly impact how diseases are treated worldwide, paving the way for a healthier future.

References

1. Kumari S, Verma R, Chauhan A, Raja V, Kumari S, Kulshrestha S. Biogenic approach for synthesis of nanoparticles via plants for biomedical applications: a review. Mater Today Proc. 2023. doi:10.1016/j.matpr.2023.04.242.
2. Nguyen NTT, Nguyen LM, Nguyen TTT, Nguyen DH, Nguyen DTC, Van Tran T. Recent advances on biogenic nanoparticles for detection and control of plant pathogens in sustainable agriculture: a review. Ind Crop Prod. 2023;198:116700. doi:10.1016/j.indcrop.2023.116700.
3. Kumar A, Shah SR, Jayeoye TJ, Kumar A, Parihar A, Prajapati B, et al. Biogenic metallic nanoparticles: biomedical, analytical, food preservation, and applications in other consumable products. Front Nanotechnol. 2023;5:1175149. doi:10.3389/fnano.2023.1175149.
4. Mughal B, Zaidi SZJ, Zhang X, Hassan SU. Biogenic nanoparticles: synthesis, characterisation and applications. Appl Sci. 2021;11(6):2598. doi:10.3390/app11062598.
5. Chatzidaki MD, Papavasileiou KD, Papadopoulos MG, Xenakis A. Reverse micelles as antioxidant carriers: an experimental and molecular dynamics study. Langmuir. 2017;33(20):5077–5085. doi:10.1021/acs.langmuir.7b00213.
6. Garibo D, Borbón-Nuñez HA, de León JND, García Mendoza E, Estrada I, Toledano-Magaña Y, et al. Green synthesis of silver nanoparticles using *Lysiloma acapulcensis* exhibit high-antimicrobial activity. Sci Rep. 2020;10(1):12805. doi:10.1038/s41598-020-69606-7.
7. Dash SS, Samanta S, Dey S, Giri B, Dash SK. Rapid green synthesis of biogenic silver nanoparticles using *Cinnamomum tamala* leaf extract and its potential antimicrobial application against clinically isolated multidrug-resistant bacterial strains. Biol Trace Elem Res. 2020;198:681–696. doi:10.1007/s12011-020-02107-w.
8. Pechyen C, Ponsanti K, Tangnorawich B, Ngernyuang N. Biogenic synthesis of gold nanoparticles mediated by *Spondias dulcis* (Anacardiaceae) peel

extract and its cytotoxic activity in human breast cancer cell. Toxicol Rep. 2022;9:1092–1098. doi:10.1016/j.toxrep.2022.04.031.

9 Karimi F, Tiri RNE, Aygun A, Gulbagca F, Özdemir S, Gonca S, et al. One-step synthesized biogenic nanoparticles using Linum usitatissimum: application of sun-light photocatalytic, biological activity and electrochemical H2O2 sensor. Environ Res. 2023;218:114757. doi:10.1016/j.envres.2022.114757.

10 Hosny M, Fawzy M, El-Fakharany EM, Omer AM, Abd El-Monaem EM, Khalifa RE, et al. Biogenic synthesis, characterization, antimicrobial, antioxidant, antidiabetic, and catalytic applications of platinum nanoparticles synthesized from Polygonum salicifolium leaves. J Environ Chem Eng. 2022;10(1):106806. doi:10.1016/j.jece.2021.106806.

11 Essien ER, Atasie VN, Okeafor AO, Nwude DO. Biogenic synthesis of magnesium oxide nanoparticles using *Manihot esculenta* (Crantz) leaf extract. Int Nano Lett. 2020;10(1):43–48. doi:10.1007/s40089-019-00290-w.

12 Iqbal J, Abbasi BA, Ahmad R, Shahbaz A, Zahra SA, Kanwal S, et al. Biogenic synthesis of green and cost-effective iron nanoparticles and evaluation of their potential biomedical properties. J Mol Struct. 2020;1199:126979. doi:10.1016/j.molstruc.2019.126979.

13 Sharifi-Rad M, Pohl P. Synthesis of biogenic silver nanoparticles (AgCl-NPs) using a *Pulicaria vulgaris* Gaertn. aerial part extract and their application as antibacterial, antifungal, and antioxidant agents. Nanomaterials. 2020;10(4):638. doi:10.3390/nano10040638.

14 Balasubramanian S, Kala SMJ, Pushparaj TL. Biogenic synthesis of gold nanoparticles using *Jasminum auriculatum* leaf extract and their catalytic, antimicrobial, and anticancer activities. J Drug Deliv Sci Technol. 2020;57:101620. doi:10.1016/j.jddst.2020.101620.

15 Gharbavi M, Mousavi M, Pour-Karim M, Tavakolizadeh M, Sharafi A. Biogenic and facile synthesis of selenium nanoparticles using *Vaccinium arctostaphylos* L. fruit extract and anticancer activity against in vitro model of breast cancer. Cell Biol Int. 2022;46(10):1612–1624. doi:10.1002/cbin.11852.

16 Tyavambiza C, Meyer M, Wusu AD, Madiehe AM, Meyer S. The antioxidant and in vitro wound healing activity of Cotyledon orbiculata aqueous extract and the synthesized biogenic silver nanoparticles. Int J Mol Sci. 2022;23(24):16094. doi:10.3390/ijms232416094.

17 Ul Haq MN, Shah GM, Gul A, Foudah AI, Alqarni MH, Yusufoglu HS, et al. Biogenic synthesis of silver nanoparticles using Phagnalon niveum and its in vivo anti-diabetic effect against alloxan-induced diabetic Wistar rats. Nanomaterials. 2022;12(5):830. doi:10.3390/nano12050830.

18 Saod WM, Al-Janaby MS, Gayadh EW, Ramizy A, Hamid LL. Biogenic synthesis of iron oxide nanoparticles using *Hibiscus sabdariffa* extract: potential for antibiotic development and antibacterial activity against multidrug-resistant bacteria. Curr Res Green Sustain Chem. 2024;8:100397. doi:10.1016/j.crgsc.2024.100397.

19 Sher N, Ahmed M, Mushtaq N, Khan RA. Synthesis of biogenic silver nanoparticles from the extract of Heliotropium eichwaldi L. and their effect as antioxidant, antidiabetic, and anti-cholinesterase. Appl Organomet Chem. 2023;37(2):e6950. doi:10.1002/aoc.6950.

20 Lubis FA, Malek NANN, Sani NS, Jemon K. Biogenic synthesis of silver nanoparticles using *Persicaria odorata* leaf extract: antibacterial, cytocompatibility, and in vitro wound healing evaluation. Particuology. 2022;70:10–19. doi:10.1016/j.partic.2022.01.001.

21 Jain A, Bhattacharya S. Recent advances in nanomedicine preparative methods and their therapeutic potential for colorectal cancer: a critical review. Front Oncol. 2023;13. doi:10.3389/fonc.2023.1211603.

22 Gurnule WB, Koinkar P, Bhagat DS, Bumbrah GS, Chawla PA. Recent advances in biomedical applications of biogenic nanomaterials. Curr Pharm Biotechnol. 2023;24:86–100. doi:10.2174/1389201023666220513101628.

23 Goel M, Sharma A, Sharma B. Recent advances in biogenic silver nanoparticles for their biomedical applications. Sustain Chem. 2023;4:61–94. doi:10.3390/suschem4010007.

24 Martins P, Rosa D, Fernandes A, Baptista PV. Nanoparticle drug delivery systems: recent patents and applications in nanomedicine. Recent Pat Nanomed. 2014;3:105–118. doi:10.2174/1877912304666140304000133.

25 US Patent Application US11617719B1. Systems and methods for secure communication and data transmission. Google Patents. Published March 28, 2023. Available from: https://patents.google.com/patent/US11617719B1/en.

26 Korean Patent Application KR101840928B1. Display device and method for controlling the same. Google Patents. Published March 20, 2018. Available from: https://patents.google.com/patent/KR101840928B1/en.

27 Joseph T, Kar Mahapatra D, Esmaeili A, Piszczyk Ł, Hasanin M, Kattali M, et al. Nanoparticles: taking a unique position in medicine. Nanomaterials. 2023;13:574. doi:10.3390/nano13030574.

28 World Intellectual Property Organization, World Intellectual Property Indicators Report: Global Patent Filings Reach Record High in 2023, PR/2024/927. [Internet] Geneva (CH): World Intellectual Property Organization; 2024. Available from: https://www.wipo.int/pressroom/en/articles/2024/article_0015.html.

29 WIPO Patent Application WO2024192242A2. Novel compositions and methods for treating neurodegenerative diseases. Google Patents. Published October 3, 2024. Available from: https://patents.google.com/patent/WO2024192242A2/en.

30 US Patent Application US20240376159A1. Heme-binding small peptide. Published October 3, 2024. Available from: https://patents.google.com/patent/US20240376159A1/en?oq=US20240376159A1.

31 US Patent Application US20140080698A1. Biotemplated inorganic materials. Published March 20, 2014. Available from: https://patents.google.com/patent/US20140080698A1/en?oq=US20140080698A1.

32 Australian Patent Application AU2024205089A1. Compositions and methods for the treatment of hemoglobinopathies. Published December 18, 2014. Available from: https://patents.google.com/patent/AU2024205089A1/en?oq=AU2024205089A1.

33 Australian Patent Application AU2016313439B2. Therapeutic nanoparticles comprising a therapeutic agent and methods of making and using same. Published May 6, 2021. Available from: https://patents.google.com/patent/AU2016313439B2/en?oq=AU2016313439B2.

34 US Patent Application US10125092B2. Lipids and lipid compositions for the delivery of active agents. Published November 13, 2018. Available from: https://patents.google.com/patent/US10125092B2/en?oq=US10125092.

35 Australian Patent Application AU2022201223C1. Devices and the use thereof in methods for ablation therapy. Published June 22, 2022. Available from: https://patents.google.com/patent/AU2022201223C1/en?oq=AU2022201223C1.

36 WIPO Patent Application WO2017122224A1. A nanobiocomposite formulation for wound healing and a process for the preparation thereof. Published July 20, 2017. Available from: https://patents.google.com/patent/WO2017122224A1/en.

37 CSIR-Institute of Himalayan Bioresource Technology. Patents [Internet]. Palampur (India): CSIR-IHBT. Available from: https://www.ihbt.res.in/en/about-us/patents.

38 Ali Hazis NU, Aneja N, Rajabalaya R, David SR. Systematic patent review of nanoparticles in drug delivery and cancer therapy in the last decade. Recent Adv Drug Deliv Formul. 2021;15:59–74. doi:10.2174/1872211314666210521105534.

39 US Patent Application US10059601B1. Synthesis of silver nanoparticles from *abelmoschus esculentus* extract. Published August 28, 2018. Available from: https://patents.google.com/patent/US10059601B1/en?oq=US10059601.

40 WIPO Patent Application WO2023038517A1. Method for *azadirachta indica* leaves extract assisted biogenic synthesis of cobalt oxide nanoparticles and its antimicrobial efficacy. Published March 16, 2023. Available from: https://patentscope.wipo.int/search/en/detail.jsf?docId=IN375171538&_cid=P21-M4IB2Q-37108-1.

41 Ajaykumar AP, Mathew A, Chandni AP, Varma SR, Jayaraj KN, Sabira O, et al. Green synthesis of silver nanoparticles using the leaf extract of the medicinal plant, Uvaria narum and its antibacterial, antiangiogenic, anticancer and catalytic properties. Antibiotics. 2023;12:564. doi:10.3390/antibiotics12030564.

42 Bankar A, Joshi B, Ravi Kumar A, Zinjarde S. Banana peel extract mediated synthesis of gold nanoparticles. Colloids Surf B: Biointerfaces. 2010;80:45–50. doi:10.1016/j.colsurfb.2010.05.029.

43 Buarki F, AbuHassan H, Al Hannan F, Henari FZ. Green synthesis of iron oxide nanoparticles using Hibiscus rosa sinensis flowers and their antibacterial activity. J Nanotechnol. 2022, 2022:1–6. doi:10.1155/2022/5474645.

44 Indian Patent Application IN202141049992. A process for extraction of copper oxide nanoparticles using green synthesis. WIPO. Published December 3, 2021. Available from https://patentscope.wipo.int/search/en/detail.jsf?docId=IN346038286.

45 Chinese Patent Application CN111515384A. Nano-silver material mediated and synthesized by moringa seed extract. Published on August 14, 2020. Available from: https://patents.google.com/patent/CN111515384A/en.

46 WIPO Patent Application WO2022070198A1. Antiviral active cinnamon extract and process. April 7, 2022. Available from: https://patents.google.com/patent/WO2022070198A1/en.

47 Srinath BS, Namratha K, Byrappa K. Eco-friendly synthesis of gold nanoparticles by *Bacillus subtilis* and their environmental applications. Adv Sci Lett. 2018;24:5942–5946. doi:10.1166/asl.2018.12224.

48 US Patent Application US9809476B2. Biological composite material loaded with magnetic nanoparticles with core-shell structure, the preparation therefore and the application. Published November 7, 2017. Available from: https://patents.google.com/patent/US9809476B2/en.

49 WIPO Patent Application WO2021232336A1. Development of antimicrobial silver nanoparticle dispersed gel for wound healing. Published March 12, 2014. Available from: https://patentscope.wipo.int/search/en/detail.jsf?docId=IN211622336&_cid=P21-M4IAZ4-32539-1.

50 WIPO Patent Application WO2019036777A1. Process for producing silver biosynthetic nanoparticles, the silver biosynthetic nanoparticles produced and use thereof. Published August 24, 2018. Available from: https://patentscope.wipo.int/search/en/detail.jsf?docId=WO2019036777&_cid=P21-M4IAZ4-32539-1.

51 Morowvat MH, Kazemi K, Jaberi MA, Amini A, Gholami A. Biosynthesis and antimicrobial evaluation of zinc oxide nanoparticles using *Chlorella vulgaris* biomass against multidrug-resistant pathogens. Materials. 2023;16:842. doi:10.3390/ma16020842.

52 Taghizadeh SM, Lal N, Ebrahiminezhad A, Moeini F, Seifan M, Ghasemi Y, et al. Green and economic fabrication of Zinc oxide (ZnO) nanorods as a broadband UV blocker and antimicrobial agent. Nanomaterials. 2020;10:530. doi:10.3390/nano10030530.

53 Thenarasu A, Chai MK, Tan YH, Wong LS, Rajamani R, Djearamane S. Study of *Chlorella vulgaris* from different growth phases as biosensor for detection of titanium and silver nanoparticles in water. Nat Environ Pollut Technol. 2024;23:959–969. doi:10.46488/NEPT.2024.v23i02.030.

54 Japanese Patent Application JP2024150519A. Devices, cartridges, and sensors for controlling fluid transfer and analyzing biological samples. Published October 23, 2024. Available from: https://patents.google.com/patent/JP2024150519A/en?oq=JP2024150519.

55 Hyrslova I, Kana A, Nesporova V, Mrvikova I, Doulgeraki AI, Lampova B, et al. in vitro digestion and characterization of selenized *Saccharomyces cerevisiae, Pichia fermentans* and probiotic *Saccharomyces boulardii*. J Trace Elem Med Biol. 2024;83:127402. doi:10.1016/j.jtemb.2024.127402.

56 Japanese Patent JP7228059B2. Delivery, engineering and optimization of systems, methods and compositions for sequence engineering and therapeutic applications. Published February 22, 2023. Available from: https://patents.google.com/patent/JP7228059B2/en?oq=JP7228059B2.

57 WIPO Patent Application WO2006084276A3. Enzyme catalyzed metallic nanoparticle synthesis. Published March 1, 2007. Available from: https://patents.google.com/patent/WO2006084276A3/en.

58 Rose GK, Thakur B, Soni R, Soni SK. Biosynthesis of silver nanoparticles using nitrate reductase from *Aspergillus terreus* N4 and their potential use as a non-alcoholic disinfectant. J Biotechnol. 2023;373:49–62. doi:10.1016/j.jbiotec.2023.07.002.

59 Ovais M, Khalil AT, Ayaz M, Ahmad I, Nethi SK, Mukherjee S. Biosynthesis of metal nanoparticles via microbial enzymes: a mechanistic approach. Int J Mol Sci. 2018;19:4100. doi:10.3390/ijms19124100.

60 Samrot AV, Jane Cypriyana PJ, Saigeetha S, Jenifer Selvarani A, Purayil SK, Ponnaiah P. Microbially synthesized silver nanoparticles: mechanism and advantages—a review. Green Synth Silver Nanomater. 2022:439–478. doi:10.1016/B978-0-12-824508-8.00015-0.

61 Lee KZ, Basnayake Pussepitiyalage V, Lee Y, Loesch-Fries LS, Harris MT, Hemmati S, et al. Engineering tobacco mosaic virus and its virus-like-particles for synthesis of biotemplated nanomaterials. Biotechnol J. 2020;16. doi:10.1002/biot.202000311.

62 US Patent Application US11654204B2. Rod-shaped plant virus nanoparticles as imaging agent platforms. Published May 23, 2023. Available from: https://patents.google.com/patent/US11654204B2/en?oq=US11654204.

63 US Patent Application US20090130186A1. Wound dressing assemblies, systems, and methods formed from hydrophilic polymer sponge structures such as chitosan and incorporating silver nanoparticles. Published May 21, 2009. Available from: https://patents.google.com/patent/US20090130186A1/en.

64 US Patent Application US20140377331A1. Antimicrobial wound-covering material and method for manufacturing same. Published December 25, 2014. Available from: https://patents.google.com/patent/US20140377331A1/en.

65 US Patent Application US20120265001A1. Composite magnetic nanoparticle drug delivery system. Published October 18, 2012. Available from: https://patents.google.com/patent/US20120265001A1/en 67.

66 US Patent Application US20070190160A1. Nanoparticles for drug-delivery. Published August 16, 2007. Available from: https://patents.google.com/patent/US20070190160A1/en.

67 WIPO Patent Application WO2020210805A1. Nanoparticles for drug delivery to brain. Published October 15, 2020. Available from: https://patents.google.com/patent/WO2020210805A1/en?oq=WO2020210805A1.

68 US Patent Application US20100216804A1. Long circulating nanoparticles for sustained release of therapeutic agents. Published August 26, 2010. Available from: https://patents.google.com/patent/US20100216804A1/en.

69 US Patent Application US10857236B2. Sustained release nanocomposite, a process for producing the same and use thereof. Published December 1, 2020. Available from: https://patents.google.com/patent/US10857236B2/en.

70 WIPO Patent Application WO2016102613A1. Nanoparticles and their use in cancer therapy. Published June 30, 2016. Available from: https://patents.google.com/patent/WO2016102613A1/ru.

71 US Patent Application US9763891B2. Therapeutic nanoparticles and methods of use thereof. Published September 19, 2017. Available from: https://patents.google.com/patent/US9763891B2/en?oq=US9763891B2.

72 US Patent Application US20190151504A1. Antimicrobial coating compositions. Published May 23, 2019. Available from: https://patents.google.com/patent/US20190151504A1/en?oq=US20190151504A1.

73 US Patent Application US20070218555A1. Anti-microbial activity of biologically stabilized silver nano particles. Published September 20, 2007. Available from: https://patents.google.com/patent/US20070218555A1/en?oq=US20070218555A1.

74 Japanese Patent Application JP2020143125A. Method and composition for wound healing. September 3, 2020. Available from: https://patents.google.com/patent/JP2020143125A/en.

75 Siddiqi NJ, Abdelhalim MAK, El-Ansary A, Alhomida AS, Ong W. Identification of potential biomarkers of gold nanoparticle toxicity in rat brains. J Neuroinflammation. 2012;9. doi:10.1186/1742-2094-9-123.

76 Ramasamy M, Lee J. Recent nanotechnology approaches for prevention and treatment of biofilm-associated infections on medical devices. Biomed Res Int. 2016;2016:1. doi:10.1155/2016/1851242.

77 Akhtar N, Pathak K. Probiotics as a tool to biosynthesize metallic nanoparticles: research reports and patents survey. Recent Pat Drug Deliv Formul. 2017;11(1):5–18. doi:10.2174/1872211311666170313124335.

78 Padhiary M, Roy D, Dey P. Mapping the landscape of biogenic nanoparticles in bioinformatics and nanobiotechnology: AI-driven insights. In: Synthesizing and Characterizing Plant-Mediated Biocompatible Metal Nanoparticles. 2025. IGI Global, p. 337–376. doi: 10.4018/979-8-3693-6240-2.ch014.

79 Zambonino MC, Quizhpe EM, Mouheb L, Rahman A, Agathos SN, Dahoumane SA. Biogenic selenium nanoparticles in biomedical sciences: properties, current trends, novel opportunities and emerging challenges in theranostic nanomedicine. Nanomaterials. 2023;13(3):424. doi:10.3390/nano13030424.

80 Petrovic S, Bita B, Barbinta-Patrascu ME. Nanoformulations in pharmaceutical and biomedical applications: green perspectives. Int J Mol Sci. 2024;25(11):5842. doi:10.3390/ijms25115842.

81 Yeung AW, Souto EB, Durazzo A, Lucarini M, Novellino E, Tewari D, et al. Big impact of nanoparticles: analysis of the most cited nanopharmaceuticals and nanonutraceuticals research. Current Res Biotechnol. 2020;2:53–63. doi:10.1016/j.crbiot.2020.04.002.

82 Jamkhande PG, Ghule NW, Bamer AH, Kalaskar MG. Metal nanoparticles synthesis: an overview on methods of preparation, advantages and disadvantages, and applications. J Drug Deliv Sci Technol. 2019;53:101174. doi:10.1016/j.jddst.2019.101174.

83 Grant E, Van den Hof M, Gold ER. Patent landscape analysis: a methodology in need of harmonized standards of disclosure. World Patent Inf. 2014;39:3–10. doi:10.1016/j.wpi.2014.09.005.

84 Ashique S, Upadhyay A, Hussain A, Bag S, Chaterjee D, Rihan M, et al. Green biogenic silver nanoparticles, therapeutic uses, recent advances, risk assessment, challenges, and future perspectives. J Drug Deliv Sci Technol. 2022;77:103876. doi:10.1016/j.jddst.2022.103876.

85 Venditto VJ, Szoka FC Jr. Cancer nanomedicines: so many papers and so few drugs! Adv Drug Deliv Rev. 2013;65(1):80–88. doi:10.1016/j.addr.2012.09.038.

86 Raghuwanshi N, Srivastava AK, Yadav TC, Gupta S, Khatri K, Pruthi V, Pruthi V. Biogenic nanoparticles as theranostic agents: prospects and challenges. In: Integrating Green Chemistry and Sustainable Engineering. 2019. Wiley, p. 647–684. doi: 10.1002/9781119509868.ch19.

87 Bhandari M, Raj S, Kumar A, Kaur DP. Bibliometric analysis on exploitation of biogenic gold and silver nanoparticles in breast, ovarian and cervical cancer therapy. Front Pharmacol. 2022;13:1035769. doi:10.3389/fphar.2022.1035769.

88 Wang X, Liu Z, Jin R, Cai B, Liu S, Bai Y, Chen X. Multifunctional hierarchical nanohybrids perform triple antitumor theranostics in a cascaded manner for effective tumor treatment. Acta Biomater. 2021;128:408–419. doi:10.1016/j.actbio.2021.04.019.

89 Ameis SH, Szatmari P. Imaging-genetics in autism spectrum disorder: advances, translational impact, and future directions. Front Psychiatry. 2012;3:46. doi:10.3389/fpsyt.2012.00046.

90 Outterson K, Samora JB, Keller-Cuda K. Will longer antimicrobial patents improve global public health? Lancet Infect Dis. 2007;7(8):559–566. doi:10.1016/S1473-3099(07)70188-3.

91 Zhang Y, Davis DA, AboulFotouh K, Wang J, Williams D, Bhambhani A, et al. Novel formulations and drug delivery systems to administer biological solids. Adv Drug Deliv Rev. 2021;172:183–210. doi:10.1016/j.addr.2021.02.01194.

92 Saraf S. Applications of novel drug delivery system for herbal formulations. Fitoterapia. 2010;81(7):680–689. doi:10.1016/j.fitote.2010.05.001.

93 United States patent US10507249B2. Lipid nanoparticles for mRNA delivery. December 17, 2019. Available from: https://patents.google.com/patent/US10507249B2/en.

94 World Intellectual Property Organization patent WO2023058037A1. Biomolecule mixture for biogenic synthesis of metal nanoparticles. April 13, 2023. Available from: https://patents.google.com/patent/WO2023058037A1/en.

95 United States patent application US20230301930A1. Nanoparticle cancer therapy. September 28, 2023. Available from: https://patents.google.com/patent/US20230301930A1/en.

96 United States patent US10828328B1. Method of preparing biogenic silver nanoparticles. November 10, 2020. Available from: https://patents.google.com/patent/US10828328B1/en.

97 Thirugnanasambandan T. Patenting issues in the development of nanodrugs. In: Intellectual Property Issues in Nanotechnology. 2020. CRC Press, p. 247–276. doi: 10.1201/9781003052104-16.

98 Ghosh S, Bhagwat T, Chopade BA, Webster TJ. Patents, technology transfer, and commercialization aspects of biogenic nanoparticles. In: Nanobiotechnology. 2021. Elsevier, p. 323–339. doi: 10.1016/B978-0-12-822878-4.00020-1.

99 Voelkl B, Altman NS, Forsman A, Forstmeier W, Gurevitch J, Jaric I, et al. Reproducibility of animal research in light of biological variation. Nat Rev Neurosci. 2020;21(7):384–393. doi:10.1038/s41583-020-0313-3.

100 Sinha J. Nanotechnology's significance in the treatment of viral illness and their future prospective with emerging patents. Int J Pharm Sci. 2024;2(10):92–110. doi:10.5281/zenodo.13882190.

101 Stawarska A, Bamburowicz-Klimkowska M, Runden-Pran E, Dusinska M, Cimpan MR, Rios-Mondragon I, Grudzinski IP. Extracellular vesicles as next-generation diagnostics and advanced therapy medicinal products. Int J Mol Sci. 2024;25(12):6533. doi:10.3390/ijms25126533 (108).

102 Keat CL, Aziz AA, Eid A, Elmarzugi NA. Biosynthesis of nanoparticles and silver nanoparticles. In: Bioresources and Bioprocessing, Vol. 2. 2015. Springer Science+Business Media. doi: 10.1186/s40643-015-0076-2.

103 Pal K, Chakroborty S, Nath N. Limitations of nanomaterials insights in green chemistry sustainable route: review on novel applications. In: Green Processing and Synthesis, Vol. 11. 2022. De Gruyter, p. 951. doi: 10.1515/gps-2022-0081.

104 Rayzah M, Elderdery AY, Alzerwi NAN, Alzahrani B, Alsrhani A, Alsultan A, et al. *Syzygium cumini* (L.) extract-derived green titanium dioxide nanoparticles induce caspase-dependent apoptosis in hepatic cancer cells. Plants. 2023;12:3174. doi:10.3390/plants12183174.

105 Rana A, Yadav K, Jagadevan S. A comprehensive review on green synthesis of nature-inspired metal nanoparticles: mechanism, application and toxicity. J Clean Prod. 2020;272:122880. doi:10.1016/j.jclepro.2020.122880.

106 Mishra S, Keswani C, Abhilash PC, Fraceto LF, Singh HB. Integrated approach of agri-nanotechnology: challenges and future trends. Front Plant Sci. 2017;8. doi:10.3389/fpls.2017.00471.

14

Green Toxicology and Safety Assessment of Biogenic Nanoparticles

Punam Talukdar[1,2], Ratul Chakraborty[1,2], Ayan Goswami[3], Chinmoyee Deori[4], and Asis Bala[1,2]

[1] *Pharmacology and Drug Discovery Research Laboratory, Division of Life Sciences; Institute of Advanced Study in Science and Technology (IASST), An Autonomous Institute under – Department of Science & Technology (Govt. of India) Vigyan Path, Guwahati, Assam, India*
[2] *Academy of Scientific and Innovative Research (AcSIR), AcSIR (an Indian Institute of National Importance), Sector 19, Ghaziabad, Uttar Pradesh, India*
[3] *Division of Pharmacology, Guru Nanak Institute of Pharmaceutical Science and Technology, Kolkata, West Bengal, India*
[4] *Department of Pharmacology, Lakhimpur Medical College and Hospital, North Lakhimpur, Assam, India*

14.1 Introduction

Nanotechnology combines essential principles from biology, physics, and chemistry, operating at the nanoscale level [1]. At this small scale, physical dimensions decrease, altering chemical properties and bonding behaviors. Additionally, biological processes, such as targeted drug delivery and molecular bonding, take place more effectively [1, 2]. The notion of nanotechnology was initially proposed by physicist Richard Feynman in 1959 during a presentation about the assembly of materials at the atomic and molecular scales [3]. Nanoparticles (NPs) are a diverse class of nanomaterials (NMs) that have been extensively researched for their unique properties and innovative applications in various fields, including biomedicine, pharmaceuticals, biosensing, electronics, and environmental technologies [4]. As the demand for NPs continues to rise, their production has experienced a substantial increase in recent years [5]. Traditionally, NPs have been synthesized using different physical and chemical methods, each with its own advantages and drawbacks [6]. However, recent trends indicate a significant shift towards more sustainable and environmentally friendly production techniques [6]. These innovative processes not only prioritize reducing immediate environmental impact but also take into account the long-term implications for

Biogenic Synthesis of Nanoparticles and Sustainable Nanotherapeutics,
First Edition. Edited by Malay K. Das.

our planet's climate, promoting a more responsible approach to nanoparticle manufacturing [6]. Biological methods for nanoparticle synthesis are exceptionally valuable as they leverage the inherent processes found in nature. The discovery of biological materials' ability to facilitate reduction in the nineteenth century was instrumental in the evolution of green synthesis techniques [7]. These innovative biological methods offer numerous advantages over conventional physical and chemical approaches, they are eco-friendly, minimizing environmental impact, and they are significantly more energy-efficient than physical processes [8].

Moreover, these methods are not only cost-effective but also scalable, making them highly suitable for extensive production (Iravani et al., 2014). Their connection to renewable resources introduces a diverse range of natural reducing agents, which enhances the versatility of the synthesis process [9]. Additionally, the high efficiency of these biological processes further solidifies their position as an appealing alternative for synthesizing NPs, promising both sustainability and effectiveness in various applications [10].

14.2 Biogenic Nanoparticles (BNPs)

Green nanotechnology represents a cutting-edge approach that leverages biological methods to create NPs using natural agents [11]. This technique provides a cost-effective and sustainable alternative to traditional nanoparticle synthesis and greatly improves environmental safety [12]. Nanotechnology is a dynamic and fast-evolving discipline that centers on the manipulation of materials at the nanoscale, generally ranging from 1 to 100 nm [13]. It encompasses a diverse range of applications across critical sectors, including energy production, food security and agriculture, bioengineering, national security and defense, medicine and healthcare, cosmetics, paints, and biotechnology [14]. Given its expansive scope, the large-scale synthesis of NMs remains a pressing requirement for the advancement of this technology [13]. Currently, the methods employed to synthesize NMs can be classified into physical, chemical, biological, or hybrid techniques, often combining multiple approaches [15]. While physical and chemical methods have been widely utilized in industrial settings, they are not without their drawbacks, including potential environmental harm and safety issues [16]. In this context, biogenic nanoparticles (BNPs) synthesized from organic biomatter or living organisms emerge as a more promising alternative [12]. These naturally derived NPs are viewed as superior due to their long-term sustainability, reusability, and biocompatibility, which align better with modern environmental and health standards compared to their conventionally synthesized counterparts [17].

The biological production of NPs is a fascinating process that utilizes different organisms, such as bacteria, fungi, and plants, to create these small materials using

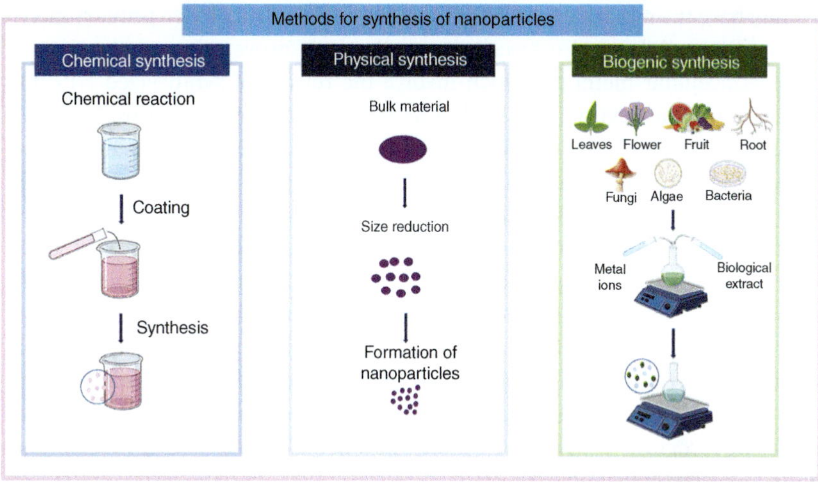

Figure 14.1 A comprehensive overview of various methods for synthesizing nanoparticles, accompanied by detailed schematic representations illustrating each technique available.

a bottom-up method [18] as shown in Figure 14.1 schematically [12]. Remarkably, microorganisms like bacteria, yeast, fungi, and actinomycetes possess the unique ability to thrive in extreme environmental conditions. This resilience allows them to withstand harsh environments and effectively convert metals into their ionic forms [19]. A notable example of this process is the synthesis of the first bacterial gold NPs, achieved using the organism *Bacillus subtilis* [20]. In this intriguing biosynthesis process, microorganisms take up metal ions from their environment and convert them into elemental metals through complex enzymatic activity [21]. Biosynthesis can be categorized into two distinct mechanisms: intracellular and extracellular. In the intracellular method, metal ions are absorbed into the microbial cell, where they are processed and transformed into NPs [22]. Conversely, in the extracellular approach, metal ions are captured on the surface of the microbial cell. Here, they undergo reduction to form NPs, demonstrating the remarkable capability of these organisms to engage in nanoscale manufacturing [19].

Fungi and bacteria employ both internal and external methods for the synthesis of NPs, with the specific pathways differing across various microorganisms [23]. In intracellular synthesis, positively charged metal ions are transported through negatively charged cell walls and internalized through specific mechanisms, such as gold nanoparticle synthesis using *Lactobacillus kimchicus* [24]. On the other hand, extracellular synthesis refers to the process where metal ions accumulate on the surface of cells, as demonstrated by the production of gold NPs by the bacterium *Paracoccus haeundaensis* [25].

Fungi have the remarkable ability to synthesize NPs through two primary mechanisms: intracellular and extracellular synthesis. However, they predominantly favor extracellular methods, which involve the release of synthesized NPs into their surrounding environment [26]. A notable example of this process is the use of the yeast species *Candida glabrata* for the production of silver nanoparticles (AgNPs) [27]. Recently, there has been an increasing interest in producing NPs using plant materials. This method is valued for its scalability, the vast biodiversity it encompasses, and the rich variety of phytochemicals available from different plant sources [12]. Although the yields of phytochemicals can fluctuate with the seasons, plant-based synthesis is highly regarded for its efficiency in terms of time and cost, as well as its safety profile [28]. While this approach often results in polydisperse NPs, meaning the particles vary in size and distribution, it is still considered an effective and practical method for the creation of stable NPs. With increasing recognition, phytofabrication is now being embraced as a key technique in the field of nanotechnology [29].

Green NPs represent a promising and sustainable alternative to conventional synthesis methods by harnessing the rich potential of bioresources such as plants and microbes [30]. Although microbes have demonstrated effectiveness in nanoparticle synthesis, their practical application is often hindered by several factors, including high operational costs, complexities in handling and scaling processes, and challenges related to downstream processing [31]. In contrast, plants offer a more straightforward and accessible solution for nanoparticle production; they are naturally abundant in a variety of bioactive compounds, such as phenols, polyphenols, and flavonoids, which play a vital role in the reduction and stabilization of NPs. These compounds not only facilitate the synthesis process but also contribute to the unique properties of the NPs themselves [17]. Therefore, focusing research efforts on plant-based synthesis of NPs holds great potential to overcome the existing challenges presented by microbial methods, ultimately leading to enhanced commercial applications and the development of greener technologies in various fields [32].

14.2.1 Green Toxicology Principles

Green toxicology is an innovative field that integrates essential toxicological principles into the design of safer chemical products (Figure 14.2). Its main goal is to significantly lessen the potential effects on human health and the environment [33]. The foundational principles of green toxicology revolve around several key strategies, including the creation of chemicals that are inherently nontoxic, conducting early safety assessments, minimizing exposure requirements, promoting sustainable testing methodologies, and employing progressive testing approaches that are still in development for regulatory use [34].

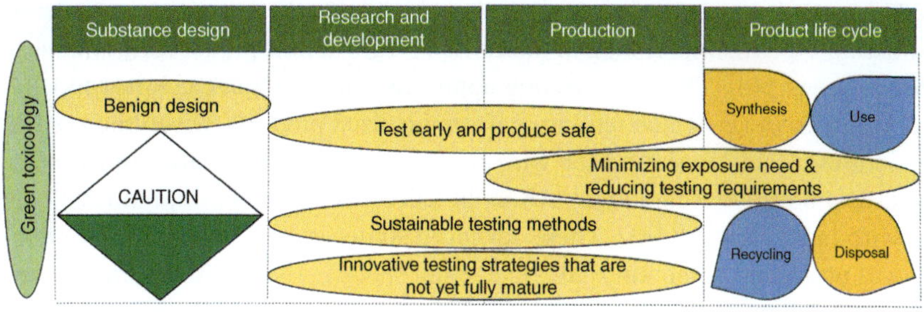

Figure 14.2 The principles of green toxicology: This figure provides a detailed schematic representation illustrating the core principles of green toxicology, as established in scientific literature. Green toxicology focuses on understanding the impact of chemicals on human health and the environment while promoting safer, more sustainable alternatives. The schematic offers a visual guide to these principles, highlighting key concepts and relationships in the field.

One of the core principles emphasizes the necessity of designing chemicals that are not only safe for people and the environment but also possess minimal potential for long-term ecological persistence [35]. An effective method for mitigating hazards during the design phase is to proactively identify and eliminate structural alerts specific to molecular features known to correlate with toxic risks. This approach necessitates a thorough evaluation of the entire chemical entity, be it a molecule, nanoparticle, or mixture, including its physical properties under various exposure conditions [36]. The strategy of prioritizing safety in chemical design must be integrated at the very outset of product development, ensuring that safety considerations are deeply embedded in the creation process. Chemists often rely on sophisticated tools such as in silico modeling and predictive methods, including Quantitative Structure–Activity Relationship models, to anticipate the toxicological profiles of new chemicals [37].

A notable example of this approach can be seen in the collaborative research undertaken by BASF SE and Metanomics GmbH. They showcased how metabolomics: a field that examines the unique chemical fingerprints left behind by cellular processes—can support the grouping of chemicals and the "read-across" method [38]. This was exemplified when they utilized biological data to successfully waive a traditional 90-day toxicity test for methylchlorophenoxypropionic acid (MCPP), a commonly used herbicide [39]. By comparing the metabolomic profiles of MCPP with those of structurally similar compounds, they were able to identify an analogous study substance and predict MCPP's toxicity profile accurately, all without the need for further animal testing. This illustrates the potential of green toxicology to innovate and improve chemical safety while also respecting ethical considerations in research [34].

The second principle of green toxicology highlights the critical importance of initiating experimental testing at the earliest stages of new product and chemical development [40]. This early testing approach stands in stark contrast to conventional methods, which tend to evaluate finalized products. Given the limited quantities of substances and the wide range of candidates available, researchers turn to innovative techniques such as in silico screening and high-throughput in vitro methodologies [41]. These advanced strategies play a vital role in prioritizing promising compounds and extracting specific insights into their mechanisms of action. A noteworthy example of such initiatives is the United States Environmental Protection Agency's (US EPA's) ToxCast program, which employs cutting-edge technologies to predict toxicity outcomes [42].

The third principle emphasizes the necessity of minimizing toxic exposure during the manufacturing process. This goal is achieved through collaborative efforts between chemists and toxicologists, who assess the product's entire lifecycle, from inception to disposal [33]. Key techniques employed include one-pot synthesis—where multiple reactions occur in a single vessel—utilization of nontoxic catalysts, and strategies aimed at reducing waste and environmental discharge [40]. While exposure data traditionally focuses on occupational settings, this approach often overlooks consumer exposure, leading to limitations in exposure-driven testing methods. In response to these challenges, RISK21 was developed, an integrated risk assessment framework that promotes collaborative design of data packages with regulators [43]. This framework features a tiered, exposure-driven methodology that visually represents both exposure and hazard data, aiding in comprehensive risk evaluations [44, 45].

The fourth principle advocates for a shift toward more sustainable practices in toxicity testing by aiming to reduce the usage of resources, chemicals, and biohazardous waste. A primary focus of this principle is decreasing reliance on animal studies, which have historically been the gold standard in toxicology [46]. Instead, green toxicology emphasizes the adoption of alternative methods, such as in vitro and in silico testing, aligning with the 3Rs principle- replace, reduce, and refine in vivo testing [33, 40]. Despite the promise of these innovative tools, the validation process can be sluggish, sometimes taking up to ten years due to stringent regulatory requirements. This lag hinders the integration of groundbreaking methods in swiftly evolving fields like NMs [47].

The fifth principle encourages the exploration of innovative testing strategies that are still in their developmental phases and not yet fully established for regulatory purposes. Toxicologists and risk assessors are pioneering integrated predictive methodologies, such as lazy structure–activity relationships (LAZAR) and read-across structure–activity relationships (RASAR) [48]. These techniques combine various strategies, including read-across and quantitative structure–activity relationships (QSAR), to estimate toxicological endpoints based on structural similarities. LAZAR establishes links through structural similarity, while RASAR enhances predictive accuracy by leveraging big data analytics, achieving remarkable accuracy rates of up to 95%. This approach not only provides consistency that often surpasses traditional animal testing but also holds significant potential to reduce the need for conventional testing methods, albeit requiring thorough validation before widespread application [49].

14.3 Safety Assessment of Biogenic Nanoparticles

Assessing the safety of BNPs has emerged as a vital research focus, particularly in light of their increasing application in both medicine and environmental science. These innovative NPs are produced through biological agents, such as fungi or plant extracts, offering a sustainable and eco-friendly alternative to traditional synthetic methods [50]. However, it is imperative to conduct comprehensive safety evaluations to confirm their compatibility with human health and ensure environmental sustainability.

Key factors to consider when evaluating the safety of BNPs include their size, shape, and surface characteristics. While smaller NPs can provide enhanced functionality and improved effectiveness in various applications, they can also present greater safety risks due to their increased reactivity and bioavailability [13]. The specific synthesis methods and concentrations employed during their production are influential in determining their ultimate size and morphology. Despite their encouraging attributes, including powerful antibacterial and anticancer effects, there remains a concerning potential for toxicity that could jeopardize the health of living organisms [51].

To determine the biocompatibility of BNPs, researchers utilize a combination of in vitro and in vivo methodologies. In vitro studies involve experimenting with various cell lines to gauge the NPs' effects on cell viability, membrane integrity, and the potential for DNA damage [52]. These preliminary assays are crucial for identifying harmful effects at the cellular level. Conversely, in vivo studies, which encompass experiments on animal models, provide valuable insights into the systemic implications of nanoparticle exposure. Such studies reveal how NPs interact with different organ systems, affect immune responses, and influence metabolic processes [53].

A growing body of evidence has highlighted the kidneys as a primary target for damage caused by nanoparticle exposure. When NPs accumulate in the renal system, they can induce nephric inflammation, necrosis of renal cells, and impair the kidneys' overall function [54]. AgNPs, in particular, have been found to disrupt energy metabolism while inflicting damage on mitochondria and cell membranes in the kidneys of rats, which may significantly contribute to the nephrotoxic effects associated with AgNP exposure This underscores the critical need for ongoing research to fully understand the implications of BNPs on health and the environment [55].

The interplay between cellular components and BNPs significantly influences their toxicity. Understanding the mechanisms through which these NPs exert harmful effects is critical. Two primary pathways of concern are

- **Oxidative stress:** BNPs have the capability to promote the production of reactive oxygen species (ROS) within cells. This surge in ROS can lead to oxidative stress, which can subsequently alter the structure and function of proteins, disrupt cellular membranes, cause lipid peroxidation—a process where lipids in cell membranes become oxidized—result in damage to DNA, and ultimately lead to mutations and genomic instability. Such cellular damage not only compromises individual cell health but can also contribute to broader disease processes [56].
- **Inflammation:** When immune cells encounter BNPs, they may become activated and trigger an inflammatory response [57]. This process involves the release of cytokines, which are signaling proteins that can lead to tissue damage and foster an environment of chronic inflammation. This inflammatory response can exacerbate tissue injury and contribute to various health issues [58].

Beyond their implications for human health, BNPs raise significant ecological concerns. Ecotoxicological research is dedicated to exploring how these NPs interact with various organisms within ecosystems, including plants, microorganisms, and aquatic species [59]. A pressing example of this is seen in aquatic environments, where BNPs can accumulate and have detrimental effects on fish and invertebrate populations. Such NPs can induce oxidative stress in these organisms or disrupt their reproductive functions, potentially leading to declines in population and biodiversity [60].

To mitigate these environmental implications, researchers are focused on evaluating the bioaccumulation potential and biodegradability of BNPs. A key model organism in these ecotoxicity studies is the zebrafish (*Danio rerio*), which has become widely used for assessing the toxicity of various chemicals and NPs [61]. By understanding the fate of BNPs in ecosystems, scientists aim to prevent their persistence and minimize the risk of long-term ecological harm, emphasizing the necessity for ongoing research and regulatory measures [59].

14.4 Factors Influencing BNP Toxicity

As the field of nanotechnology rapidly advances, increasing concerns have begun to surface regarding the potential toxicity of NPs. While scientists and toxicologists have dedicated efforts to assess the safety of various conventional NPs, the research surrounding the toxicity of green NPs—those derived from natural sources—remains surprisingly underexplored [62]. Furthermore, the investigation into the biocompatibility of biological metal NPs within animal models is limited, creating a gap in our understanding of their safety profile. The cytotoxic effects of these biological NPs are influenced by a multitude of factors [63]. Key among these is the capacity of plant molecules to effectively reduce metals, which plays a crucial role in the nanoparticle's properties. Other important considerations include the number of surface atoms on the NPs, their specific morphology or shape, their capping status—referring to any coatings that might stabilize or modify the NPs' behavior—and the presence of any residual unreduced metallic ions. Each of these characteristics can significantly affect toxicity assessments, revealing the complexity of evaluating the safety of these NMs [64]. Considering the complex and varied nature of the factors involved, there is a pressing necessity to create a swift, effective, and economical approach for carrying out toxicity studies on biological NPs. Developing such methodologies is crucial as they would significantly deepen our comprehension of the safety implications associated with these NPs [65].

The toxicity of NPs is largely determined by their potential to produce ROS, which may cause damage to cells, neurons, and genetic material. Various traits are significant in influencing the degree of their toxic impact [66]. These include the NPs' shape, which can affect how they interact with biological systems; surface area, which influences the amount of interaction; and dissolution rate, which governs how quickly they break down in biological environments. Furthermore, the aspect ratio, size, surface coating, agglomeration tendencies, and crystallinity of the NPs also contribute to their overall toxicity profile [67].

Together, these factors create a complex interplay that ultimately determines the safety and risk associated with the use of NPs in several applications. The comprehensive review on factor affecting the risk associated with NPs are shown in Figure 14.3 and described below:

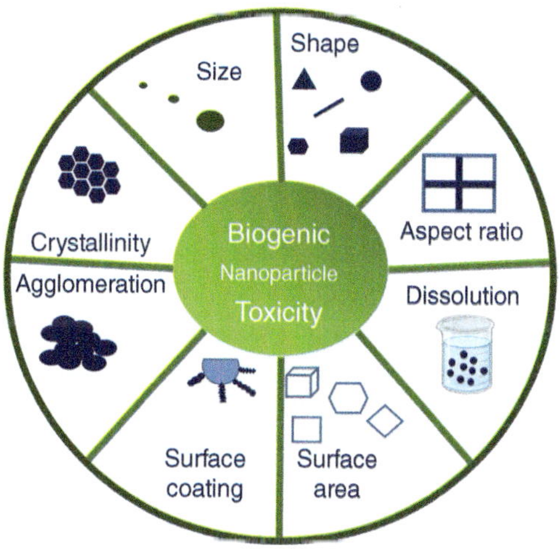

Figure 14.3 Factors influencing the toxicity of biogenic NPs play a crucial role in determining their safety and efficacy in various applications. This schematic representation highlights the numerous elements that can directly or indirectly impact the toxicological profile of NPs.

14.4.1 Size and Surface Area

The decrease in nanoparticle size leads to a dramatic increase in the surface area-to-volume ratio, which significantly enhances their biological and chemical reactivity. In contrast to larger particles, NPs possess a much greater surface area and a higher number of particles per unit mass. This unique characteristic enables them to engage in distinct interactions with biological systems [68]. The engineered nature of these NPs, combined with their augmented surface reactivity and increased surface area, can amplify the production of ROS. This elevation in reactive species may ultimately contribute to harmful effects at the cellular level, inducing cytotoxic and genotoxic responses [69].

14.4.2 Shape

The shape of NPs significantly influences their biological interactions, cellular uptake, and overall toxicity. Research has shown that spherical NPs are taken up by cells in much larger quantities compared to other shapes, which notably affects their reactivity in biological systems [70]. For instance, when examining the internalization of carbon NPs, both spherical and tubular forms have been observed to penetrate gut and gill epithelial tissues effectively, while cube-shaped particles fail to do so.

In another study, silver nanoplates were found to exhibit a markedly higher rate of endocytic uptake and increased adhesion properties, leading to greater toxicity levels than their spherical counterparts when tested on zebrafish embryos [71]. Similarly, when exploring gold NPs, it was noted that gold nanorods accumulated significantly fewer autophagosomes than gold nanospheres, suggesting a difference in their biological processing. Interestingly, star-shaped gold NPs were found to possess the highest cytotoxicity against human cells among various gold nanoparticle shapes. Additionally, needle-shaped NPs demonstrated a higher toxicity profile compared to their spherical equivalents, revealing a clear relationship between nanoparticle shape and both their harmful effects and potential applications, such as in the targeted delivery of oral medications [72].

14.4.3 Aspect Ratio

The toxicity of biogenic NMs is significantly influenced by their aspect ratio, which indicates the ratio of the NPs' width to height [73]. Nanoparticles that have an aspect ratio of 1 are perfectly spherical in shape, whereas nanotubes exhibit an aspect ratio that approaches zero, indicating a slender, elongated form. Research suggests that as the aspect ratio increases, the particles become longer and thinner, and the potential for toxicity also rises [74]. This correlation highlights the importance of considering the shape and dimensions of NMs when evaluating their safety and impact on biological systems.

14.4.4 Surface Coating

Nanoparticles often have surface coatings applied to them to modify and enhance their intrinsic properties. These coatings play a crucial role in improving various characteristics, including the stability of the NPs, their wettability in different environments, their rate of dissolution, and overall functionality in various applications. Importantly, the nature of these surface coatings can significantly influence the toxicity levels of the NPs [75].

14.4.5 Crystallinity

The crystallinity of biogenic NMs, which essentially reflects their level of structural organization, significantly influences their toxicological effects. Specifically, materials that possess higher degrees of crystallinity tend to demonstrate enhanced surface reactivity. This increased reactivity can trigger the formation of detrimental ROS, which, in turn, may induce oxidative stress within cellular environments [76]. Such oxidative stress can disrupt cellular functions and contribute to various health issues.

14.4.6 Dissolution

The dissolution of NPs is a crucial factor that affects their biological safety, uptake, and mechanisms of toxicity. Even NPs with the same size and composition can exhibit different behavior in terms of dissolution due to variations in their surface modifications. Media-soluble NPs, before being taken up by cells, tend to release ions, which can trigger oxidative stress and increase cytotoxic effects. This process influences their bioavailability and the resulting toxic response [77].

14.4.7 Agglomeration

Agglomeration is the clustering of NPs, which affects their reactivity and potential toxicity. Due to their high free surface energy, NPs tend to form larger aggregates unless stabilized by protective coatings. This phenomenon of agglomeration can lead to increased toxicity, which has been particularly noted in materials such as carbon nanotubes and various oxide NPs [78].

14.5 Regulatory Framework and Safety Guidelines

Despite the remarkable progress made in the field of nanotechnology, lingering concerns continue to surround the potential health and environmental risks related to green NMs. These innovative materials possess the capacity to deliver

Table 14.1 Regulatory framework for biogenic nanoparticles: A tabulated overview of the regulatory bodies and their key functions.

Regulatory body	Region/scope	Functions and key points	References
Food and Drug Administration (FDA)	United States (Global Outreach)	Flexible, adaptive regulatory approach; assesses NM risks and safety based on scientific evidence. Focuses on NM interactions with biological systems, product quality, and post-marketing safety.	[82]
European Medicines Agency (EMA)	European Union	Regulates NMs under medicinal product legislation. Created the European Nanomedicines Expert Group (2009). Collaborates with FDA and other agencies for biomedical NM regulation.	[83]
Scientific Committee on Emerging and Newly-Identified Health Risks (SCENIHR)	European Union	Evaluates health and environmental risks of NMs, particularly in cosmetics. Provides comprehensive risk assessments for multidisciplinary issues.	[84]
National Nanotechnology Initiative (NNI) and National Science Foundation (NSF)	United States	NNI funds nanotechnology research, while NSF raises awareness of NM risks through workshops, reports, and discussions.	[85, 86]
Ministry of Health, Labor and Welfare (MHLW) and Pharmaceuticals and Medical Devices Agency (PMDA)	Japan	Regulates NMs case-by-case under medicinal legislation. Collaborates with EMA and other agencies for data sharing and policy development.	[87]
Action Group on Erosion, Technology, and Concentration (ETC) and Health Canada	Canada	ETC oversees NM commercialization; Health Canada Evaluates NMs and nanomedicines based on regulatory principles. Provides guidance on nanotechnology-based food and health products.	[86]

both significant benefits and potential harms, underscoring the urgent need for comprehensive research to establish safe exposure levels and toxicity thresholds [79]. The development of advanced and sensitive analytical techniques is essential for the early detection of toxicity induced by NMs, allowing for timely intervention and mitigation of risks [80]. Furthermore, thorough risk assessments are critical; they must carefully consider not only the substances that might be released during the application of these materials but also those that may be present in the final products. This multifaceted approach is vital for effectively addressing safety concerns and ensuring the responsible development and use of nanotechnology in the future [81]. Table 14.1 describes the regulatory framework for BNPs.

In recent studies, various BNPs have been evaluated for their properties and biocompatibility across different biological models. Notably, urea NPs exhibited excellent biocompatibility, showing no significant harmful effects in both in vitro and in vivo conditions [88], which utilized WI-38 human fibroblast cells and rabbit models. Another significant finding involved AgNPs derived from *Malva crispa Linn*. These NPs were observed to cause cytological changes and intrahepatic localization that led to oxidative stress and immunotoxicity in zebrafish during in vivo experiments [89]. Furthermore, silver NPs synthesized from *Althaea officinalis* revealed concerns regarding cellular toxicity and oxidative stress when tested on both cell lines and zebrafish models [51].

Despite the promising applications of BNPs, the green synthesis process remains fraught with challenges that hinder its advancement. Key limitations include the heterogeneity of particle size and shape, polydispersity during production, and stability issues of the NPs [11]. Critical variables influencing the size, shape, and stability of NPs, including reaction time, temperature, pH, the concentration of natural catalysts, salt formation, and reduction processes, require careful control. Adjusting the reaction pH and temperature, in particular, plays a vital role in enhancing the quality and functionality of the synthesized NPs, paving the way for more effective applications in biomedical and environmental fields [90].

14.6 Case Studies

Investigations into biogenic NMs reveal a diverse range of biological effects that are significantly influenced by factors such as their origins and the methods used for their synthesis. Some biogenic NMs exhibit reversible toxicity, which may manifest as harmful effects like the accumulation of substances in organs, the induction of oxidative stress, and alterations in metabolic processes [66]. In contrast, other types of biogenic NMs demonstrate remarkable stability and biocompatibility, causing few to negligible adverse effects on living organisms. These varying outcomes highlight the critical need for comprehensive toxicity assessments and

Table 14.2 Studies of biogenic nanoparticle toxicity: Tabulated form of several available studies their finding and model/methods used.

Study	Nanoparticle type	Findings	Model	Type	References
Sargassum siliquosum-mediated silver NPs	Silver NPs	No mortality at 2000 mg/kg. Minimal increase in serum creatinine and blood urea nitrogen.	Rats	In vivo	[92]
Ficus religiosa-mediated silver NPs	Silver NPs	Increased serum AST, ALT, LDH, TNF-α, and IL-6 on day 29, normalized by day 89. Silver accumulation in liver, brain, and lungs (4.77, 3.94, and 3.043 μg/g tissue). Liver histological changes reversed by day 89.	Rats	In vivo	[93]
Mulberry, olive and fig-mediated sulfur and silver NPs	Silver and Sulfur NPs	Increased mortality in larvae, pupae, and adults; reduced larval longevity.	Insects	In vivo	[94]
Solanum nigrum-mediated silver NPs	Silver NPs	Improved body weight in diabetic rats; restored total cholesterol and triglyceride levels after 21 days of administration.	Rats	In vivo	[95]
Nephrotoxicity study with *Desmodium gangeticum*	Silver NPs (chemical and green synthesis)	Renal architecture changes, supported by urine and blood chemistry. Toxic similarities observed between chemically and green-synthesized NPs.	Wistar rats, renal cells	Both	[12]
Curcuma mangga-mediated gold NPs	Gold NPs	Stable in physiological media for 24 hours; cytocompatible with CCD-18Co and MRC-5 cells. Hemocompatibility with <10% hemolysis and no erythrocyte aggregation.	Cell lines	In vitro	[96]
Bacillus sp.	Urea NPs	Nanoparticles showed good biocompatibility without in vitro and In vivo damage	WI-38 and Rabbit	In vitro and In vivo	[88]
Malva crispa Linn.	Silver NPs	Cytological changes and intrahepatic localization caused oxidative stress and immunotoxicity	Zebrafish	In vivo	[89]
Althaea officinalis	Silver NPs	Cellular toxicity and oxidative stress	Cell lines and Zebra Fish	In vitro and In vivo	[51]

the establishment of standardized protocols. Such measures are essential to ensure the safe and effective application of biogenic NMs in numerous fields, including medicine, environmental science, and technology [91]. Table 14.2 describes the studies of BNP toxicity.

14.7 Challenges and Future Perspectives

BNPs have shown significant promise in a wide range of applications, yet the green synthesis of these NPs still faces several challenges that require attention [30]. One of the primary limitations stems from the heterogeneity in particle size and shape, as well as challenges related to polydispersity during the production process. Additionally, the stability of these NPs can be compromised, which poses further complications [97]. Researchers have been actively developing targeted methodologies and refining processing parameters to tackle these hurdles, as these factors have a direct correlation with the size, shape, polydispersity, and overall stability of the resulting NPs [11].

It is crucial to control various variables during the synthesis process, such as reaction time, temperature, pH levels, and the concentration of natural catalysts. The formation of salts and the reduction processes involved also play pivotal roles [98]. Among these variables, pH and temperature stand out as particularly influential; they significantly affect the size distribution, reduction rate, and morphology of BNPs [99]. For instance, adjustments in pH can alter the ionization state of the phytocompounds present in the solvent system, which subsequently influences the stability, surface functionalization, and efficiency of capping around the NPs. Establishing optimal synthesis conditions is essential to ensure the production of BNPs that are stable and functionally effective [100].

There is a critical demand for a more affordable, eco-friendly, and economically feasible approach to producing BNPs. Many potential natural reducing agents capable of facilitating this production have yet to be thoroughly investigated [101]. Critical factors such as particle size, shape, concentration, and the intended applications have profound implications on the perceived cytotoxicity of these NPs [17]. Addressing these gaps is vital for the advancement of green-synthesized NPs and for their safe and effective integration across various fields of application [102, 103].

Acknowledgments

The authors sincerely thank the Institute of Advanced Study in Science and Technology (IASST) and the Department of Science and Technology, Government of

India. Ms. Punam Talukdar and Mr. Ratul Chakraborty acknowledge the IASST and CSIR UGC for providing the I-JRF and NET-JRF fellowships, respectively.

References

1 Bhushan B (ed.). Springer Handbook of Nanotechnology. 2017. Springer, Berlin, Heidelberg.

2 Malik S, Muhammad K, Waheed Y. Nanotechnology: a revolution in modern industry. Molecules. 2023;28(2):661.

3 Haleem A, Javaid M, Singh RP, Rab S, Suman R. Applications of nanotechnology in medical field: a brief review. Glob Health J. 2023;7(2):70–77.

4 Singh RP, Rai AR, Abdala A, Chaudhary RG. Biogenic Sustainable Nanotechnology: Trends and Progress. 2022. Elsevier.

5 Fu Y, Liu T, Wang H, Wang Z, Hou L, Jiang J, Xu T. Applications of nanomaterial technology in biosensing. J Sci: Adv Mater Devices. 2024;9:100694.

6 Verma A, Gautam SP, Bansal KK, Prabhakar N, Rosenholm JM. Green nanotechnology: advancement in phytoformulation research. Medicine. 2019;6(1):39.

7 Kumar P, Singh PK, Hussain M, Kumar Das A. Synthesis of silver metal nanoparticles through electric arc discharge method: a review. Adv Sci Lett. 2016;22(1):3–7.

8 Iravani S, Korbekandi H, Mirmohammadi SV, Zolfaghari B. Synthesis of silver nanoparticles: chemical, physical and biological methods. Res Pharm Sci. 2014;9(6):385–406.

9 Naghdi M, Taheran M, Brar SK, Verma M, Surampalli RY, Valéro JR. Green and energy-efficient methods for the production of metallic nanoparticles. Beilstein J. Nanotechnol. 2015;6(1):2354–2376.

10 Álvarez-Chimal R, Arenas-Alatorre JÁ. Green Synthesis of Nanoparticles. A Biological Approach. 2023. IntechOpen.

11 Osman AI, Zhang Y, Farghali M, Rashwan AK, Eltaweil AS, Abd El-Monaem EM, et al. Synthesis of green nanoparticles for energy, biomedical, environmental, agricultural, and food applications: a review. Environ Chem Lett. 2024;22(2):841–887.

12 Patil S, Chandrasekaran R. Biogenic nanoparticles: a comprehensive perspective in synthesis, characterization, application and its challenges. J Genet Eng Biotechnol. 2020;18(1):67.

13 Khan I, Saeed K, Khan I. Nanoparticles: properties, applications and toxicities. Arab J Chem. 2019;12(7):908–931.

14 Lugani Y, Sooch BS, Singh P, Kumar S. Nanobiotechnology applications in food sector and future innovations. In: Microbial Biotechnology in Food and Health. 2021. Elsevier, p. 197–225.

15 Altammar KA. A review on nanoparticles: characteristics, synthesis, applications, and challenges. Front Microbiol. 2023;14:1155622.

16 Singh KR, Nayak V, Singh J, Singh AK, Singh RP. Potentialities of bioinspired metal and metal oxide nanoparticles in biomedical sciences. RSC Adv. 2021;11(40):24722–24746.

17 Kulkarni D, Sherkar R, Shirsathe C, Sonwane R, Varpe N, Shelke S, et al. Biofabrication of nanoparticles: sources, synthesis, and biomedical applications. Front Bioeng Biotechnol. 2023;11:1159193.

18 Adeleke BS, Olowe OM, Ayilara MS, Fasusi OA, Omotayo OP, Fadiji AE, et al. Biosynthesis of nanoparticles using microorganisms: a focus on endophytic fungi. Heliyon. 2024;10(21):e39636.

19 Li X, Xu H, Chen ZS, Chen G. Biosynthesis of nanoparticles by microorganisms and their applications. J Nanomater. 2011;2011(1):270974.

20 Southam G, Beveridge TJ. The in vitro formation of placer gold by bacteria. Geochim Cosmochim Acta. 1994;58(20):4527–4530.

21 Singh NA, Narang J, Garg D, Jain V, Payasi D, Suleman S, Swami RK. Nanoparticles synthesis via microorganisms and their prospective applications in agriculture. Plant Nano Biol. 2023;5:100047.

22 Lahiri D, Nag M, Sheikh HI, Sarkar T, Edinur HA, Pati S, Ray RR. Microbiologically-synthesized nanoparticles and their role in silencing the biofilm signaling cascade. Front Microbiol. 2021;12:636588.

23 Bhardwaj K, Sharma A, Tejwan N, Bhardwaj S, Bhardwaj P, Nepovimova E, et al. *Pleurotus* macrofungi-assisted nanoparticle synthesis and its potential applications: a review. J Fungi. 2020;6(4):351.

24 Mughal B, Zaidi SZJ, Zhang X, Hassan SU. Biogenic nanoparticles: synthesis, characterisation and applications. Appl Sci. 2021;11(6):2598.

25 Cherian T, Maity D, Rajendra Kumar RT, Balasubramani G, Ragavendran C, Yalla S, et al. Green chemistry based gold nanoparticles synthesis using the marine bacterium Lysinibacillus odysseyi PBCW2 and their multitudinous activities. Nanomaterials. 2022;12(17):2940.

26 Guilger-Casagrande M, Lima RD. Synthesis of silver nanoparticles mediated by fungi: a review. Front Bioeng Biotechnol. 2019;7:287.

27 Ghosh S, Ahmad R, Zeyaullah M, Khare SK. Microbial nano factories: synthesis and biomedical applications. Front Chem. 2021;9:626834. doi:10.3389/fchem.2021.626834.

28 Ying S, Guan Z, Ofoegbu PC, Clubb P, Rico C, He F, Hong J. Green synthesis of nanoparticles: current developments and limitations. Environ Technol Innov. 2022;26:102336.

29 Zhang X, Zhao X, Li H, Hao X, Xu J, Tian J, Wang Y. Detection methods of nanoparticles synthesized by gas-phase method: a review. Front Chem. 2022;10:845363.

30 Hosseingholian A, Gohari SD, Feirahi F, Moammeri F, Mesbahian G, Moghaddam ZS, Ren Q. Recent advances in green synthesized nanoparticles: from production to application. Mater Today Sustain. 2023;24:100500.

31 Chopra H, Bibi S, Singh I, Hasan MM, Khan MS, Yousafi Q, et al. Green metallic nanoparticles: biosynthesis to applications. Front Bioeng Biotechnol. 2022;10:874742.

32 Fahim M, Shahzaib A, Nishat N, Jahan A, Bhat TA, Inam A. Green synthesis of silver nanoparticles: a comprehensive review of methods, influencing factors, and applications. JCIS Open. 2024;16:100125.

33 Crawford SE, Hartung T, Hollert H, Mathes B, van Ravenzwaay B, Steger-Hartmann T, et al. Green toxicology: a strategy for sustainable chemical and material development. Environ Sci Eur. 2017;29:1–16.

34 Krebs J, McKeague M. Green toxicology: connecting green chemistry and modern toxicology. Chem Res Toxicol. 2020;33(12):2919–2931.

35 De Marco BA, Rechelo BS, Tótoli EG, Kogawa AC, Salgado HRN. Evolution of green chemistry and its multidimensional impacts: a review. Saudi Pharm J. 2019;27(1):1–8.

36 Khan FH. Chemical hazards of nanoparticles to human and environment (a review). Orient J Chem. 2013;29(4):1399–1408.

37 Benigni R, Bossa C. Data-based review of QSARs for predicting genotoxicity: the state of the art. Mutagenesis. 2019;34(1):17–23.

38 Ramirez T, Daneshian M, Kamp H, Bois FY, Clench MR, Coen M, et al. Metabolomics in toxicology and preclinical research. ALTEX. 2013;30(2):209–225.

39 Van Ravenzwaay B, Sperber S, Lemke O, Fabian E, Faulhammer F, Kamp H, et al. Metabolomics as read-across tool: a case study with phenoxy herbicides. Regul Toxicol Pharmacol. 2016;81:288–304.

40 Maertens, A., Anastas, N., Spencer, P.J., Stephnes, M., Goldberg, A. and Hartung, T., 2014. Food for thought: green toxicology.

41 Szymański P, Markowicz M, Mikiciuk-Olasik E. Adaptation of high-throughput screening in drug discovery—toxicological screening tests. Int J Mol Sci. 2012;13(1):427–452.

42 Kavlock R, Chandler K, Houck K, Hunter S, Judson R, Kleinstreuer N, et al. Update on EPA's ToxCast program: providing high throughput decision support tools for chemical risk management. Chem Res Toxicol. 2012;25(7):1287–1302.

43 Ball N, Bars R, Botham PA, Cuciureanu A, Cronin MT, Doe JE, et al. A framework for chemical safety assessment incorporating new approach methodologies within REACH. Arch Toxicol. 2022;96(3):743–766.

44 Sewell F, Aggarwal M, Bachler G, Broadmeadow A, Gellatly N, Moore E, et al. The current status of exposure-driven approaches for chemical safety assessment: a cross-sector perspective. Toxicology. 2017;389:109–117.

45 Turley AE, Isaacs KK, Wetmore BA, Karmaus AL, Embry MR, Krishan M. Incorporating new approach methodologies in toxicity testing and exposure assessment for tiered risk assessment using the RISK21 approach: case studies on food contact chemicals. Food Chem Toxicol. 2019;134:110819.

46 Anastas PT, Warner JC. Green Chemistry: Theory and Practice. 1998. Oxford University Press, New York, p. 30.

47 Mehta M, Bui TA, Yang X, Aksoy Y, Goldys EM, Deng W. Lipid-based nanoparticles for drug/gene delivery: an overview of the production techniques and difficulties encountered in their industrial development. ACS Mater Au. 2023;3(6):600–619.

48 Maunz A, Gütlein M, Rautenberg M, Vorgrimmler D, Gebele D, Helma C. Lazar: a modular predictive toxicology framework. Front Pharmacol. 2013;4:38.

49 Luechtefeld T, Marsh D, Rowlands C, Hartung T. Machine learning of toxicological big data enables read-across structure activity relationships (RASAR) outperforming animal test reproducibility. Toxicol Sci. 2018;165(1):198–212.

50 Senthil Rathi B, Ewe LS, Yew WK, Tiong SK. Recent trends and advancement in metal oxide nanoparticles for the degradation of dyes: synthesis, mechanism, types and its application. Nanotoxicology. 2024;18:272–298.

51 Rheder DT, Guilger M, Bilesky-José N, Germano-Costa T, Pasquoto-Stigliani T, Gallep TBB, et al. Synthesis of biogenic silver nanoparticles using Althaea officinalis as reducing agent: evaluation of toxicity and ecotoxicity. Sci Rep. 2018;8(1):12397.

52 Sharma D, Kanchi S, Bisetty K. Biogenic synthesis of nanoparticles: a review. Arab J Chem. 2019;12(8):3576–3600.

53 Mitchell MJ, Billingsley MM, Haley RM, Wechsler ME, Peppas NA, Langer R. Engineering precision nanoparticles for drug delivery. Nat Rev Drug Discov. 2021;20(2):101–124.

54 Boudreau MD, Imam MS, Paredes AM, Bryant MS, Cunningham CK, Felton RP, et al. Differential effects of silver nanoparticles and silver ions on tissue accumulation, distribution, and toxicity in the Sprague Dawley rat following daily oral gavage administration for 13 weeks. Tox Sci. 2016;150(1):131–160.

55 Ma W, Jing L, Valladares A, Mehta SL, Wang Z, Li PA, Bang JJ. Silver nanoparticle exposure induced mitochondrial stress, caspase-3 activation and cell death: amelioration by sodium selenite. Int J Biol Sci. 2015;11(8):860.

56 Maqsood MF, Shahbaz M, Khalid F, Rasheed Y, Asif K, Naz N, et al. Plant-based nanoparticles in ROS mitigation and abiotic stress tolerance: a review. Plant Stress. 2023:100281.

57 Cronin JG, Jones N, Thornton CA, Jenkins GJ, Doak SH, Clift MJ. Nanomaterials and innate immunity: a perspective of the current status in nanosafety. Chem Res Toxicol. 2020;33(5):1061–1073.

58 Liu J, Liu Z, Pang Y, Zhou H. The interaction between nanoparticles and immune system: application in the treatment of inflammatory diseases. J Nanobiotechnol. 2022;20(1):127.

59 Tran TK, Nguyen MK, Lin C, Hoang TD, Nguyen TC, Lone AM, et al. Review on fate, transport, toxicity and health risk of nanoparticles in natural ecosystems: emerging challenges in the modern age and solutions toward a sustainable environment. Sci Total Environ. 2024;912:169331.

60 Olawade DB, Wada OZ, Fapohunda O, Egbewole BI, Ajisafe O, Ige AO. Nanoparticles for microbial control in water: mechanisms, applications, and ecological implications. Front Nanotechnol. 2024;6:1427843.

61 Luis AI, Campos EV, Oliveira JL, Vallim JH, Proença PL, Castanha RF, et al. Ecotoxicity evaluation of polymeric nanoparticles loaded with ascorbic acid for fish nutrition in aquaculture. J Nanobiotechnol. 2021;19(1):163.

62 Wang Q, Huang JY, Li HQ, Chen Z, Zhao AZJ, Wang Y, et al. TiO2 nanotube platforms for smart drug delivery: a review. Int J Nanomedicine. 2016;11:4819–4834.

63 Patel J, Kumar GS, Roy H, Maddiboyina B, Leporatti S, Bohara RA. From nature to nanomedicine: bioengineered metallic nanoparticles bridge the gap for medical applications. Discover Nano. 2024;19(1):1–24.

64 Francis DV, Abdalla AK, Mahakham W, Sarmah AK, Ahmed ZF. Interaction of plants and metal nanoparticles: exploring its molecular mechanisms for sustainable agriculture and crop improvement. Environ Int. 2024;190:108859.

65 Sharma A, Madhunapantula SV, Robertson GP. Toxicological considerations when creating nanoparticle-based drugs and drug delivery systems. Expert Opin Drug Metab Toxicol. 2012;8(1):47–69.

66 Egbuna C, Parmar VK, Jeevanandam J, Ezzat SM, Patrick-Iwuanyanwu KC, Adetunji CO, et al. Toxicity of nanoparticles in biomedical application: nanotoxicology. J Toxicol. 2021;2021(1):1–21.

67 Abbasi R, Shineh G, Mobaraki M, Doughty S, Tayebi L. Structural parameters of nanoparticles affecting their toxicity for biomedical applications: a review. J Nanopart Res. 2023;25(3):43.

68 Johnston HJ, Hutchison G, Christensen FM, Peters S, Hankin S, Stone V. A review of the in vivo and in vitro toxicity of silver and gold particulates: particle attributes and biological mechanisms responsible for the observed toxicity. Crit Rev Toxicol. 2010;40(4):328–346.

69 Oberdörster G, Oberdörster E, Oberdörster J. Nanotoxicology: an emerging discipline evolving from studies of ultrafine particles. Environ Health Perspect. 2005;113(7):823–839.

70 Bacchetta R, Santo N, Valenti I, Maggioni D, Longhi M, Tremolada P. Comparative toxicity of three differently shaped carbon nanomaterials on Daphnia magna: does a shape effect exist? Nanotoxicology. 2018;12(3):201–223.

71 Abramenko NB, Demidova TB, Abkhalimov EV, Ershov BG, Krysanov EY, Kustov LM. Ecotoxicity of different-shaped silver nanoparticles: case of zebrafish embryos. J Hazard Mater. 2018;347:89–94.

72 Carnovale C, Bryant G, Shukla R, Bansal V. Identifying trends in gold nanoparticle toxicity and uptake: size, shape, capping ligand, and biological corona. ACS Omega. 2019;4(1):242–256.

73 Lippmann M. Effects of fiber characteristics on lung deposition, retention, and disease. Environ Health Perspect. 1990;88:311–317.

74 Muller J, Huaux F, Moreau N, Misson P, Heilier JF, Delos M, et al. Respiratory toxicity of multi-wall carbon nanotubes. Toxicol Appl Pharmacol. 2005;207(3):221–231.

75 Malvindi MA, De Matteis V, Galeone A, Brunetti V, Anyfantis GC, Athanassiou A, et al. Toxicity assessment of silica coated iron oxide nanoparticles and biocompatibility improvement by surface engineering. PLoS One. 2014;9(1):e85835.

76 Wang Q, Huang JY, Li HQ, Chen Z, Zhao AZJ, Wang Y, et al. TiO2 nanotube platforms for smart drug delivery: a review. *Int. J. Nanomed.* 2016:11:4819–4834.

77 Misra SK, Dybowska A, Berhanu D, Luoma SN, Valsami-Jones E. The complexity of nanoparticle dissolution and its importance in nanotoxicological studies. Sci Total Environ. 2012;438:225–232.

78 Kumar CS, Hormes J, Leuschner C (eds). Nanofabrication towards biomedical applications: techniques, tools, applications, and impact. 2006. John Wiley & Sons.

79 Iavicoli I, Leso V, Ricciardi W, Hodson LL, Hoover MD. Opportunities and challenges of nanotechnology in the green economy. Environ Health. 2014;13:1–11.

80 Yu H, Fu PP. Toxicity and environmental risks of nanomaterials: challenges and future needs. J Environ Sci Health C Environ Carcinog Ecotoxicol Rev. 2009;27(1):1–35.

81 McCall MJ. Nanoparticles in the real world. Nat Nanotechnol. 2011;6(10):613–614.

82 Wiedermann CJ, Eisendle K. Comparison of hydroxyethyl starch regulatory summaries from the Food and Drug Administration and the European Medicines Agency. J Pharm Policy Pract. 2017;10:1–6.

83 Pita R, Ehmann F, Papaluca M. Nanomedicines in the EU—regulatory overview. AAPS J. 2016;18:1576–1582.

84 Thomas K, Aguar P, Kawasaki H, Morris J, Nakanishi J, Savage N. Research strategies for safety evaluation of nanomaterials, part VIII: international efforts to develop risk-based safety evaluations for nanomaterials. Toxicol Sci. 2006;92(1):23–32.

85 Barhoum A, García-Betancourt ML, Jeevanandam J, Hussien EA, Mekkawy SA, Mostafa M, et al. Review on natural, incidental, bioinspired, and engineered nanomaterials: history, definitions, classifications, synthesis, properties, market, toxicities, risks, and regulations. Nanomaterials. 2022;12(2):177.

86 Jeevanandam J, Barhoum A, Chan YS, Dufresne A, Danquah MK. Review on nanoparticles and nanostructured materials: history, sources, toxicity and regulations. Beilstein J Nanotechnol. 2018;9(1):1050–1074.

87 Ullah A, Lim SI. Plant extract-based synthesis of metallic nanomaterials, their applications, and safety concerns. Biotechnol Bioeng. 2022;119(9):2273–2304.

88 Priyam A, Seth P, Mishra J, Manna PK, Singh PP. Occupational safety assessment of biogenic urea nanofertilisers using in vitro pulmonary, and in vivo ocular models. Heliyon. 2023;9(11):e21623.

89 Krishnaraj C, Harper SL, Yun SI. In vivo toxicological assessment of biologically synthesized silver nanoparticles in adult Zebrafish (Danio rerio). J Hazard Mater. 2016;301:480–491.

90 Azad A, Zafar H, Raza F, Sulaiman M. Factors influencing the green synthesis of metallic nanoparticles using plant extracts: a comprehensive review. Pharmaceut Fronts. 2023;5:e117–e131.

91 Kyriakides TR, Raj A, Tseng TH, Xiao H, Nguyen R, Mohammed FS, et al. Biocompatibility of nanomaterials and their immunological properties. Biomed Mater. 2021;16(4):042005.

92 Vasquez RD, Apostol JG, de Leon JD, Mariano JD, Mirhan CMC, Pangan SS, et al. Polysaccharide-mediated green synthesis of silver nanoparticles from Sargassum siliquosum JG Agardh: assessment of toxicity and hepatoprotective activity. OpenNano. 2016;1:16–24.

93 Manivasagan P, Oh J. Production of a novel fucoidanase for the green synthesis of gold nanoparticles by Streptomyces sp. and its cytotoxic effect on HeLa cells. Mar Drugs. 2015;13(11):6818–6837.

94 Araj SEA, Salem NM, Ghabeish IH, Awwad AM. Toxicity of nanoparticles against *Drosophila melanogaster* (Diptera: Drosophilidae). J Nanomater. 2015;2015(1):758132.

95 Sengottaiyan A, Aravinthan A, Sudhakar C, Selvam K, Srinivasan P, Govarthanan M, et al. Synthesis and characterization of Solanum nigrum-mediated silver nanoparticles and its protective effect on alloxan-induced diabetic rats. J Nanostruct Chem. 2016;6:41–48.

96 Foo YY, Periasamy V, Kiew LV, Kumar GG, Malek SNA. *Curcuma mangga*-mediated synthesis of gold nanoparticles: characterization, stability, cytotoxicity, and blood compatibility. Nanomaterials. 2017;7(6):123.

97 Mahesh B. A comprehensive review on current trends in greener and sustainable synthesis of ferrite nanoparticles and their promising applications. Results Eng. 2024;21:101702.

98 Kumari S, Raturi S, Kulshrestha S, Chauhan K, Dhingra S, András K, et al. A comprehensive review on various techniques used for synthesizing nanoparticles. J Mater Res Technol. 2023;27:1739–1763.

99 Pechyen C, Tangnorawich B, Toommee S, Marks R, Parcharoen Y. Green synthesis of metal nanoparticles, characterization, and biosensing applications. Sens Int. 2024;5:100287.

100 Ashique S, Upadhyay A, Hussain A, Bag S, Chaterjee D, Rihan M, et al. Green biogenic silver nanoparticles, therapeutic uses, recent advances, risk assessment, challenges, and future perspectives. J Drug Deliv Sci Technol. 2022;77:103876.

101 Singh J, Dutta T, Kim KH, Rawat M, Samddar P, Kumar P. Green' synthesis of metals and their oxide nanoparticles: applications for environmental remediation. J Nanobiotechnol. 2018;16:1–24.

102 Nyabadza A, McCarthy É, Makhesana M, Heidarinassab S, Plouze A, Vazquez M, Brabazon D. A review of physical, chemical and biological synthesis methods of bimetallic nanoparticles and applications in sensing, water treatment, biomedicine, catalysis and hydrogen storage. Adv Colloid Interf Sci. 2023;321:103010.

103 Malvindi MA, De Matteis V, Galeone A, Brunetti V, Anyfantis GC, Athanassiou A, et al. Toxicity assessment of silica coated iron oxide nanoparticles and biocompatibility improvement by surface engineering. PLoS One. 2014;9(1):e85835.

15

Market Outlook for Biogenic Nanopharmaceutical Products

Himakshi Baruah[1], *Anupam Sarma*[1], *Tapash Chakraborty*[1], *and Malay K. Das*[2]

[1] *Advanced Drug Delivery Laboratory, Department of Pharmaceutics, School of Pharmaceutical Sciences, Girijananda Chowdhury University, Guwahati, Assam, India*
[2] *Drug Delivery Research Laboratory, Department of Pharmaceutical Sciences, Dibrugarh University, Dibrugarh, Assam, India*

15.1 Introduction

A new era in materials science was introduced by the development of nanotechnology in the 1960s. Nanotechnology has brought together numerous academic fields, including physics, biology, chemistry, materials engineering, and medicine. Nanoparticles are defined as materials with dimensions smaller than 100 nm [1]. In the vast field of novel technology, nanoparticles are used to encapsulate active pharmaceutical ingredients in the inside core or outer surface of nanoparticles (NPs). The encapsulation of bioactive compounds (BCs) in nanocarriers results in the enhancement of their bioactivity and shields the BCs from pH-, enzyme-, and temperature-induced destruction. The biophysical-chemical characteristics that are essential for drug delivery via nanodelivery systems or nanocarriers are as follows: (i) Stability and chemical inertness of the drug under physiological conditions; (ii) enhanced encapsulation efficiency of the drug; (iii) controlled plasma drug concentration; (iv) does not elicit any immunogenic reaction and is compatible with the physiological environment; (v) lower cost, highly scalable, and a targeted function at the disease site with no or minimal effect on healthy tissues; and (vi) the drug releases in a sustained manner from the nanocarrier to the disease site. Nanoparticles improve the pharmacokinetic characteristics such as half-life and bioavailability (or bioaccessibility) of the drug after administration into the body, leading to enhanced *in vivo* functional and clinical efficacy of the drug [2]. Many types of nanoparticles (NPs) are synthesized, characterized,

Biogenic Synthesis of Nanoparticles and Sustainable Nanotherapeutics,
First Edition. Edited by Malay K. Das.

and used in nanotechnology. Numerous chemical and physical methods are frequently employed to develop NPs from different noble metals like gold, silver, or platinum; however, these procedures are harmful to the environment. These chemical and physical methods that are employed to formulate nanoparticles show several limitations, such as high cost, decreased productivity, and toxicity, and therefore, the biological production of nanomaterials is rapidly gaining attention [3]. There is an urgent need for the development of non-toxic, ecologically safe technologies for producing NPs. Several green synthesis techniques for NPs that are safe, straightforward, inexpensive, reproducible, scalable, and motivated by the safety-by-design philosophy have recently been developed. Also, currently a number of biological systems – including bacteria, yeast, fungi, and plant extracts – are widely being employed in green synthesis techniques to formulate nanoparticles [4]. These biologically derived nanoparticles are called biogenic nanoparticles, which have garnered a lot of attention due to their unique properties and wide variety of applications in numerous fields. The many characteristics of biogenic nanoparticles such as size, shape, surface chemistry, and functionality, make them ideal for use in environmental remediation, agriculture, medicine, and other fields [5]. Biogenic nanocarriers are delivery vehicles made from natural or biologically derived substances, entities obtained from biology or living organisms such as whole cells, or cell components and waste products. They have superior bioactivities, such as little or no toxicity, biocompatibility, bioavailability, and biodegradability. Biogenic nanocarriers enhance the biological characteristics of bioactives [2]. Biogenic nanoparticles offer potential uses in biomedicine, including targeted drug delivery, imaging, diagnostics, and treatments, due to their nontoxicity and biocompatibility [5]. Some researchers have produced gold-coated iron nanoparticles using coconut oil, liquorice root (*Glycyrrhiza glabra*), and olive oil. Biogenic Silver nanoparticles and Gold nanoparticles play a significant role in other domains outside cancer. When *Ajuga bracteosa* aqueous extract-derived biogenic Gold nanoparticles and Silver nanoparticles are utilized in mice, studies reveal *in vivo* antidiabetic and anticancer benefits in addition to anti-hypoglycemia and hepatic-protective capabilities [3]. Engineers and scientists have long understood the importance of looking to nature for inspiration [6]. In biomaterials, there are two main approaches – the first uses natural molecules that have the ability to influence the formulation of nanomaterials [6]. Compared to physical and chemical nanoparticle development methods, this approach offers significant advantages. The following are some of the benefits of producing NPs biologically: Biogenic NPs are environment-friendly because they do not require harmful chemicals or high temperatures. They are sustainable and more cost-effective compared to chemical and physical approaches. Biogenic NPs are less expensive because they can be produced from easily accessible sources like plant and microbial extracts.

Biogenic nanoparticles are biocompatible and useful in a range of biological processes, including medication administration, tissue engineering, and imaging. The size and shape of the biogenic nanoparticles is influenced by different parameters like temperature, pH, concentration of the substrate and exposure time to the substrate [6]. Because of their several benefits, the development of biogenic nanoparticles for the treatment and diagnosis of several diseases is increasing extensively. Several reports on biogenic nanoparticles have been published, but the present scenario of these biogenic nanocarriers has not been fully explored yet. This article aims to provide in-depth information on the current scenario of biogenic nanoparticles in the market, along with industry trends, a global forecast, the regulatory aspects of biogenic nanoparticles, a report on clinical developments, and records of marketed products of biogenic nanocarriers.

15.2 Classification

Biogenic nanoparticles are classified based on their genesis into two main categories: nature-made and man-made (Figure 15.1). Each category has unique biosynthetic pathways, characteristics, uses in nanomedicine, delivery of therapeutics, and environmental science.

15.2.1 Nature-Made Biogenic Nanoparticles

15.2.1.1 Outer Membrane Vesicles

Nanoparticles have the potential to quickly bind with immune system cells and move through lymphatic and blood channels. They may have the potential to operate as immunostimulatory agents. Biogenic nanoparticles, which are made by spontaneously vacuolating bacteria (outer membrane vesicles, or OMVs) or extracellular vesicles (exosomes), or by expressing viral structural proteins – virus-like particles (VLPs) – have garnered increasing attention in the field of tumor vaccines among the numerous carriers assessed for their display of

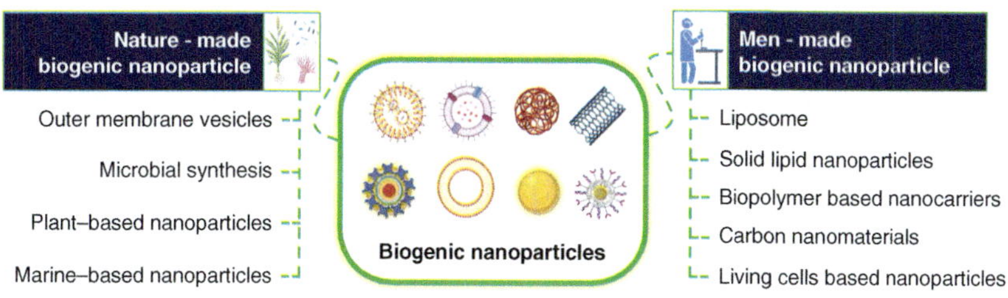

Figure 15.1 Types of biogenic nanoparticles.

cancer antigens [7]. In tumor immunotherapy, biogenic nanoparticles are crucial because they have the capacity to control the immune system of the body. OMVs' strong biocompatibility is demonstrated by their ability to be employed as an adjuvant in clinical treatment. OMVs are easily obtained and large-scale production can be realized due to the ease of bacterial culture and its rapid growth. The vaccine industry has made extensive use of VLPs (e.g. HEPTAVAX – B, Cervarix, Anti-Acnes). They can elicit both innate and adaptive immune responses from the host by preserving the normal shape of viral antigen proteins [8]. The role of exosomes in tumor immunomodulation has also been extensively researched. On the other hand, immune cell-derived exosomes have the ability to stimulate the immune system and enhance the effects of immunotherapy. However, because exosomes are essential for tumor cells to transmit information, they are becoming a major target for the therapy of malignant tumors [9].

15.2.1.2 Microbial Synthesis of Biogenic Nanoparticles

Many distinct nanoparticles can be produced by microorganisms. Because of this, scientists are now more interested in using these microorganisms to create nanoparticles for a range of uses. It is possible to use nanoparticles for a range of applications in the pharmaceutical field. Inorganic chemicals can be produced by fungi and bacteria through physiologically mediated and induced synthesis [10]. More and more microorganisms, including bacteria, fungi, and algae, are being used to create nanoparticles. A diverse range of microorganisms create nanoparticles primarily through distinct reactions with metal precursors. For example, fungi and bacteria are capable of both exterior and intracellular synthesis and the pathways for each process differ according to the kind of microbe [11]. The fact that bacteria are known to have reductases and other enzymes that change metal ions into metal nanoparticles comes from this. Biomolecules that are reducing agents including enzymes, carbohydrates, and amino acids, and a large concentration of functional groups, including carboxyl, thiol, and amino groups, are also present in the bacterial cell wall. Using an appropriate growth media combined with a metal precursor, microorganisms are cultured in this comparatively simple approach [12, 13]. The antibacterial and antifungal properties of silver nanoparticles, which are produced by *Bacillus subtilis,* make them useful in biomedical and textile applications. *Pseudomonas aeruginosa* gold nanoparticles are perfect for biosensing and drug delivery because they are stable and biocompatible. *Fusarium oxysporum* generates selenium nanoparticles that are useful in nutraceuticals and medicine administration because of their antioxidant and anticancer properties. Due to their superparamagnetism, iron oxide nanoparticles *Magnetospirillum magnetotacticum* are used as MRI contrast agents and in hyperthermia treatment. *Aspergillus fumigatus* zinc oxide nanoparticles have both photocatalytic and antibacterial properties [5]. The production of fungal nanoparticles has a number of benefits over bacterial techniques, such as environmental friendliness, cost-effectiveness, and scalability [14]. Silver is the main target of research because fungi have enzymes that convert metal ions into nanoparticles. Research has looked into using fungi such as *F. oxysporum* for producing different metal and metal oxide nanoparticles and *Cladosporium cladosporioides* for gold nanoparticle synthesis [15, 16]. Both intracellular and extracellular formations of nanoparticles are possible in fungi, with the internal synthesis of frequently producing nanoparticles. By making it easier to remove depleted metals, this method has the potential to clear environmental contamination [17]. Bacterial nanocellulose nanoparticles are designed to enhance their properties for specific applications. *Acetobacter xylinum* generates hydrophilic carboxylated nanocellulose that can be used to heal wounds and deliver medications. Phosphorylated nanocellulose from *Gluconeobacter hansenii* is beneficial for tissue engineering and bone regeneration due to its high phosphate content and bioactivity. Because of its anti-inflammatory and anticoagulant properties, the sulphated nanocellulose of *A. xylinum* is ideal for tissue scaffolds and medical devices. *A. xylinum* produces highly reactive oxidized nanocellulose for tissue engineering and drug administration. *Komagataeibacter xylinus* bacterial cellulose nanocrystals are used in biomedical applications and composite materials due to their great strength and crystallinity [5].

15.2.1.3 Plant-Based Biogenic Nanoparticles

There has been a lot of interest in the biogenic production of metal/metal oxide nanoparticles (NPs) such as Ag-NPs, Au-NPs, and iron oxide NPs, employing intracellular living plants, extracellular plant extracts, and phytochemicals, as alternative to traditional techniques. A variety of plant-mediated synthesis techniques are frequently used for this objective. For the synthesis of metal/metal oxide nanoparticles, researchers are currently looking into biogenic methods. In particular, they are interested in using particular phytoconstituents from plants, like proteins, polyphenols, and organic acids, to improve accuracy and control size and shape during nanoparticle manufacturing. There are three main methods for producing biogenic nanoparticles: extracellular or plant extract-mediated synthesis, phytochemical-based synthesis, and intracellular or plant biomass-based synthesis [18]. Gold nanoparticles from Tulsi that are stable and uniform respond effectively against bacteria and cancer. Neem silver nanoparticles are used for wound healing and water purification due to their potent antibacterial activity. *Aloe vera* zinc oxide nanoparticles' photocatalytic and UV-blocking properties make them ideal for usage in sunscreens and cosmetics. Iron oxide nanoparticles from green tea exhibit superparamagnetism and are used as MRI contrast agents for drug delivery. Because of their antifungal, antioxidant, and catalytic properties, eucalyptus copper nanoparticles can be used in biological applications [19].

Existing studies revealed that *Argemone mexicana L., Tribulus terrestris, Medicago sativa, Olax scandens, Hubertia ambavilla, Hypericum lanceolatum,* and liquorice root (*G. glabra*), marine algae, *Eucalyptus tereticornis* leaves, *Artemisia Sieberi Besser* leaves, *Russelia equisetiformis* and *Leucophyllum frutescens* leaves, *Carissa carandas* leaves, *Asparagus bracteosa, Gelidiella acersaon, Curcumae kwangsiensis* leaves, *Spondias dulcis (Anacardiaceae)* peels, *Vigna radiate, botanical extracts of Garcinia mangostana, Caesalpinia pulcherrima* stem, *Callistemon viminalis* flower extracts, *Zeylanicum* leaves of *Salvia officinalis,* blooms of *Tilia cordata, Aegle martelo* are used in the synthesis of biogenic gold and silver nanoparticles for anti-cancer application [3].

15.2.1.4 Marine-Based Nanoparticles

About 70% of the surface of the globe is made up of the ecologically varied marine environment, which is yet mostly undiscovered. According to recent research, a number of marine plants can function as biofactories to produce nanoparticles. Several types of marine algae have the capacity to synthesize a range of nanoparticles, which has sparked interest in creating novel and strong medications from natural marine plant sources. Seaweed and algae produce marine-based nanoparticles, which are environmentally safe and have unique properties. It has been discovered that several of these nanoparticle forms are potent antibacterial agents [20]. Research on seaweeds indicates a variety of useful qualities. These include environmental functions like heavy metal accumulation and anti-fouling activity, additionally medical uses offer anti-inflammatory and anticancer benefits. Notably, research shows that seaweeds can act as agents for the synthesis of stable metal nanoparticles, as well as reduce metal ions. Additionally, different studies highlight the possibility for comparable uses by extrapolating these results to specific seagrass species [20]. Biogenic synthesis of nanoparticles using marine algae and seagrasses is a non-toxic and environmentally friendly green chemistry method that may produce a diverse range of particle sizes, morphologies, contents, and characteristics [21]. Currently, there is limited understanding of how seaweeds and seagrasses produce nanoparticles. Seaweed cell walls include biomolecules that may function as biocatalysts, producing metal and metal oxide nanoparticles and decreasing precursor metal salts, according to research. The development of particles can also be directed and controlled by bigger amphiphilic biomolecules acting as surfactants. In the room-temperature biogenic production of nanoparticles, an aqueous solution containing seaweed or seagrass extract is combined with a metal salt solution. A color shift in the reaction mixture signifies the creation of nanoparticles since reduction proceeds swiftly. *Arockiya* et al. used brown algae (*Stoechospermum marginatum*) to synthesize gold nanoparticles [22]. Antimicrobial and environmentally safe silver nanoparticles are produced by *Sargassum muticum* and are utilized in biological and coating applications. *Ulva lactuca* gold nanoparticles are perfect for imaging and drug administration since they are stable and biocompatible. *Gelidium amansii* zinc oxide nanoparticles' photocatalytic and UV-blocking properties make them ideal for usage in sunscreen and cosmetics. *Gracilaria corticata* iron oxide nanoparticles exhibit superparamagnetism. These particles are employed in MRI contrast and drug delivery. Copper nanoparticles from *Kappaphycus alvarezii* have antifungal and antioxidant properties [5].

15.2.1.5 Virus-Like Particles

Protein-based nanoparticles, including viruses and protein cages, are a new class of innovative drug carriers. Self-assembling supramolecular structures made of separate protein monomers are known as protein cages. By definition, their components are not viral and VLPs are formed exclusively from the natural viruses' coat proteins. Although, they are robust structures that can resist environmental stresses and avoid disintegration, viruses are also sensitive enough to recognize cues in the cellular environment and release their genome in response to commands. Viral nanoparticles (VNP) and virus-like particles (VLPs) are the two subcategories of viruses. VLPs are thought to be non-infectious and are the genome-free variants of their VNP counterparts. The presence or absence of viral genomes may also cause them to impart distinct immunostimulatory profiles [23, 24]. Because of their high absorption efficiency, biocompatibility, water solubility, morphological homogeneity, and ease of functionalization, natural delivery carriers – especially those based on VNP offer certain special benefits. Utilizing biological behaviors, optimal nanomedical techniques for drug administration or imaging seek to create efficient nanoparticle size cages with enhanced rigidity, suitable pharmacokinetic characteristics, targeting ability to the specific cell, and effective cellular penetrability. Since tissue tropisms are typically absent from plant viruses, these characteristics can be added to a formulation of nanoparticles, whether they are natural or synthetic, to provide them with features like penetration, membrane crossing, cell surface receptor binding, and target capabilities [25]. Viruses have the potential to be both uniform and programable. Tobacco mosaic virus nanoparticles from the tobacco mosaic virus, influenza virus nanoparticles from the influenza A virus, bacteriophage nanoparticles from the influenza A virus, bacteriophage nanoparticles from the M13 bacteriophage, and plant virus nanoparticles from the cowpea mosaic virus (CPMV) are a few examples. These nanoparticles are used in the creation of vaccines, gene delivery, and nanocarriers and they are ideal for biosensing and drug delivery [5].

15.2.2 Man-Made Biogenic Nanoparticles

15.2.2.1 Liposomes

Vesicular nanoliposomes, also known as nano-lipid vesicles, are the earliest and most effective kind of delivery systems. These vesicular nanoparticles are assembled by themselves and the aqueous compartments separate the monophospholipid bilayer or several phospholipid bilayers (membranes). They have garnered a lot of interest from the scientific community since Bangham, and his associates discovered them in 1965. They are employed as transporters for both hydrophilic and hydrophobic substances in a variety of sectors, particularly the biomedical sector [26]. Liposomes have a monophospholipid bilayer in their unilamellar structure, but their multilamellar structure resembles an onion. Since multiple unilamellar vesicles are created inside bigger liposomes, multivesicular vesicles (MVV) form a multilamellar configuration with concentric phospholipid spheres. For hydrophilic substances exclusively, liposome encapsulation effectiveness rises with the number of bilayers. One crucial element influencing the circulation half-life of liposomes is their size. The amount of drug enclosed depends on the size and quantity of bilayers. When liposomes are used to deliver drugs, the intended vesicles typically measure between 50 nm and 150 nm [27, 28]. Liposomes are an example of bioinspiration and are being utilized as cell membrane models because their layers are quite similar to those of natural cells. Although liposomes have many benefits, including high encapsulation efficiency, biocompatibility, biodegradability, transportability through cell membranes, and the capacity to carry both hydrophilic and hydrophobic compounds, they also have drawbacks, including weak chemical and physical stability, quick removal from systemic circulation, degradation in the acidic environment of GIT, great loss of loaded drugs, and inefficient response to stimulus [26]. Because of their content and coating, liposomes today exhibit a remarkable molecular diversity, including ufasomes, phytosomes, terpesomes, bilosomes, and aspasomes. For instance, liposomes exhibit a long-circulating activity when coated with polyethylene glycol liposome (PEGylated liposomes) or a combination PEG and polyvinyl alcohol, which is preferable in the latter scenario. Cell membranes can be applied to liposomes to improve them and produce biomimetic nanocarriers. By allowing for targeted distribution, extending circulation duration, and simulating biological processes, this method enhances medication delivery. Utilizing hybrid membranes from other cell types, such as fusing the membranes of cancer and red blood cells, can improve these "cell membrane-coated liposomes" even further. Tumor targeting, immunological evasion, and biocompatibility are all enhanced by this combination strategy [29].

15.2.2.2 Solid Lipid Nanoparticles (SLN)

Nanostructured lipid carriers (NLCs) are composed of both liquid and solid lipids, while SLNs are made up of lipids that remain in the solid state at body temperature and emulsifying agents induce their stability in aqueous dispersions [30, 31]. Both NLCs and SLNs originate from oil/water emulsions. These nanocarriers can carry both hydrophilic and lipophilic medications and exhibit biocompatibility [32]. SLNs are classified into three primary groups according to the location of the drug composition: (i) A homogenous matric model that evenly distributes medication within a lipid matrix; (ii) the concentration of the drug in the liquid shell of the outer portion identifies the drug enrich shell model; and (iii) the liquid core component contains drug aggregations in the drug-enriched core model [33]. The drug's location affects how quickly it is released from SLNs. While type-2 SLNs are suitable for burst release, types 1 and 3 are suitable for controlled release. Furthermore, SLNs are designed to preserve the release profile of the drug, which lessens the requirement for rapid dosing and boosts therapeutic benefits [34]. Myricetin, a flavonoid with anticancer effects, was added to SLNs by Khatamian and co-workers. Folic acid (FA) was used as an active target and chitosan (CS) was used to modify SLNs [32]. When tested against human breast carcinoma MCF-7 cells, the produced myrcetin-SLN-CS-FA showed strong antioxidant and anticancer properties. NLCs are enhanced SLN systems that exhibits greater stability and encapsulation efficiency for a variety of bioactive substances. Additionally, during storage, NLCs stop the active ingredients from leaking out of the nanocarrier. *Coc* et al. synthesized NLCs encapsulated with the chemical caffeic acid which is hydrophilic in nature and linseed oil which is lipophilic in nature and showed a synergistic effect that increased the antioxidant activity of both natural actives [35].

15.2.2.3 Biopolymer-Based Nanocarriers

Biopolymers, such as nucleic acids, peptides, proteins, and polysaccharides, are effective and efficient alternatives to polymers that are synthetically made for the development of biodegradable and biocompatible nano-vehicles for the delivery of bioactives. The main natural sources of biopolymers are waste products from the agricultural and food industries, marine sources (such as sponges, lobsters, shrimp, corals, and fish), microorganisms (yeasts, fungi, and algae), plants (such as maize, rice, tapioca, barley, soybean, wheat, corn, banana), and animals (cattle are the most prevalent animal sources) [36]. Naturally occurring and polymeric proteins and peptides that are found in living things make excellent options for drug nanocarriers. When taken orally and digested *in vivo,* they are biodegradable, biocompatible, and do not emit any toxic chemicals [37, 38]. Because of their increased selectivity, bioactivity, capacity to pass across cell membranes (particularly for peptides), and little to zero damage to normal cells, protein and peptide-based nanoformulations have made significant strides in recent years, particularly in the treatment of cancer [39]. Another class of natural polymers employed in drug delivery system development are polysaccharides, which are chosen for their mucoadhesiveness, hydrophilicity, biocompatibility, high availability, and potential antibacterial and anti-inflammatory qualities. Starch, chitosan, cellulose, dextran, hyaluronic acid, alginate, and pectin are polysaccharides that are most frequently utilized in the formulation of drug delivery systems [40]. DNA is a sequence-specific programmable molecule that uses the complementary base pairing idea to allow nanostructures based on DNA to self-assemble [41]. This method yields useful DNA nanocarriers for the delivery of bioactives. As a result, *Yuan* et al. created DNA nanocarrier encapsulated curcumin (DNA-CUR), a new nanovector that improves the stability and biostability of curcumin delivery. DNA/chitosan nanoparticles were produced by *Wang* et al. for the delivery of astaxanthin (ATX); these nanocarriers demonstrated improved cellular absorption efficiency and antioxidant activity [42].

15.2.2.4 Carbon Nanomaterials

All the nanoparticles that are made of carbon atoms are known as carbon-based nanomaterials (CBNs) and they might be spherical, planar, tubular, or horn-shaped. Because of their structural diversity, bioactivities (antioxidant, antimicrobial, and antitumoral qualities) and chemical, thermal, structural, electrical, optical, and mechanical characteristics, fullerenes, carbon nano-onions (CNOs), nano-diamonds (NDs), carbon nanotubes (CNTs), and carbon quantum dots (CQDs) have gained immense focus from researchers in the biomedical field, particularly as drug carriers. By functionalizing CBNs with biologically derived materials, their toxicity is reduced, and their biocompatibility is improved. However, CBNs must undergo biodegradations in cells after serving their purpose, particularly in nano-delivery applications. CBNs' biological degradability remains a problem [43, 44]. Their use is restricted by possible health and environmental hazards, which makes it difficult to make an informed choice despite their promise. Their possible toxicity, including effects on the heart and platelet activation, is a matter of concern. Research is still being conducted to improve their biocompatibility and target drug delivery to particular cells and tissues by modifying and functionalizing their surfaces, despite these obstacles. This includes creating highly effective drug-loading graphene-based nanocarriers and investigating the application of nano-diamonds for bioimaging and targeted drug delivery. To guarantee the safe and successful transition of CBNs into clinical applications, more study is essential to completely comprehend the hazards and advantages of these drugs [45].

15.2.2.5 Drug Delivery Using Living Cells and Cell Derivatives

Primary benefits of cell-derived biological vehicles are their inherent stealth, biocompatibility, minimal immunogenicity, and targeting capabilities. One intriguing technique for biomimetic particle engineering is to encapsulate nanoparticles by biological cellular membranes from immune cells, macrophages, platelets, red blood cells (RBCs), cancer cells, and stem cells. These cell membrane-camouflaged nanoparticles are utilized in targeted drug delivery and are biocompatible, retaining many of the characteristics of their parent cells. *Wu* et al. formulated a new bioinspired drug delivery system based on RBC membrane-loaded nanoparticles (UrNAPs). UrA is an antitumoral pentacyclic triterpenoid that occurs naturally. The resulting UrNAPs exhibited enhanced water solubility, high stability and anticancer impact by causing non-small-cell lung cancer cells to undergo autophagy and apoptosis [2]. Yeast cells are another intriguing carrier that can improve stability and facilitate the delivery of a variety of bioactive chemicals because they can load both hydrophilic and hydrophobic molecules. Diatoms' unique properties, such as their large surface area, porous structure, and ease of modification, make them suitable scaffolds for efficient nanocarriers for the transport of therapeutics to the targeted site of action. Plant cells are being investigated as carriers as well because they provide advantages including effective encapsulation of therapeutic proteins and antigens, low cost, and excellent safety. These biological carriers offer a viable path toward the creation of sophisticated and more efficient drug delivery systems [2].

15.3 Limitations of Biogenic Nanoparticles

Researchers still do not fully understand how NPs are perceived, despite their existence for more than 50 years. As an environmentally benign substitute for traditional nanoparticle synthesis techniques, the green chemistry concept supports the synthesis of nanoparticles. Furthermore, the green chemistry method of synthesizing nanoparticles is based on the idea that the procedure should be safe, use natural resources instead of harmful once, and be toxicity-free and reasonably priced. The manufacture of metal/metal oxide nanoparticles (NPPs) using plants, bacteria, fungi, yeast, and actinomycetes has been informed in a variety of research studies so far. Plants are a common natural resource used for the synthesis of NPs because of their ethnobiological significance, active components, ease of availability, cheap cost, and streamlined procedure. Despite this, there are still many important problems and technical difficulties that researchers must overcome in order to successfully develop biogenic nanoparticles [46]. Although biogenic synthesis has several benefits, polydispersity makes it difficult to control the size of the nanoparticles. To improve the capacity to control particle size and shape as well as to fine-tune particle size distribution, significant work is also needed. This is why a lot of biogenic research has concentrated on creating techniques that can enable monodisperse nanoparticles production [47]. One way to influence the morphology of nanoparticles is to modify the process parameters in order to determine the ideal synthesis parameter. In the process of nanoparticle formation, variables including PH, temperature, reaction duration, and reagent concentration, all have an equal impact on the nucleation process [48]. Another obstacle is determining which microbial candidates are optimal based on their intrinsic characteristics, such as growth rate, replicability, and biochemical activity. Adapting to adapt the biogenic synthesis to large-scale production requires consideration of both the yield and the total biosynthesis rate. Because the yields of biosynthesized nanoparticles are significantly lower than those of chemical synthesis, their large-scale manufacture presents difficulties. Commercial or scale-up methods for synthesizing nanoparticles also vary widely. The overall cost would be lower, nevertheless, because of the enormous quantity, accessibility, and robust development of bacteria. These procedures also save money because they do not require costly processes, organic solvents, or thermal stabilizers [10].

15.4 Current Industry Trends

Researchers have identified a number of benefits of nanotechnology, which is currently in widespread use. Nanoparticles' exceptionally small size makes them essential in a range of reactions. The research in nanotechnology and nanomedicine is broad and varied. With notable medicinal outcomes, drugs have reached a new level due to impressive developments in nanomedicine. Research on the enormous potential of nanoparticles for medical applications is essential. Many studies are being carried out in the medical area to look into the best methods and strategies, such as cardiovascular disease therapeutic genes, nephrology, and cancer treatment. While the quality of nanotechnology and nanoparticles have both improved and yielded encouraging results, traditional medicine has made great strides [49]. Due to their improved electric mobility, stimulated texture, and stability, nanomaterials are especially helpful in the domains of medicine, safety, textiles, space exploration, skincare goods, food production, and agriculture [50]. The advantages of nanomaterials over traditional biotechnological methods include high stability, adaptability, and target selectivity. Nanotechnology can contribute to a sustainable future and the global economy due to its various uses in a different range of areas. Drug delivery, therapeutics, and diagnostics are just a few of the medical and healthcare uses for nanomaterials [51]. It is essential to comprehend the structure–function relationship of nanomaterials in order to accomplish a successful application. It will be challenging to choose the best nanomaterial for a particular application, as well as to produce nanomaterials in a way that is sustainable, scalable and affordable. Treatments with improved biocompatibility and retention duration that are either water-soluble or insoluble could be administered via biodegradable nanoparticles. By developing materials like nanoadditives and nanozymes, the industrial sector has the ability to broaden the use of nanobiotechnology. It is possible to construct medical devices and technology to produce extremely specific subcellular or molecular interactions [52]. The practical application of biogenic nanoparticles requires the development of a strategy to produce biogenic nanoparticles on a larger scale. Officially, they are referred to as intracellular or extracellular biogenic processes. The biological entity employed for biogenic nanomaterials may be reusable in extracellular production methods. However, because the biological entity must be disrupted or killed by destroying its living structure, separation of the generated nanoparticles will be a little more challenging in the case of intracellular synthesis. Due to excessive waste generation and the lack of biological entity reuse following destruction; such intracellular situations might not be suitable for large-scale production [53]. It can be difficult to scale up the biosynthesis of biogenic nanomaterials in terms of their dependability, productivity, and economy. The biosynthesis of biogenic nanomaterials (BBN) is often reaction-rate-limited for rapid reactions, and reactor size has a significant impact on the production rate. The majority of BBN schemes, if not all of them, are only appropriate for large-scale batch procedures. Therefore, appropriate design of consistent and appropriate for large-scale batch procedures. Therefore, appropriate design, consistent and appropriate temperature, PH, and salinity control, as well as agitation/mixing of the solid-liquid mixture in the batch system are necessary scale-up parameters. The majority of BBN processes lack detailed response rate data, making it difficult to create correlations for process control applications. When designing BBN large-scale batch reactors, unit dependability must also be taken into consideration [53]. Several biogenic nanoparticles are there which are available in the market and approved by the FDA such as (i) liposomes like AmBisome® for systemic fungal infections (FDA1997), DaunoXome® for HIV-related KS (FDA 1996), DepoCyt® for lymphomatous malignant meningitis (FDA1997/2007), DepoDur® for the treatment of chronic pain (FDA2004), Doxil® for AIDS-related KS, multiple myeloma, ovarian cancer (FDA1995), Visudyne® for the treatment of wet age-related macular degeneration, pathological myopia, ocular histoplasmosis syndrome (FDA2000), (ii) lipid-based (nonliposomal) formulations like Abelcet®, Amphotec® for the treatment of systemic fungal infections (FDA1995/1996), (iii) PEGylated proteins, aptamers, polypeptides like Adagen® for adenosine deaminase deficiency disease (FDA1990), Neulasta® for Febrile neutropenia (FDA2002), Oncaspar® for acute lymphoblastic leukemia (FDA2002), Somavert® for acromegaly (FDA2003), Macugen® for intravitral neovascular degeneration (FDA2004), Mircera® for the treatment of anemia (FDA2007), (iv) protein-drug conjugates like Abraxane® for metastatic breast cancer and non-small-cell lung cancer (FDA2005), Kadcyla® for metastatic breast cancer (FDA2013), Ontak® for primary cutaneous lymphoma (FDA1994/2006), (v) metal-based nanoformulations like Feridex® for liver/spleen lesion (FDA1996), Feraheme™ for the treatment of iron deficiency anemia (FDA2009), (vi) virosomes like Gendicine® for head and neck squamous-cell carcinoma and Rexin-G® for the treatment of all solid tumors [54]. Table 15.1 lists a few biogenic nanoparticles that are marketed or undergoing clinical trials.

Table 15.1 List of clinical trials of biogenic nanoparticles.

Study Title	Source	Conditions	Interventions	Clinical Trial Identifier	References
Use of Autologous Plasma Rich in Platelets and Extracellular Vesicles in the Surgical Treatment of Chronic Middle-ear Infections	Plasma	Middle-ear Infections	Plasma Rich in Platelets and Extracellular Vesicles	NCT04761562	[55]
Isolation and Characterization of the Extracellular Vesicles Secreted by the Human Endometrium	Human Endometrium	Healthy, Fertile, Normal Cycling Women	Extracellular Vesicles	NCT02797834	[56]
Exosome – based Liquid Biopsies for Upper Gastrointestinal Cancers Diagnosis	Plasma	Upper Gastrointestinal Cancer	Exosome	NCT06278064	[57]
Venezuelan Equine Encephalitis Monovalent Virus-like Particle Vaccine (VEEV)	Virus	Encephalitis	Virus-like Particle Vaccine	NCT03776994	[58]
Immunogenicity of a Quadrivalent Virus-like Particles (VLP) Influenza Vaccine in Healthy Adults	Virus	Respiratory Tract Disease	Quadrivalent Virus-like Particles (VLP)	NCT02768805	[59]
Safety and Immunogenicity of Norovirus Bivalent Virus – like Particle Vaccine in Healthy Adults	Virus	Norovirus Prevention	Norovirus Bivalent Virus-like Particle Vaccine	NCT02142504	[60]
Antimicrobial Efficacy of Biogenic Gold Nano Particle From *Pelargonium Graveolens* Leaves Extract Mouthwash for Children	*Pelargonium Graveolens* Leaves	Gingivitis, Periodontitis, Dental Caries	Biogenic Gold Nanoparticle	NCT05816512	[61]
Efficacy of Novel Titania Nanoparticles-enriched Mouth Wash in Treating Dentine Hypersensitivity	Metal Oxide	Dentine Hypersensitivity	Titania Nanoparticles	NCT06790953	[62]
Effect of Propolis Nanoparticles on Postoperative Pain and Bacterial Reduction in Mandibular Premolars	Propolis	Necrotic Pulp	Propolis Nanoparticles	NCT05146713	[63]
The Effects of Nanocurcumin on Treg Cells and Th17 Cells Responses in Ankylosing Spondylitis Patients	Curcumin	Ankylosing Spondylitis	Nanocurcumin	NCT03140657	[64]
T4N5 Liposome Lotion Compared with Placebo Lotion for Preventing Actinic Keratoses in Patients with Xeroderma Pigmentosum	Phospholipids	Precancerous Condition	T4N5 Liposome Lotion	NCT00002811	[65]

15.5 Regulatory Aspects of Biogenic Nanoparticles

Although a number of approved nanopharmaceuticals have been developed, these nanopharmaceuticals' potential to have an impact on clinical practice in the future has been hampered by the absence of certain generic procedures for preclinical research and characterization. As an alternative, conventional methods for developing medications have frequently been altered to evaluate the compatibility, safety, and toxicity of nanopharmaceuticals. Regulatory authorities claim that the specifications that need to be observed within the regulatory framework are determined by the Active Pharmaceutical Ingredients (APIs) in nanoparticles [66]. The US Food and Drug Administration (US FDA) and the European Medical Agency (EMA), two international regulatory agencies, do not yet have any particular requirements for the market approval of nanopharmaceuticals. The US FDA formed a Nano Technology Interest Group to foster effective communication and educate staff and policymakers about scientific developments in the field of nanoparticles with the goal of improving awareness of the regulatory aspects of products containing nanoparticles. For drug delivery systems based on nanopharmaceuticals, safety and toxicity are crucial factors that must be satisfied for a product to be approved by regulators. Nanoparticles are complex systems made up of a variety of excipients, including stabilizers, co-surfactants, and surfactants. As recommended by international regulatory agencies, the excipients should be selected from the recognized list, i.e. Generally Regarded as Safe (GRAS). The ingredients listed as GRAS have undergone safety and toxicity testing [67]. Additionally, these excipients must be used at the regulatory-accepted concentration; toxicity studies must be conducted to show the acceptability and safety at the previously mentioned concentration if larger concentrations are to be utilized [66]. The GRAS list also shows the safe acceptable concentration range of the excipients in different types of administration methods [67]. Because nanoparticles are complex and varied, controlling them is a major task. Their properties and the mechanisms that nanoparticles follow to interact with the body can be drastically changed by even slight modifications to their manufacturing or design. Because of this, developing standardized procedures and tests to evaluate their efficacy and safety is challenging. The structure, biological characteristics, and biodistribution pattern of nanoparticles can all be greatly impacted by minor modifications to the raw ingredients or manufacturing procedures. Additionally, since researchers frequently attach drug and prodrugs, targeting molecules, tracking and imaging entities, and other entities to nanoparticles, new quality control assays and reliable techniques must be developed in order to effectively monitor and characterize not only the physicochemical properties of nanoparticles, such as size and size variability, morphology, and charge, but also their performance, such as drug release, metabolism assessment, protein binding, and specific cellular uptake. Additionally, these methods can make it possible to link the alterations that lead to broad physicochemical changes and affect biological traits, biocompatibility, and treatment outcomes. Because of the varied properties of nanoparticles, it can be challenging to produce them reliably and in large quantities. It's important to identify and manage critical qualities in any manufacturing process. The use of "quality-by-design" concepts, like process analytical technology (PAT), will guarantee an online/at-line quality assessment technique [68]. It is easier to implement automated processes to address problems as they emerge along the way when the most important manufacturing attributes are identified and anticipated. These innovative concepts have led to the development and implementation of the new pharmaceutical development regulations, International Conference on Harmonization (ICH) Q8, Q9, and Q10. They intend to significantly influence already existing drugs and promote the future development and manufacturing of nanoparticles. Another barrier to nanoparticle regulation is the need for *in vivo* and clinical studies, as well as the type of information that must be supplied both before and during the product life cycle [69–71]. Beginning with the initial stages of research and development, applicants for Marketing Authorisation Applications (MAA) are permitted by the European regulatory framework to receive "scientific counselling" from regulators. This could swiftly result in the coordinated formation of cutting-edge nanopharmaceuticals by lessening the influence of significant preventable obstacles throughout the procedure. This can help foster an integrated approach to the development of novel pharmaceuticals and mitigate the impact of major obstacles in that process [68]. By developing new regulatory standards and principles with active participation from academia and industry, as it has frequently been in the United States and Europe, a better regulatory guideline will be promoted progressively and transparently. The Innovative Medicines Initiative in Europe and the National Center for Advancing Translational Sciences (NCTAS/NIH) in the United States are two significant platforms that are going to help improve regulatory science in innovative technologies, precision medicine, and nanoparticles in the future [68, 72].

15.6 Global Forecast

The Global market for nanomaterials was valued at $12.42 billion in 2023 and is expected to expand at a compound annual growth rate (CAGR) of 15.0% between 2024 and 2030 [73]. The market for nanoparticles used in biotechnology, drug development, and drug delivery systems is projected to reach $102.7 billion in 2023 and $156.8 billion in 2028, globally, at a CAGR of 8.8%. The growing need for targeted therapeutic delivery systems and the application of nanoparticles in diagnostic imaging are the main factors propelling the growth of nanoparticles in the biotechnology, drug development, and drug delivery systems markets. Diagnostic nanoparticles aim to improve understanding of the key (patho-) physiological principles underlying a variety of diseases and disease treatments, as well as to depict disorders. A new field of application in biotechnology and diagnostics using particle materials in the nanometer size range has also emerged as a result of the development of nanoparticles and their quick integration into research and development [74]. Due to the product's high superparamagnetic qualities and enhanced surface area at the time of application, the market is anticipated to rise as a result of increased investment from research labs and biomedical industries in targeted drug delivery, gene therapy, and the treatment of malignant tumors. Depending on the manufacturing procedures used, different raw materials are needed for the creation of nanomaterials. Composites, polymers, rare-earth metals, alloys, and metals are examples of such raw materials that are either provided to commercial producers or research facilities [73]. Among the leading suppliers in the market who offer the necessary raw ingredients are Bernd Kraft GmbH and AppliChem GmbH. Growth in North America and Europe is expected to be impeded by strict rules. The possible negative effects of nanoparticles on the environment and human health are being evaluated by regulatory bodies like the Food and Drug Administration, the US Environmental Protection Agency and the Health and Consumer Protection Directorate. The market is expanding at a high rate and at a faster rate [73]. There is a moderate level of innovation because many nanomaterial businesses are constantly working on research and development projects and implementing new terminologies to obtain a competitive edge and meet the needs of different applications. The market for nanomaterials is thought to be quite competitive due to the vast number of manufacturers. American elements, US Research Nanomaterials, Inc., Nanoshel LLC, Nanocomposix Inc., Sigma Aldrich Inc., and Strem Chemicals, Inc. are some of the major companies in the market [73]. Standard requirements and end-user design specifications serve as the foundation for the items that manufacturers produce. In the United States, nanomaterials are not currently governed by any federal or state regulations. The current regulatory framework for food, cosmetic, pharmaceutical, chemical, and pesticide products also applies to nanomaterials. The Food and Drug Administration (FDA) and the Environmental Protection Agency (EPA) are two of the main regulatory agencies involved, which results in a high influence of regulations [73]. A significant portion of the global revenue in 2023 – roughly 31.2% – came from the medical application segment because the drugs are highly effective in treating cancer and other malignant tumors. Additionally, nanoparticles are utilized for a number of specialized tasks in the medical sector, such as targeted drug delivery, imaging, surgical robots, nano-diagnostics, cell repair, and nanobiosensors. During the forecast period 2024–2030, the Indian nanomaterials market is expected to expand at the highest

CAGR of 19.0%. One potential way to close the gap between India's rich and diverse natural resources and its growing demand for things like electronics, medical care, clean food and water, and premium fabrics is through nanomaterials [73].

15.7 Conclusion

Various nanotechnology techniques demonstrate the ability to develop drug carriers at the nanoscale with particular optical, chemical, and physical characteristics. For the manufacturing of nanoparticles, a variety of physical and chemical techniques have been developed, nevertheless, the various limitations associated with these methods have prompted researchers to search for other approaches. An effective strategy for the formation of nanoparticles has been found in the biological or natural sources. Compared to conventional chemical synthesis techniques, biogenic nanoparticles – which are produced utilizing biological organisms such as bacteria, fungi, plants, or algae, offer a number of benefits. These biogenic nanoparticles are easier to formulate, biocompatible, less toxic, less stable, more affordable, and environmentally friendly. However, the precise process of developing biogenic nanoparticles must be developed. A lot of effort is needed to enhance morphology, particle size control, and synthesis performance. Particle size and monodispersity are two main issues when evaluating nanoparticle synthesis. Therefore, a deeper comprehension of the stability and safety of biodegradable nanoparticles is required. They are used in different industries, such as electronics, agriculture, environmental clean-up, and medicine. To sum up, biogenic nanoparticles have a lot of potential to develop technology in an environmentally responsible and sustainable way. Their environmentally friendly synthesis techniques are very appealing since they lessen the negative effects of producing nanoparticles on the environment. There are still some drawbacks to be resolved, nevertheless, such increasing manufacturing rates are guaranteeing uniformity in nanoparticle characteristics and handling regulatory issues. Further developments in nanotechnology could result from more creative ideas and wider applications brought about by ongoing research and development in this area.

References

1 Sharma VK, Yngard RA, Lin Y. Silver nanoparticles: green synthesis and their antimicrobial activities. Adv Colloid Interf Sci. 2009;145(1–2):83–96.

2 Petrovic SM, Barbinta-Patrascu ME. Organic and biogenic nanocarriers as bio-friendly systems for bioactive compounds' delivery: state-of-the art and challenges. Materials. 2023;16(24):7550.

3 Bhandari M, Raj S, Kumar A, Kaur DP. Bibliometric analysis on exploitation of biogenic gold and silver nanoparticles in breast, ovarian and cervical cancer therapy. Front Pharmacol. 2022;13(December):1–13.

4 Hano C, Abbasi BH. Plant-based green synthesis of nanoparticles: production, characterization and applications. Biomolecules. 2022;12(1):1–9.

5 Padhiary M, Roy D, Dey P. Mapping the Landscape of Biogenic Nanoparticles in Bioinformatics and Nanobiotechnology. 2024. IGI Global.

6 Hassan AA, Sayed-Elahl RMH, Hamaky AME, Mansour MK, Oraby NH, Barakat MH. Green synthesis of metallic nanoparticles and their biomedical applications. Handb Green Sustain Nanotechnol Fundam Dev Appl. 2023;1–4(1):47–71.

7 Petousis-Harris H, Paynter J, Morgan J, Saxton P, McArdle B, Goodyear-Smith F, et al. Effectiveness of a group B outer membrane vesicle meningococcal vaccine against gonorrhoea in New Zealand: a retrospective case-control study. Lancet. 2017;390(10102):1603–1610.

8 Donaldson B, Lateef Z, Walker GF, Young SL, Ward VK. Virus-like particle vaccines: immunology and formulation for clinical translation. Expert Rev Vaccines. 2018;17(9):833–849.

9 Syn NL, Wang L, Chow EKH, Lim CT, Goh BC. Exosomes in cancer nanomedicine and immunotherapy: prospects and challenges. Trends Biotechnol. 2017;35(7):665–676.

10 Moitra P, Alafeef M, Alafeef M, Alafeef M, Dighe K, Frieman MB, et al. Selective naked-eye detection of SARS-CoV-2 mediated by N gene targeted antisense oligonucleotide capped plasmonic nanoparticles. ACS Nano. 2020;14(6):7617–7627.

11 Mathivanan K, Selva R, Chandirika JU, Govindarajan RK, Srinivasan R, Annadurai G, et al. Biologically synthesized silver nanoparticles against pathogenic bacteria: synthesis, calcination and characterization. Biocatal Agric Biotechnol. 2019;22:101373.

12 Ibrahim S, Ahmad Z, Manzoor MZ, Mujahid M, Faheem Z, Adnan A. Optimization for biogenic microbial synthesis of silver nanoparticles through response surface methodology, characterization, their antimicrobial, antioxidant, and catalytic potential. Sci Rep. 2021;11(1):1–18.

13 Salem SS, Ali OM, Reyad AM, Abd-Elsalam KA, Hashem AH. *Pseudomonas indica*-mediated Silver nanoparticles: antifungal and antioxidant biogenic tool for suppressing mucormycosis fungi. J Fungi. 2022;8(2):126.

14 Syed A, Ahmad A. Extracellular biosynthesis of platinum nanoparticles using the fungus *Fusarium oxysporum*. Colloids Surf B: Biointerfaces. 2012;97:27–31.

15 Bhargava A, Jain N, Khan MA, Pareek V, Dilip RV, Panwar J. Utilizing metal tolerance potential of soil fungus for efficient synthesis of gold nanoparticles

with superior catalytic activity for degradation of rhodamine B. J Environ Manag. 2016;183:22–32.

16 Pham ND, Duong MM, Le MV, Hoang HA, Pham LKO. Preparation and characterization of antifungal colloidal copper nanoparticles and their antifungal activity against Fusarium oxysporum and Phytophthora capsici. Comptes Rendus Chim. 2019;22(11–12):786–793.

17 Crane RA, Dickinson M, Popescu IC, Scott TB. Magnetite and zero-valent iron nanoparticles for the remediation of uranium contaminated environmental water. Water Res. 2011;45(9):2931–2942.

18 Al-Zahrani FAM, Salem SS, Al-Ghamdi HA, Nhari LM, Lin L, El-Shishtawy RM. Green synthesis and antibacterial activity of Ag/Fe2O3 nanocomposite using buddleja lindleyana extract. Bioengineering. 2022;9:452.

19 Bodade RG, Kumar R, Kutty R. Crop plant-mediated nanoparticle synthesis and applications. In: Nanoparticles Green Org Synth Strateg Towar Sustain, Vol. January. 2023, p. 351–399.

20 Fawcett D, Verduin JJ, Shah M, Sharma SB, Poinern GEJ. A review of current research into the biogenic synthesis of metal and metal oxide nanoparticles via marine algae and seagrasses. J Nanosci. 2017;2017(4):1–15.

21 Smit AJ. Medicinal and pharmaceutical uses of seaweed natural products: a review. J Appl Phycol. 2004;16(4):245–262.

22 Arockiya Aarthi Rajathi F, Parthiban C, Ganesh Kumar V, Anantharaman P. Biosynthesis of antibacterial gold nanoparticles using brown alga, *Stoechospermum marginatum* (kützing). Spectrochim Acta A Mol Biomol Spectrosc. 2012;99:166–173.

23 Steinmetz NF. Biological and evolutionary concepts for nanoscale engineering. EMBO Rep. 2019;20(8):1–6.

24 Wang C, Beiss V, Steinmetz NF. Cowpea Mosaic Virus nanoparticles and empty virus-like particles show distinct but overlapping immunostimulatory properties. J Virol. 2019;93(21):1–14.

25 Ferrer-Miralles N, Rodríguez-Carmona E, Corchero JL, García-Fruitós E, Vázquez E, Villaverde A. Engineering protein self-assembling in protein-based nanomedicines for drug delivery and gene therapy. Crit Rev Biotechnol. 2015;35(2):209–221.

26 Tan C, Wang J, Sun B. Biopolymer-liposome hybrid systems for controlled delivery of bioactive compounds: recent advances. Biotechnol Adv. 2021;48(2):107727.

27 Maja L, Željko K, Mateja P. Sustainable technologies for liposome preparation. J Supercrit Fluids. 2020;165:104984.

28 Ong SGM, Ming LC, Lee KS, Yuen KH. Influence of the encapsulation efficiency and size of liposome on the oral bioavailability of griseofulvin-loaded liposomes. Pharmaceutics. 2016;8(3):25.

29 Petrovic S, Bita B, Barbinta-Patrascu ME. Nanoformulations in pharmaceutical and biomedical applications: green perspectives. Int J Mol Sci. 2024;25(11):5842.

30 Jagaran K, Singh M. Lipid nanoparticles: promising treatment approach for Parkinson's disease. Int J Mol Sci. 2022;23(16):9361.

31 Khosa A, Reddi S, Saha RN. Nanostructured lipid carriers for site-specific drug delivery. Biomed Pharmacother. 2018;103:598–613.

32 Khatamian N, Motavalizadehkakhky A, Homayouni Tabrizi M, Mehrzad J, Zhiani R. Preparation and characterization of the myricetin-loaded solid lipid nanoparticles decorated with folic acid-bound chitosan and evaluation of its antitumor and anti-angiogenic activities *in vitro* and *in vivo* in mice bearing tumor models. Cancer Nanotechnol. 2023;14(1):1–22.

33 Prastiwi N, Ervina LK. Application of solid lipid nanoparticles preparation in infection caused by antibiotic-resistant bacteria. Indones J Pharmacol Ther. 2021;2(3):104–111.

34 Yadav N, Khatak S, Singh Sara UV. Solid lipid nanoparticles- a review. Int J Appl Pharm. 2013;5(2):8–18.

35 Coc LMC, Lacatusu I, Badea N, Barbinta-Patrascu ME, Meghea A. Effective lipid nanocarriers based on linseed oil for delivery of natural polyphenolic active. J Nanomater. 2021;2021(3):1–9.

36 Jaya B, Brajesh B, Antonella F, Giovanna LD, Amit K. Biopolymer: a sustainable material for food and medical applications. Polymers. 2022;14(5):983.

37 Tavakoli N, Divsalar A, Haertlé T, Sawyer L, Saboury AA, Muronetz V. Milk protein-based nanodelivery systems for the cancer treatment. J Nanostructure Chem. 2021;11(4):483–500.

38 Kianfar E. Protein nanoparticles in drug delivery: animal protein, plant proteins and protein cages, albumin nanoparticles. J Nanobiotechnol. 2021;19(1):1–32.

39 Sorolla A, Sorolla MA, Wang E, Ceña V. Peptides, proteins and nanotechnology: a promising synergy for breast cancer targeting and treatment. Expert Opin Drug Deliv. 2020;17(11):1597–1613.

40 Kurczewska J. Recent reports on polysaccharide-based materials for drug delivery. Polymers. 2022;14(19):4189.

41 Guo Y, Cao X, Zheng X, Abbas SJ, Li J, Tan W. Construction of nanocarriers based on nucleic acids and their applications in nanobiology delivery systems. Natl Sci Rev. 2022;9(5).

42 Wang Q, Zhao Y, Guan L, Zhang Y, Dang Q, Dong P, et al. Preparation of astaxanthin-loaded DNA/chitosan nanoparticles for improved cellular uptake and antioxidation capability. Food Chem. 2017;227:9–15.

43 de Carvalho Lima EN, Barros Martins GL, Diaz RS, Schechter M, Piqueira JRC, Justo JF. Effects of carbon nanomaterials and Aloe vera on Melanomas—where are we? Recent Updates. Pharmaceutics. 2022;14(10):1–34.

44 Rahmati M, Mozafari M. Biological response to carbon-family nanomaterials: interactions at the nano-bio interface. Front Bioeng Biotechnol. 2019;7:1–22.

45 Bhardwaj AK, Singh V, Rehalia A, Pandit AK, Sharma R. A review on nanomaterials for drug delivery systems and application of carbon-based nanomaterials. ES Mater Manuf. 2023;21:1–20.

46 Patil S, Chandrasekaran R. Biogenic nanoparticles: a comprehensive perspective in synthesis, characterization, application and its challenges. J Genet Eng Biotechnol. 2020;18(1):67.

47 Karunakaran G, Jagathambal M, Gusev A, Torres JAL, Kolesnikov E, Kuznetsov D. Rapid biosynthesis of AgNPs using soil bacterium Azotobacter vinelandii with promising antioxidant and antibacterial activities for biomedical applications. JOM. 2017;69(7):1206–1212.

48 El Domany EB, Essam TM, Ahmed AE, Farghali AA. Biosynthesis physico-chemical optimization of gold nanoparticles as anti-cancer and synergetic antimicrobial activity using *Pleurotus ostreatus* fungus. J Appl Pharm Sci. 2018;8(5):119–128.

49 Malik M, Aamir Iqbal M, Iqbal Y, Malik M, Bakhsh S, Irfan S, et al. Biosynthesis of silver nanoparticles for biomedical applications: a mini review. Inorg Chem Commun. 2022;145:109980.

50 Mohammadzadeh V, Barani M, Amiri MS, Taghavizadeh Yazdi ME, Hassanisaadi M, Rahdar A, et al. Applications of plant-based nanoparticles in nanomedicine: a review. Sustain Chem Pharm. 2022;25:100606.

51 Jafarizadeh-Malmiri H, Sayyar Z, Anarjan N, Berenjian A. Nanobiotechnology in Food: Concepts, Applications and Perspectives. 2019. Springer, p. 1–155.

52 Abdel-Megeed RM. Biogenic nanoparticles as a promising drug delivery system. Toxicol Rep. 2025;14:1.

53 Mohammadinejad R, Mansoori GA. Large-scale production/biosynthesis of biogenic nanoparticles. In: Biog Nano-Particles their Use Agro-ecosystems. 2020, p. 67–83.

54 Weissig V, Pettinger TK, Murdock N. Nanopharmaceuticals (part 1): products on the market. Int J Nanomedicine. 2014;9:4357–4373.

55 Vozel D, Božič D, Jeran M, Jan Z, Pajnič M, Pađen L, et al. Treatment with platelet- and extracellular vesicle-rich plasma in otorhinolaryngology-a review and future perspectives. Adv Biomembr Lipid Self-Assem. 2021;33:119–153.

56 Isolation and Characterization of the Extracellular Vesicles Secreted by the Human Endometrium. ClinicalTrials.gov. Available from: https://clinicaltrials.gov/study/NCT02797834?term=outer membrane vesicle&rank=8&tab=table.

57 Exosome-based Liquid Biopsies for Upper Gastrointestinal Cancers Diagnosis. ClinicalTrials.gov. Available from: https://clinicaltrials.gov/study/NCT06278064?term=outer membrane vesicle&rank=9.

58 Venezuelan Equine Encephalitis Monovalent Virus-Like Particle Vaccine. ClinicalTrials.gov. Available from: https://clinicaltrials.gov/study/NCT03776994?term=Virus like particles&rank=1.

59 Immunogenicity of a Quadrivalent Virus-Like Particles (VLP) Influenza Vaccine in Healthy Adults.ClinicalTrials.gov. Available from: https://clinicaltrials.gov/study/NCT02768805?term=Virus like particles&rank=3.

60 Safety and Immunogenicity of Norovirus Bivalent Virus-Like Particle Vaccine in Healthy Adults. ClinicalTrials.gov. Available from: https://clinicaltrials.gov/study/NCT02142504?term=Virus like particles&rank=6.

61 Antimicrobial Efficacy of Biogenic Gold Nano Particle From *Pelargonium Graveolens* Leaves Extract Mouthwash for Children. ClinicalTrials.gov. [cited 2025 Feb 8]. Available from: https://clinicaltrials.gov/study/NCT05816512?term=gold nanoparticles&rank=3.

62 Efficacy of Novel Titania Nanoparticles Enriched Mouth Wash in Treating Dentine Hypersensitivity. ClinicalTrials.gov. Available from: https://clinicaltrials.gov/study/NCT06790953?term=biogenic&rank=11.

63 Effect of Propolis Nanoparticles on Postoperative Pain and Bacterial Reduction in Mandibular Premolars. ClinicalTrials.gov. Available from: https://clinicaltrials.gov/study/NCT05146713?term=PLANT, NANOPARTICLE&rank=1.

64 The Effects of Nanocurcumin on Treg Cells and Th17 Cells Responses in Ankylosing Spondylitis Patients. ClinicalTrials.gov. Available from: https://clinicaltrials.gov/study/NCT03140657?term=PLANT, NANOPARTICLE&rank=3.

65 T4N5 Liposome Lotion Compared With Placebo Lotion for Preventing Actinic Keratoses in Patients With Xeroderma Pigmentosum. ClinicalTrials.gov. Available from: https://clinicaltrials.gov/study/NCT00002811?term=LIPOSOME&rank=2.

66 Sainz V, Conniot J, Matos AI, Peres C, Zupančič E, Moura L, et al. Regulatory aspects on nanomedicines. Biochem Biophys Res Commun. 2015;468(3):504–510.

67 Liu M, Wen J, Sharma M. Solid lipid nanoparticles for topical drug delivery: mechanisms, dosage form perspectives, and translational status. Curr Pharm Des. 2020;26(27):3203–3217.

68 Chhikara BS, Mandal D, Parang K. Synthesis, anticancer activities, and cellular uptake studies of lipophilic derivatives of doxorubicin succinate. J Med Chem. 2012;55(4):1500–1510.

69 Singh J. International conference on harmonization of technical requirements for registration of pharmaceuticals for human use. J Pharmacol Pharmacother. 2015;6(3):185–187.

70 International Council for Harmonisation. Q9(R1) - Quality Risk Management. Int Conf Harmon Geneva. 2015;44(1):1–20.

71 Pharmaceutical Quality Systems. Pharm Qual Syst. 2007 (September). Available from: https://picscheme.org/docview/3462.

72 Adhikari P, Pal P, Das AK, Ray S, Bhattacharjee A, Mazumder B. Nano lipid-drug conjugate: an integrated review. Int J Pharm. 2017;529(1–2):629–641.

73 Nanomaterials Market Size, Share and Growth Report, 2030. Available from: https://www.grandviewresearch.com/industry-analysis/nanotechnology-and-nanomaterials-market.

74 Global Nanoparticles in Biotechnology & Drug Development Market. Available from: https://www.bccresearch.com/market-research/biotechnology/nanoparticles-biotechnology-drug-development-drug-delivery-report.html.

Index

Biogenic Synthesis of Nanoparticles and Sustainable Nanotherapeutics,
First Edition. Edited by Malay K. Das.

Z